Solar Sailing

Technology, Dynamics and Mission Applications

Springer

London
Berlin
Heidelberg
New York
Barcelona
Hong Kong
Milan
Paris
Santa Clara
Singapore
Tokyo

Colin Robert McInnes

Solar Sailing

Technology, Dynamics and Mission Applications

Springer

Published in association with

Praxis Publishing
Chichester, UK

Colin Robert McInnes BSc, PhD, CEng, FRAS, FRAes
Reader in Space Systems Engineering
Department of Aerospace Engineering
University of Glasgow, Glasgow, Scotland

SPRINGER–PRAXIS SERIES IN SPACE SCIENCE AND TECHNOLOGY
SERIES EDITOR: John Mason B.Sc., Ph.D.

ISBN 1-85233-102-X Springer-Verlag Berlin Heidelberg New York

British Library Cataloguing in Publication Data
   McInnes, Colin Robert
      Solar sailing : technology, dynamics and mission applications.
      — (Springer-Praxis series in space science and technology)
      1. Solar sails. I. Title
      629.4'75'4
      ISBN 1-85233-102-X

Library of Congress Cataloging-in-Publication Data
   McInnes, Colin Robert, 1968–
      Solar sailing : technology, dynamics, and mission applications / Colin Robert McInnes
      p. cm. — (Springer-Praxis series in space science and technology)
      "Published in association with Praxis Publishing, Chichester, UK."
      Includes index.
      ISBN 1-85233-102-X (alk. paper)
      1. Solar sails. 2. Orbital dynamics. I. Title. II. Series.
      TL783.9.M39   1999   98-43763 CIP
      629.47'54—dc 21

© Praxis Publishing Ltd, Chichester, UK, 1999
Printed by MPG Books Ltd, Bodmin, Cornwall, UK

Cover design: Jim Wilkie
Typesetting: Originator, Gt. Yarmouth, Norfolk, UK

Printed on acid-free paper supplied by Precision Publishing Papers Ltd, UK

For Karen and Calum

Our traveller knew marvellously the laws of gravitation, and all the attractive and repulsive forces. He used them in such a timely way that, once with the help of a ray of sunshine, another time thanks to a co-operative comet, he went from globe to globe, he and his kin, as a bird flutters from branch to branch.

VOLTAIRE: *Micromegas*, 1752

Contents

List of illustrations and tables

ILLUSTRATIONS

TABLES

Foreword

Dear Reader

You are holding in your hands *the* reference book on Solar Sailing. There have been other books on various aspects of solar sails (you will find them appropriately listed on page 31 of this book), but whereas the other books have concentrated on one aspect (mathematically rigorous solar sail astrodynamics for mathematicians) or another (strut and film solar sail construction techniques for hardware engineers) or another (fun missions using solar sails for space enthusiasts), you will find in this book that Colin McInnes, rigorous mathematician, practical aerospace engineer and inspiring writer, has covered all the aspects.

This book not only contains all the right mathematical formulas that you need to design your own 'pole-sitter' solar sail spacecraft, whether the 'pole' is that of the Sun, Earth, Mars or Mercury, but it also describes in detail how to design and build the sails. Finally, the book inspires you to get busy doing so by outlining all the interesting missions that solar sails can do that no other propulsion system can do, from 'hanging' between here and the Sun to warn of impending solar flares about to black out entire continents, to multiple sample returns from a multiple asteroid mission, to round trip missions to the stars – all using that miraculous propulsion system that uses no energy, uses no propellant, and lasts forever – the solar sail.

Dr Robert L. Forward
Forward Unlimited

Author's preface

I first stumbled across the idea of solar sailing not long after I had matriculated as an undergraduate student at Glasgow University in 1984. While browsing through some popular science books in Hillhead library, not far from the main University campus in the West end of Glasgow, I came upon a wonderful colour painting of a large solar sail. I remember at the time being struck by the aesthetic beauty and sheer excitement of the idea. Several years later, in 1988, I was offered a scholarship by the Royal Society of Edinburgh to pursue postgraduate research. The scholarship was not tied to any particular topic, but was awarded for work in the general area of Astronomy. My supervisor, Professor John Brown (now the Astronomer Royal for Scotland) took a fairly liberal interpretation of the definition of Astronomy. Knowing of my interest in orbital mechanics, he suggested I follow up some work on the effect of light pressure in the classical three-body problem. John suggested that a good place to start would be to consider solar sails.

I quickly found that although solar sailing had been studied for many years, not much had been published in the technical literature. For an intending PhD student this was both good news and bad. It implied that the field was wide open for exploitation as a research topic, but also meant that I wouldn't find too many collaborators. After some initial work investigating the effect of the finite angular size of the solar disc on two-body orbits, I returned to the three-body problem. This was a topic my other supervisor Dr John Simmons had explored in great detail. John had discovered that if one or both of the masses in the problem were luminous, several new equilibrium points appeared; that is, an infinitesimal particle would remain at rest relative to the two primary bodies. I revisited this problem for the Sun–Earth system, assuming one luminous body, and a flat solar sail instead of a point mass. Rather than just a few additional equilibrium points, whole surfaces now appeared where a solar sail could remain at rest. These equilibrium surfaces extended in a bubble attached to the classical L_1 equilibrium point, 1.5 million km sunward of the Earth. At the time I was approaching the problem as a scientist (having just graduated with a degree in Physics and Astronomy) rather than an engineer, so any practical applications weren't foremost in my mind. Almost simultaneously however,

Dr Robert Forward, the well-known proponent of advanced propulsion, was investigating the use of solar sails to 'levitate' over the night side of the Earth to provide communication services to high-latitude and polar regions. For some time to come Bob was to be the only other individual I knew who was actively pursuing new orbits and mission applications for solar sails.

Several years later, in August 1996, I was put in touch with Dr Patricia Mulligan at NOAA and Dr John West at JPL. Pat and John were beginning a study to investigate the use of an inflatable solar sail to orbit sunward of the L_1 point. A large inflatable structure had been flight tested on the STS-77 mission a few months earlier and appeared to hold great promise for solar sailing. An orbit sunward of L_1 would provide enhanced warning of the energetic plasma streams from the Sun which can induce magnetic storms on Earth, leading to the disruption of satellite communications. Apart from predicting terrestrial weather, NOAA also predicts 'space weather'. The orbit for the mission, later named Geostorm, was in fact one of the family of equilibrium surfaces I had found at the Earth–Sun L_1 point many years before. It has been a privilege for me to contribute to this mission study as a consultant ever since. With the start of the Geostorm study, a long circle from my initial work on solar sailing had finally closed. At about the same time I was approached by Clive Horwood of Praxis Publishing Ltd who kindly invited me to write a technical book on solar sailing. Given that my interest in solar sailing had been rekindled by the Geostorm study, it seemed to be both a good idea and a good time to start a book. The remaining pages contain the product of my efforts. Although the title is my own, I almost opted for the much more appropriate suggestion by Dr Jean-Yves Prado, 'Solar Sailing – What Are We Waiting For?'. While this captures the spirit of what many of us believe, I'm not sure that my publisher would have approved.

This book is an attempt to bring together much of my own work on solar sailing along with that of many others into a complete volume which will form a primer for those new to the field, and a reference document for practising scientists and engineers who wish to explore solar sailing for their own purposes. Inevitably, some prior knowledge is required to access all of the chapters. For those without a background in orbital mechanics, Chapters 1, 3, 6 and parts of Chapter 7 should provide an insight into solar sailing and hopefully inspire interest in the technology. For those versed in orbital mechanics, Chapters 4 and 5 will complete the overview of the field. Again, somewhat inevitably, notation can become contorted in a book which covers a field in breadth. As far as possible I have tried to keep to the accepted notation for key variables and constants. Where duplication occurs I have used the tilde notation where appropriate. For example, σ is used to represent the solar sail mass per unit area, while $\tilde{\sigma}$ is used to represent the Stefan–Boltzmann constant.

I hope that those who read this book will find the same delight that I have at both the sheer excitement which solar sailing invokes and also the amazing range of mission applications it can enable. While I have no objection to excitement and enthusiasm, I also hope that readers will be hard-headed in their evaluation of solar sailing. I firmly believe that solar sailing can only succeed if we confront the challenges it poses and focus on what solar sails can do, rather than try to advance solar sailing for its own sake. My only other wish is that I may have the opportunity to write a second edition of this book some years from now, detailing the successful use of solar technology for some initial mission applications. As Jean-Yves says: What are we waiting for?

Colin McInnes
Glasgow, October 1998

Acknowledgements

Many individuals have contributed to this book, either by providing material, information, advice or simple encouragement. In particular my thanks go to John Brown, Kieran Carroll, Bob Forward, Craig French, Roderick Galbraith, Chuck Garner, Manfred Leipold, Esther Morrow, Pat Mulligan, Elena Poliakhova, Alex Shvartsburg, Alan Simpson, Rob Staehle, Giovanni Vulpetti, John West, Henry Wong and Jerome Wright. I would like to offer special thanks to Jean-Yves Prado, who kindly reviewed the entire book as each chapter was produced, and Bob Forward who reviewed the completed manuscript and wrote the foreword. Jean-Yves Prado kindly provided the cover picture by Lionel Bret. Clive Horwood at Praxis Publishing Ltd was an ideal to which all scientific publishers should aspire. Lastly, my thanks go to my family who have been a constant source of support and encouragement. My wife Karen and son Calum were a true inspiration, particularly through the long final weeks of preparation.

Glossary of terms

ACE	Advanced Composition Explorer
ASAP	Ariane structure for auxiliary payload
CCD	Charge-coupled device
CCL	Cambridge Consultants Ltd
CFRP	Carbon fibre-reinforced plastic
CLCB	Continuous longeron coilable boom
CME	Coronal mass ejection
CNES	Centre National d'Etudes Spatiales
DLC	Diamond-like carbon
DLR	German Aerospace Research Establishment
DSN	Deep space network
EDM	Engineering development mission
ESA	European Space Agency
GEO	Geostationary orbit
GTO	Geostationary transfer orbit
IAE	Inflatable antenna experiment
JPL	Jet Propulsion Laboratory
MEMS	Microelectromechanical systems
NASA	National Aeronautics and Space Administration
NKO	Non-Keplerian orbit
NOAA	National Oceanic and Atmospheric Administration
ODISSEE	Orbital demonstration of an innovative solar sail driven expandable structure experiment
PVDF	Polyvinyldifluoride
SSRV	Solar sail race vehicle
STEM	Storable tubular expandable member
SSUJ	Solar Sail Union of Japan
U3P	Union pour la Promotion de la Propulsion Photonique
UV	Ultra-violet
WSF	World Space Foundation

1

Introduction to solar sailing

1.1 PRINCIPLES OF SOLAR SAILING

For all of its short history, practical spacecraft propulsion has been dominated by the unaltering principles of Newton's third law. All forms of propulsion, from simple solid rocket motors to complex solar-electric ion drives, rely on a reaction mass which is accelerated into a high velocity jet by some exothermal or electromagnetic means. A unique and elegant form of propulsion which transcends this reliance on reaction mass is the solar sail. Since solar sails are not limited by a finite reaction mass they can provide continuous acceleration, limited only by the lifetime of the sail film in the space environment. Of course, solar sails must also obey Newton's third law. However, solar sails gain momentum from an ambient source, namely photons, the quantum packets of energy of which Sunlight is composed.

The momentum carried by an individual photon is vanishingly small. Therefore, in order to intercept large numbers of photons, solar sails must have a large, extended surface. Furthermore, to generate as high an acceleration as possible from the momentum transported by the intercepted photons, solar sails must also be extremely light. For a typical solar sail the mass per unit area of the entire spacecraft may be an order of magnitude less than the paper on which this text is printed. Not only must solar sails have a small mass per unit area, they must also be near perfect reflectors. Then, the momentum transferred to the sail can be almost double the momentum transported by the incident photons. At best, however, only 9 Newtons of force are available for every square kilometre of sail located at the Earth's distance from the Sun. Adding the impulse due to incident and reflected photons, it will be seen that the solar radiation pressure force is directed almost normal to the surface of the solar sail. Then, by controlling the orientation of the sail relative to the Sun, the solar sail can gain or lose orbital angular momentum. In this way the solar sail is able to tack, spiralling inwards towards the Sun, or outwards to the farthest edge of the solar system and indeed beyond to interstellar space.

The picture then is clear. A solar sail is a large shining membrane of thin reflective film held in tension by some gossamer structure. Using momentum gained only by reflecting ambient light from the Sun, the solar sail is slowly but continuously accelerated to accomplish any number of possible missions. Without the violence of reaction propulsion the solar sail is tapping a tiny fraction of the energy released through nuclear fusion at the core of the Sun. Solar sailing, with its analogies with terrestrial sailing, may seem a fanciful and romantic notion. However, as will be shown in this book, the romanticism is overshadowed by the immense practicability and quiet efficiency with which solar sails can be put to use.

1.2   PERSPECTIVES

1.2.1   Pioneers

Although solar sailing has been considered as a practical means of spacecraft propulsion only relatively recently, the fundamental ideas are by no means new. The actual concept of solar sailing has a long and rich history, dating back to the Soviet pioneers of astronautics, and indeed before. While the existence of light pressure was demonstrated in theory by the Scottish physicist James Clerk Maxwell in 1873, it was not measured experimentally until precision laboratory tests were performed by the Russian physicist Peter Lebedew in 1900. Similarly, while a number of science fiction authors, most notably the French authors Faure and Graffigny in 1889, wrote of spaceships propelled by mirrors, it was not until early this century that the idea of a practical solar sail was articulated. As early as the 1920s the Soviet father of astronautics, Konstantin Tsiolkovsky, and his co-worker, Fridrickh Tsander, both wrote of 'using tremendous mirrors of very thin sheets' and 'using the pressure of sunlight to attain cosmic velocities'. Although there is some uncertainty regarding dates, it appears that Tsander was the first to write of practical solar sailing some time late in the summer of 1924. His ideas seem to have been inspired in part by Tsiolkovsky's more general writings from 1921 on propulsion using light.

Tsiolkovsky (1857–1935) (Fig. 1.1), the Soviet father of astronautics, was by any measure a most remarkable individual. Born near Moscow, he became profoundly deaf following a childhood illness and was largely self-taught. He was a true visionary who inspired many of the scientist and engineers who would later develop rocketry and practical spaceflight in the Soviet Union. Tsiolkovsky also made notable contributions to aviation, building the first wind tunnel in the Soviet Union in 1890 and designing an all-metal monoplane aircraft as early as 1894. One of his devotees was Fridrickh Tsander (1887–1933) (Fig. 1.2), a Latvian engineer, born in Riga. Apart from solar sailing, Tsander is also remembered as a pre-war pioneer of liquid rocket prolusion. Indeed, he led early experiments with liquid prolusion in the Soviet Union, although he died of typhoid in March 1933, not long before the flight of the first Soviet liquid-fuelled rocket. It is interesting to note that Tsander was so infatuated with the notion of space travel that he named his daughter 'Astra' and his son 'Mercury'.

Fig. 1.1. Konstantin Tsiolkovsky (1857–1935).

1.2.2 Early optimism

After the initial writings of Tsiolkovsky and Tsander in the 1920s the concept of solar sailing appears to have remained essentially dormant for over thirty years. It was not until the 1950s that the concept was re-invented and published in the popular literature. The first American author to propose solar sailing appears to have been aeronautical engineer Carl Wiley, writing under the pseudonym Russell Sanders to protect his professional credibility. In his May 1951 article in *Astounding Science Fiction*, Wiley discussed the design of a feasible solar sail and strategies for orbit raising in some technical detail. In particular he noted that solar sails could be 'tacked', allowing a spiral inwards towards the Sun. Even in 1951 Wiley was optimistic about the benefits of solar sailing for interplanetary travel and saw it as ultimately more practical than rocket propulsion.

A similar, optimistic view was aired some time later in 1958 through a separate proposal and evaluation by Richard Garwin, then at the IBM Watson laboratory of Columbia University. Garwin authored the first solar sail paper in a western technical publication, the journal *Jet Propulsion*, and coined the term 'solar sailing'. Like Wiley, Garwin recognised the unique and elegant features of solar sailing; namely, that solar sails require no propellant and are continuously accelerated,

Fig. 1.2. Fridrickh Tsander (1887–1933).

therefore allowing large velocity changes over an extended period of time. Such was Garwin's enthusiasm and optimism for solar sailing that he concluded that 'there are considerable difficulties connected with space travel, but those connected with the sail appear relatively small'. The obvious fact that such early optimism has not led to the actual flight of a solar sail after some forty years will be discussed later in this chapter.

Following the discussion of solar sailing by Garwin, more detailed studies of the orbits of solar sails were undertaken during the late 1950s and early 1960s. Several authors were able to show that, for a fixed sail orientation, solar sail orbits are of the form of logarithmic spirals. Simple comparisons of solar sailing with chemical and ion propulsion systems showed that solar sails could match, and in many cases out-perform, these systems for a range of mission applications. While these early studies explored the fundamental problems and benefits of solar sailing, they lacked a specific mission to drive detailed analyses and act as a focus for future utilisation. It is useful to note that in 1963 Arthur C. Clarke first published his well-known short story *The Wind from the Sun*. The story, centring on a manned solar sail race in Earth orbit, popularised solar sailing and indeed led to the dissemination of the idea of solar sailing to many science fiction reading engineers. Even now this story is the first introduction to solar sailing for many readers.

1.2.3   Chasing a comet

In the early 1970s the development of the Space Shuttle promised the means of being able to transport and deploy large payloads in Earth orbit. In addition, the development of technologies for deployable space structures and thin films suggested that solar sailing could be considered for a specific mission. By 1973 NASA was funding low-level studies of solar sailing at the Battelle laboratories in Ohio, which gave positive recommendations for further investigation. During the continuation of this work Jerome Wright, who would later move to Jet Propulsion Laboratory, discovered a trajectory which could allow a solar sail to rendezvous with comet Halley at its perihelion in the mid-1980s. The flight time of only four years would allow for a late 1981 or early 1982 launch. This was a remarkable finding. Until then a difficult rendezvous mission was thought to be near impossible in such a short time using the technology of the day. A seven- to eight-year mission had been envisaged using solar-electric ion propulsion, requiring a launch as early as 1977. An actual rendezvous was seen by the science community as essential for a high-quality mission. These positive results prompted the then NASA Jet Propulsion Laboratory director, Bruce Murray, to initiate an engineering assessment study of the potential readiness of solar sailing. Newly appointed from a faculty position at the California Institute of Technology, Murray adopted solar sailing as one of his bold 'purple pigeon' projects, as opposed to the more timid 'grey mice' missions he believed Jet Propulsion Laboratory had been proposing for future missions. Following this internal assessment, a formal proposal was put to NASA management in September 1976. The design of a comet Halley rendezvous mission using solar sailing was initiated in November of the same year.

During the initial design study an 800×800 m three-axis stabilised square solar sail configuration was considered (Fig. 1.3), but was dropped in May 1977 owing to the high risks associated with deployment. The design work then focused on a spin-stabilised heliogyro configuration. The heliogyro, which was to use twelve 7.5 km long blades of film rather than a single sheet of sail film (Fig. 1.4), had been developed some ten years earlier by Richard MacNeal at the Astro Research Corporation and later by John Hedgepath. The heliogyro could be more easily deployed than the square solar sail by simply unrolling the individual blades of the spinning structure. As a result of this design study, the structural dynamics and control of the heliogyro were characterised and potential sail films manufactured and evaluated. Also important for NASA institutional considerations, the solar sail work had sparked public interest and enthusiasm for a comet Halley rendezvous mission.

As a result of the interest in solar sailing, proponents of solar-electric propulsion re-evaluated their performance estimates and, in the end, were competing directly with solar sailing for funding. The solar-electric propulsion system had a larger advocacy group both within NASA and in industry. As a result of an evaluation of these two advanced propulsion concepts, NASA selected the solar-electric system in September 1977, upon its merits of being a less but still considerable risk for a comet Halley rendezvous. A short time later a rendezvous mission using solar-electric propulsion was also dropped owing to escalating cost estimates. The enthusiasm of

Fig. 1.3. Comet Halley square sail configuration (NASA/JPL).

Fig. 1.4. Comet Halley heliogyro configuration (NASA/JPL).

the science community for a rendezvous mission had in fact led to the lower cost fly-past option being discounted. Therefore, when the advanced propulsion required to enable a rendezvous mission was deleted, some careful back-stepping was required to justify a lower cost, but less capable, fly-past mission. Ultimately, however, it was too late and a dedicated NASA comet Halley mission was never flown. Comet Halley *was* intercepted, however, by an armada of Soviet, Japanese and European spacecraft.

1.2.4   Celestial races

Although dropped by NASA for near-term mission applications, the design studies of the mid-1970s stimulated world-wide interest in solar sailing. Low-level European studies were taken up by CNES (Centre National d'Etudes Spatiales) in Toulouse to assess the potential of the new Ariane launch vehicle for deep space missions. Perhaps more importantly for the long-term prospects of solar sailing was the formation of the World Space Foundation (WSF) in California in 1979 and the Union pour la Promotion de la Propulsion Photonique (U3P) in Toulouse later in 1981. The WSF, formed principally by Jet Propulsion Laboratory engineer Robert Staehle and others after the termination of the Jet Propulsion Laboratory solar sail work, attempted to raise private funds to continue solar sail development and to undertake a small-scale demonstration flight. Shortly afterwards the U3P group was formed and in 1981 proposed an ambitious Moon race to promote solar sail technology. Both of these groups were joined by the Solar Sail Union of Japan (SSUJ) in 1982 and have worked tirelessly over many years to advance solar sailing and the idea of a race to the Moon (Fig. 1.5).

More recently the US Columbus Quincentennial Jubilee commission, formed to organise celebrations of the 1992 quincentenary of Columbus discovering the New World, attempted to stimulate interest in a solar sail race to Mars. The proposal generated significant international interest in the early 1990s, inspiring some of the most technically advanced and innovative solar sail designs to date. While the use of competitive races to accelerate the development of technology has a long and successful history in aviation, the concept of solar sail races to the Moon and Mars eventually foundered. Perhaps it was that the duration of the race was significantly longer than the attention span of a television audience, thus limiting opportunities for sponsorship. Or, that the Earth–Moon system is not an ideal location for the optimum use of solar sails – as will be seen, solar sails are used to their optimum potential away from the steep gravitational well of the Earth. It is clear, however, that the race proposals brought about a renaissance in solar sailing not seen since the comet Halley studies almost fifteen years before.

1.2.5   Testing times

Although a true solar sail has yet to be flown, the 1990s have seen the development and flight testing of some key technologies for future utilisation. Firstly, under the leadership of Vladimir Syromiatnikov, the Russian Space Regatta Consortium

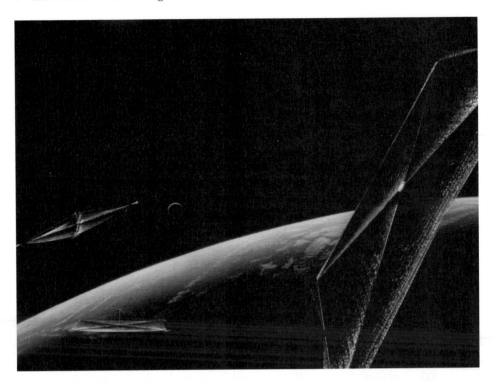

Fig. 1.5. Solar sail Moon race (U3P/Lionel Bret).

successfully deployed a spinning 20 m reflector from a Progress supply vehicle in February 1993 (Fig. 1.6). The simple deployment process was driven solely by spinning up the stowed reflector using an on-board electric motor. Observed from the MIR space station, the test demonstrated that such spin deployment can be controlled by passive means. The Znamya (translated as 'banner' or 'flag') experiment was conducted at extremely low cost, and is the first in a series of planned flight tests of deployable reflectors. While the reflectors can demonstrate technologies for solar sailing, their principal use is to illuminate northern Russian cities during dark winter months to aid economic development.

Another spectacular demonstration of a large deployable reflector was achieved in May 1996 during the STS-77 Space Shuttle mission (Fig. 1.7). The 14 m diameter Inflatable Antenna Experiment (IAE) was designed to test the deployment of a large inflatable structure, to be used principally as a radio-frequency reflector. Owing to venting of trapped air in the stowed film, however, the deployment sequence did not proceed as planned. In addition, the shape of the reflecting surface was not as precise as desired. While not achieving all of its goals, the experiment clearly demonstrated the promise of inflatable structures for robust and reliable deployment. As will be seen, this flight validation of inflatable structures has driven the design of some recent solar sail concepts.

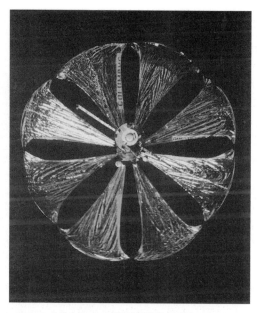

Fig. 1.6. Znamya deployment test, 4 February 1993 (SRC Energia).

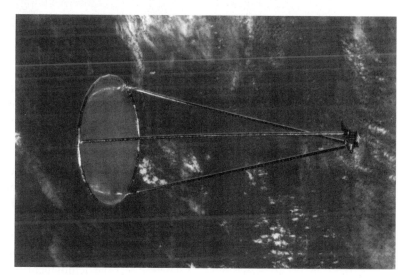

Fig. 1.7. Inflatable antenna deployment test, 20 May 1996 (NASA/JPL).

1.2.6   New millennium

Recent years have seen a major shift in thinking concerning future robotic space exploration. Both NASA and ESA have initiated programmes to develop small, low-cost but highly capable spacecraft for planetary and space science missions. The

NASA New Millennium and ESA SMART programmes will flight test new technologies to demonstrate the possibilities offered by such low-mass, high-performance spacecraft. These developments have in part been spurred on by advances in the miniaturisation of terrestrial electronics. For example, low-mass charge-coupled device (CCD) cameras, solid-state memory and high-performance processors can enable highly autonomous, self-sufficient spacecraft with a mass of order 10 kg in the relatively near term. Future advances in microelectromechanical systems (MEMS) technology may ultimately lead to spacecraft-on-a-chip with a mass of under 1 kg.

While these developments offer exciting opportunities for conventional spacecraft missions, such as large co-operating constellations, a wonderful opportunity is now at hand to capitalise on these low-mass spacecraft to reduce the size and complexity of solar sails. While the Jet Propulsion Laboratory comet Halley mission required an 800×800 m solar sail to transport a 850 kg payload, future solar sails may be more than ten times smaller. Indeed, some current solar sail concepts are even smaller than the reflective vanes to be used on the comet Halley solar sail for attitude control. This reduction in scale simplifies the manufacture, packing and deployment of solar sails. Perhaps just as importantly, it brings the scale of solar sails down to a size which can be easily visualised. No longer do solar sails need to be viewed as vast, unmanageable structures to be developed at some time in the indefinite future. The engineering required for solar sailing is now on a more human scale, thus lending greater plausibility to the whole concept.

1.2.7   Lessons of history

Since Garwin's paper initiated modern developments in solar sailing some forty years ago, the concept has inspired many individuals to devote their time and energy to advance the field. Countless technical papers have been written which demonstrate the potential advantages of solar sailing, many by graduate students who then move on to the more immediate problems of industry. Studies have been conducted which demonstrate the technical feasibility of solar sailing. However, for all of these sometimes heroic efforts an operational solar sail has yet to fly. The reasons for this failure are, of course, open to debate. It may be a stubborn reluctance on the part of some to believe that solar sailing can ever work, or that it offers benefits over conventional propulsion. Furthermore, reviewing the development of solar sailing detailed earlier, the view could be taken that solar sailing has been a technology in search of a mission. It is certainly clear that the last major effort on behalf of a space agency was due to the mission-enabling properties of solar sailing to chase and rendezvous with comet Halley. While solar sailing enabled a rendezvous, it did of course lose in the end to electric propulsion. The threat posed by solar sailing to long-standing investments in electric propulsion should not be underestimated, even now.

In order to advance solar sailing, proponents need to step back from their enthusiasm which can give the mistaken impression that it is an elegant idea which should be funded for the sake of aesthetics. A cold look at the strengths and weaknesses of the technology is required in order to build a convincing case for support. In particular, it is the weaknesses of solar sailing, either real or perceived, which need to be addressed. While the obvious advantage of potentially unlimited velocity change is perhaps the greatest benefit, it is useless if the first operational solar sail fails to deploy. Historical problems with the deployment of even modest space structures can unfortunately taint solar sailing by association. Similarly, competition from solar-electric propulsion is still a threat, although the new institutional approach to advanced technologies provides a welcome opportunity for exploitation. Given these factors it seems that what is required is a small, low-cost and low-risk solar sail mission for which there is either no feasible alternative form of propulsion or no alternative option of comparable cost. It is also a key requirement that there is an absolutely compelling mission application which will demand the development of solar sail technology to flight status. If these criteria are met, then mission planners and their political masters will be cornered into developing solar sail technology and so bring to fruition the dreams of Tsander, Tsiolkovsky and many others.

1.3   PRACTICALITIES OF SOLAR SAILING

1.3.1   Solar sail configurations

The fundamental goal for any solar sail design is to provide a large, flat reflective film which requires a minimum of structural support mass. Secondary issues, such as ease of manufacture and reliability of deployment, are also clearly of great importance. In general, the sail film must be kept flat through the application of tension forces at the edges of the film. These forces can be generated mechanically by cantilevered spars, or by centripetal force induced by spinning the sail film. While these two approaches are quite different, a synthesis of both is also possible.

The first concept is the square solar sail, the optimum design of which has four deployable spars cantilevered from a central load-bearing hub, as shown in Fig. 1.8. The hub contains the payload and spar deployment mechanisms, which may be jettisoned following deployment. Attitude control of the square solar sail can be achieved by inducing torques generated by articulated reflecting vanes attached to the spar tips, or through relative translation of the centre-of-mass and centre-of-pressure of the sail. For example, the payload can be mounted at the tip of a deployable boom, erected normal to the sail surface from the central hub. Boom rotations will then displace the solar sail centre-of-mass over a fixed centre-of-pressure. The major difficulty with the square solar sail is packing and deployment. As will be seen in Chapter 3, the square solar sail requires a large number of serial operations for deployment, thus increasing the opportunity for failure.

While the square solar sail is an appealing concept, the cantilevered spars are subject to bending loads and so must be sized accordingly. Even although the load imposed by the sail film when under pressure is small, the spars can comprise a significant mass fraction of the solar sail. An alternative concept is to use

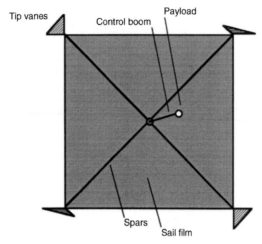

Fig. 1.8. Square solar sail configuration.

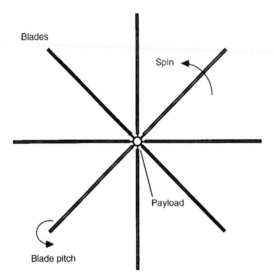

Fig. 1.9. Heliogyro configuration.

spin-induced tension. In this concept the sail film is divided into a number of long, slender blades which are again attached to a central load-bearing hub. The so-called heliogyro slowly spins to maintain a flat, uniform surface, as shown in Fig. 1.9. By rotating the blades of the heliogyro in a cyclic fashion asymmetric forces can be generated across the blade disc, inducing torques which will precess the spin axis of the heliogyro. At first the heliogyro appears more efficient since spin-induced tension is utilised rather than mechanical cantilevered spars. However, the heliogyro blades may require edge stiffeners to transmit radial loads and to provide torsional stiffness

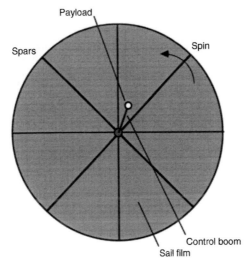

Fig. 1.10. Disc solar sail configuration.

to allow cyclic blade rotations. The principal advantage of the heliogyro is its ease of packing and deployment. The individual blades need only be rolled during manufacture and unrolled from the spinning structure for deployment.

The final concept is the disc solar sail in which a continuous film, or elements of film, are held flat, again using spin-induced tension, as shown in Fig. 1.10. The disc solar sail offers the same potential advantages as the heliogyro in reducing structural mass, but avoids extremely long slender blades. The spin axis of the disc solar sail can be precessed by again inducing torques through displacements of the centre-of-mass. In order to provide some stiffness in the sail disc during precession, radial spars may be required, although the bending loads are small owing to the spin-induced tension. An exterior hoop structure can also be used to provide tension at the edge of the sail film. Packing and deployment are perhaps less problematic than for the square solar sail since flexible radial spars can be wound around the central hub during manufacture. The elastic energy stored in the spars then drives the deployment, unfurling the sail film in the process. This passive deployment scheme is particularly attractive.

1.3.2   Performance metrics

In order to compare solar sail designs a standard performance metric is required. The most common metric is the solar sail characteristic acceleration, defined as the solar radiation pressure acceleration experienced by a solar sail facing the Sun at a distance of one astronomical unit (au), the mean distance of the Earth from the Sun. At this distance from the Sun the magnitude of the solar radiation pressure P is 4.56×10^{-6} N m^{-2}, a value which will be derived in Chapter 2. Therefore, multiplying this pressure by the sail area A yields the solar radiation pressure force exerted on the solar sail. Dividing by the sail mass m then yields the solar sail acceleration. A factor of 2 must also be added to account for the sail reflectivity since reflected photons impart a reaction of equal magnitude to incident photons. However, a finite sail efficiency η must be incorporated to allow for the non-perfect optical properties of the sail coating and billowing of the sail film. From this calculation the solar sail characteristic acceleration a_0 is then defined as

$$a_0 = \frac{2\eta P}{\sigma}, \qquad \sigma = \frac{m}{A} \tag{1.1}$$

where σ is the solar sail mass per unit area, termed the sail loading. While the actual solar sail acceleration is a function of heliocentric distance, and indeed the sail orientation, the characteristic acceleration allows a comparison of solar sail design concepts on an equal footing. Other related performance metrics will be introduced in Chapter 2. If a thin sail film, say $2\,\mu m$ thick, is available a typical solar sail characteristic acceleration may be of order $1\,mm\,s^{-2}$. While this is a useful canonical value, first-generation solar sails are likely to have somewhat lower performance. Later, however, ultra-thin sail films may enable solar sails with a characteristic acceleration of over $6\,mm\,s^{-2}$.

    The size of solar sail required to generate some desired characteristic acceleration can now be obtained from Eq. (1.1). Assuming, for example, that the mass of the payload comprises one-third of the total mass of the solar sail, and adopting a conservative efficiency of 0.85, the required sail area can be easily obtained. For a square solar sail configuration this then leads directly to the sail physical dimensions, as shown in Fig. 1.11. For example, a canonical characteristic acceleration of $1\,mm\,s^{-2}$ requires a sail loading of $7.8\,g\,m^{-2}$, assuming the above efficiency. Then, if the solar sail mass is to be divided equally between the sail film, the structure and the payload, a sail area of $387\,m^2$ per kilogram of payload is required. In order to transport a $25\,kg$ payload, a square solar sail with dimensions of $98 \times 98\,m$ is required. Alternatively, assuming a blade width of $3\,m$, ten heliogyro blades each of length $322\,m$ are required, while a disc solar sail of radius $55\,m$ will also yield the same performance. Now that a physical scale has been assigned to the design concepts discussed in section 1.3.1, the means by which solar sail orbits are generated will be investigated.

1.3.3   Solar sail orbits

There are several ways to consider the physics of solar radiation pressure, as will be discussed in Chapter 2. Perhaps the simplest to visualise is the transfer of momentum to the solar sail by photons, the quantum packets of energy of which light is composed. As a photon intercepts the surface of the sail it will impart its momentum to the sail film, thus applying an impulse to the entire solar sail. Then, when the photon is reflected a reaction impulse will also be exerted on the solar sail. The combination of these two impulses summed across the entire flux of photons incident on the sail film then leads to a force directed normal to the surface of the sail, as shown in Fig. 1.12. The orientation of the solar sail, and so the force vector, is

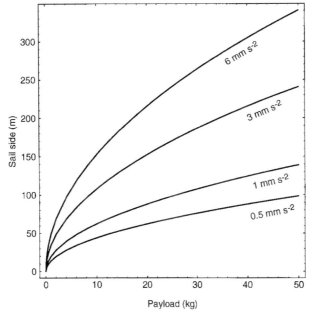

Fig. 1.11. Square solar sail dimensions as a function of payload mass for a payload mass fraction of 1/3.

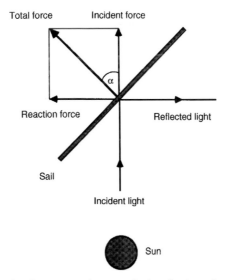

Fig. 1.12. Incidence and reaction forces exerted on a perfectly reflecting solar sail.

described relative to the Sun-line by the sail pitch angle α. Therefore, by altering the orientation of the solar sail relative to the incoming photons, the solar radiation pressure force vector can in principle be directed to any orientation within 90° of the Sun-line. As the pitch angle increases, however, the magnitude of the solar radiation

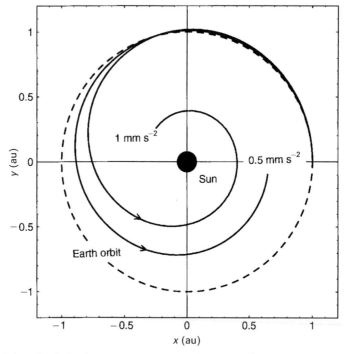

Fig. 1.13. Solar sail spiral trajectories over 300 days with $\alpha = -35°$.

pressure force decreases due to the reduction in projected sail area and the reduction in the component of the solar radiation pressure force directed normal to the sail surface.

Solar sails are continuously accelerated so that their orbits are quite different from the usual ballistic arcs followed by conventional spacecraft. By choosing either a negative or positive sail pitch angle the solar sail will either spiral inwards towards the Sun, or outwards away from the Sun. Such 'tacking' provides solar sails with the agility to enable a wide range of mission applications. The rate at which the solar sail will spiral is of course a function of its characteristic acceleration. Typical spiral trajectories are shown in Figs 1.13 and 1.14 for a fixed sail pitch angle. It can be seen that as the characteristic acceleration increases, the solar sail spirals more rapidly towards or away from the Sun. Furthermore, it will be seen later that if the solar sail is oriented such that a component of the solar radiation pressure force is directed out of the orbit plane, the inclination of the spiral orbit can also be changed. By an appropriate sequence of sail orientations, essentially any point in the solar system can be reached. However, for missions to rendezvous with planets, comets or asteroids, the sail orientation must be continuously altered along its trajectory to ensure that the solar sail matches the velocity of the target body when it is intercepted. Optimisation methods can be used in such instances to determine the best sail orientation time history to minimise the transfer time to the target body.

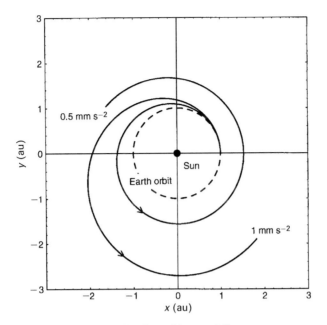

Fig. 1.14. Solar sail spiral trajectories over 900 days with $\alpha = +35°$.

1.3.4   Comparison with other propulsion systems

While characteristic acceleration may be used to evaluate competing solar sail configurations, another metric is required for comparison with other forms of propulsion. For conventional rocketry, specific impulse is used to measure the efficiency of any given propulsion system. This long-standing definition states that the specific impulse of a propulsion system is equal to the momentum gained by the rocket per unit weight of propellant consumed. While this is a useful definition for reaction rockets, it appears unsuitable for solar sails which do not consume propellant. However, an effective specific impulse can be assigned to solar sails using the rocket equation.

For some rocket of initial mass m_1, the final mass m_2 obtained after acceleration through a change in speed Δv using a propulsion system of specific impulse I_{sp} is given by the relationship

$$m_2 = m_1 \exp\left(-\frac{\Delta v}{g I_{sp}}\right) \tag{1.2}$$

where g is the standard sea-level acceleration due to gravity ($9.81\,\mathrm{m\,s^{-2}}$). This is the rocket equation, first derived by the ubiquitous Tsiolkovsky as early as 1897. Now, it can be argued from Eq. (1.2) that solar sails must have infinite specific impulse since their mass does not change. While this is true, for a finite mission duration the solar sail will only deliver a finite total impulse. In particular, if the solar sail is used to deliver a payload to some high-energy orbit, the mass of the solar sail itself becomes redundant after the payload is delivered. Therefore, if it happens that the mass of the solar sail is greater than the mass of a chemical propulsion system required to perform the same mission, the solar sail can hardly be said to be more efficient, even although it has in principle infinite specific impulse. Infinite specific impulse is only achieved in practice for an infinite mission duration.

In order to measure the effective specific impulse delivered by a solar sail the initial mass m_1 will be considered to be the mass of the solar sail and the payload, while m_2 will be the mass of the payload alone. Rearranging Eq. (1.2), the effective specific impulse may now be written as

$$I_{sp} = \frac{\Delta v}{g} \ln\left(\frac{1}{R}\right)^{-1} \tag{1.3}$$

where $R (= m_2/m_1)$ is the payload mass fraction of the solar sail. The effective Δv delivered by the solar sail can also be obtained from the solar sail characteristic acceleration and mission duration. For a mission of duration T, the effective Δv delivered by the solar sail in the vicinity of the Earth's orbit is approximately $\Delta v \sim a_0 T$ so that

$$I_{sp} \sim \frac{a_0 T}{g} \ln\left(\frac{1}{R}\right)^{-1} \tag{1.4}$$

It should be noted that for operation closer to the Sun the actual acceleration experienced by the solar sail is significantly greater than the characteristic acceleration. In addition, the useful acceleration generated by the solar sail is a function of the sail orientation which, in general, will vary during the mission. It can be seen, however, that the effective specific impulse delivered by a solar sail increases linearly with mission duration. Therefore, even a low-performance solar sail can be a competitive form of propulsion if the mission duration is long enough.

This comparison can be seen clearly in Fig. 1.15, which shows the effective specific impulse of a number of solar sails in comparison to other propulsion options, assuming a reasonable payload mass fraction R of 1/3. A higher payload mass fraction will of course yield a greater effective specific impulse, but at the expense of a thinner sail film if the total mass of the solar sail is unchanged. It can be seen that while low-performance solar sails are only efficient for long-duration missions, high-performance solar sails can rapidly out-perform other propulsion options. For example, the NASA NSTAR thruster, which represents the current state-of-the-art in flight ready electric propulsion technology, can deliver a specific impulse between 2200 and 3370 seconds, shown in Fig. 1.15. Chemical propulsion systems typically have a specific impulse above 200 seconds with an upper limit of 450 seconds for cryogenic propulsion using liquid oxygen and liquid hydrogen, again shown in Fig. 1.15. While this comparison highlights some of the benefits of solar sailing and indicates their applicability, it is an extremely simplistic analysis. Any detailed comparison between propulsion options must consider total mission cost, initial launch mass and engineering complexity in addition to single parameter metrics such as effective specific impulse. The initial launch mass of the spacecraft is of particular

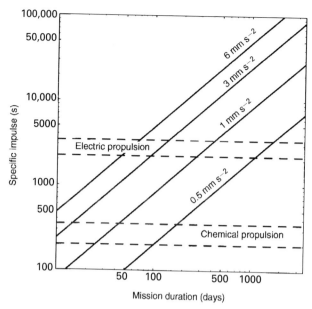

Fig. 1.15. Solar sail effective specific impulse as a function of mission duration at 1 astronomical unit for a payload mass fraction of 1/3.

importance since this drives the selection of the launch vehicle which can be a substantial fraction of the total mission cost. Even the selection of a somewhat lower capacity launch vehicle of the same family can yield significant cost savings.

1.4   SOLAR SAIL MISSION APPLICATIONS

1.4.1   Applicability

Now that the design and operation of various solar sail concepts has been discussed, possible mission applications will be considered. It has been emphasised that solar sails require no propellant and therefore offer considerable advantages, for example delivering payloads to high-energy orbits or for sample return missions. Further-more, since the solar sail has in principle unlimited Δv capability, multiple objective missions such as asteroid surveys are enabled, as are exotic non-Keplerian orbits. All of these potential applications make optimum use of the benefits of solar sails by attempting high-energy and/or long-duration missions. Indeed, the approximate specific impulse analysis of section 1.3.4 clearly highlights the need for long-duration missions for low- or moderate performance solar sails.

In operational terms, a solar sail may be used simply to augment the performance of a launch vehicle upper stage by taking the payload to escape from some initial high Earth orbit, or the solar sail may be used solely for propulsion after Earth escape. If required, escape spirals must begin from altitudes between 600 and 900 km

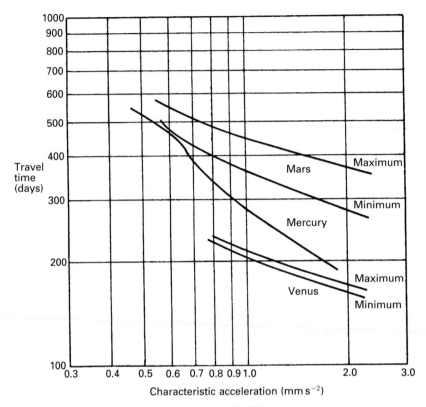

Fig. 1.16. Transfer times in the inner solar system (NASA/JPL).

owing to residual atmospheric drag acting on the sail. In any case, escape times may be of the order of months so that it is desirable to deliver the solar sail to an Earth escape trajectory by conventional means. Unlike ballistic trajectories solar sails have an essentially open launch window so that trajectories may be modified continuously to meet any required launch date. For planetary rendezvous, short spiral and capture times are possible in the inner solar system, as shown in Fig. 1.16. However, for payload delivery to the outer planets capture times are prohibitively long due to the greatly diminished solar radiation pressure. Payloads must then be inserted into planetary orbit using space storable propellants, or aerocapture if appropriate.

While solar sailing can be utilised as a form of high specific impulse propulsion for the delivery of payloads, other mission concepts centre on the continuously available solar radiation pressure force to generate families of non-Keplerian orbits. These are orbits which do not follow the usual rules of Keplerian orbital motion, as obeyed by other spacecraft and solar system bodies. While all solar sail orbits are essentially non-Keplerian owing to the perturbation exerted by solar radiation pressure, some orbits are so strongly perturbed as to merit classification in a new family. These orbits are possible if the solar radiation pressure force is of the same order of magnitude as the local gravitational force. While a high-performance solar sail may be required for some orbits, other orbits only require a moderate performance solar sail. For example, near the Lagrange balance points in the Sun–Earth system the local gravitational acceleration is so small that a moderate performance solar sail can be used to generate new orbits with quite unique mission applications.

1.4.2 Inner solar system missions

Owing to the enhanced solar radiation pressure in the inner solar system, solar sails can easily deliver payloads to close, high-inclination orbits about the Sun. Payloads can be delivered by firstly spiralling inwards to a close, circular heliocentric orbit, the radius of which is limited by the thermal tolerance of the sail film (typically between 0.2 and 0.3 au for a conventional sail). Then, the orbit inclination of the solar sail is cranked by turning the sail and alternately directing a component of the solar radiation pressure force above and below the ecliptic plane every half orbit. For example, a solar sail with a characteristic acceleration of $1\,\mathrm{mm\,s^{-2}}$ can deliver a payload to a solar polar orbit at 0.5 au with a mission duration of the order of 2.5 years starting from an Earth escape trajectory. Further manoeuvres to increase the eccentricity of the polar orbit can then allow the delivery of payloads to orbits which graze the solar atmosphere at a perihelion below 0.02 au. This is the Solar Probe mission which can only be achieved with chemical propulsion using a gravity assist at Jupiter and a high-energy launch vehicle. Since the solar sail starts from a low-energy Earth escape trajectory, a much smaller and hence lower cost launch vehicle can be used.

Other inner solar system missions, such the delivery of payloads to Mercury, offer quite spectacular opportunities. A ballistic transfer to Mercury requires a Δv of the order of $13\,\mathrm{km\,s^{-1}}$, although this can be reduced using gravity assists at the expense of increased mission duration. The same mission can be performed using only a relatively modest solar sail. A solar sail with a characteristic acceleration of $0.25\,\mathrm{mm\,s^{-2}}$ will deliver a payload to Mercury in 3.5 years, while a solar sail with double the performance will require only 1.5 years. The time for capture and spiral into a useful mission orbit at Mercury is almost negligible since the solar sail acceleration at Mercury can be over ten times the characteristic acceleration at 1 au. These inner solar system missions then make optimum use of solar sailing by utilising the increased solar radiation pressure available to enable extremely high-energy missions.

For payload delivery to Mars the outward spiral times tend to be somewhat longer than for ballistic transfers. However, solar sailing is not constrained by the synodic waiting period between ballistic launch opportunities. Again, for a characteristic acceleration of $1\,\mathrm{mm\,s^{-2}}$ the cruise time to Mars is of the order of 400 days, with an additional 100 days required for capture and inward spiral to a low planetary orbit. While solar sails can deliver a larger payload mass fraction than chemical propulsion for a given launch mass, one-way Mars missions do not make optimum use of solar sailing since the required Δv is relatively modest, particularly in comparison to solar polar-type missions.

Fig. 1.17. Solar sail payload delivery in support of the human exploration of Mars (NASA/JPL).

Although one-way Mars missions do not appear attractive, two-way sample return missions do provide opportunities. For a ballistic mission, the mass delivered to Mars must include propellant for the return leg of the trip. For a solar sail mission, however, propellant is only required for a lander to descend to and ascend from the Martian surface. Therefore, for the same launch mass solar sailing can return a greater mass of samples, perhaps from different geological regions by using a surface rover. Alternatively, for the same mass of returned samples, solar sailing can allow the use of a smaller, lower cost launch vehicle. The Δv for the sample return mission can be approximately double that for a one-way mission so that sample returns are clearly high-energy missions, thus making optimum use of solar sailing.

Other inner solar system missions for the future include the use of large solar sails to reduce the total mass in low Earth orbit (and so total cost) required for the human exploration of Mars. A solar sail with a large payload mass fraction and low characteristic acceleration can deliver logistics supplies to Mars which are not time critical, as illustrated in Fig. 1.17. A large payload mass fraction is important for such missions where the effective Δv is quite modest. This then allows the solar sail to be used efficiently with a high effective specific impulse. For example, a 2×2 km square solar sail of mass 19,200 kg can deliver a 32,000 kg payload to Mars orbit in 4.2 years from an initial Earth parking orbit. The solar sail can then return to Earth in 2 years to be loaded with another payload for delivery to Mars. This appears to be the largest solar sail which could reasonably be delivered to Earth orbit in a single section using the Space Shuttle, or a large expendable launch vehicle such as Titan IV.

1.4.3   Outer solar system missions

Owing to the diminished solar radiation pressure in the outer solar system, insertion of payloads into planetary orbit must be achieved using a conventional chemical propulsion system, or aerobraking if appropriate. Payloads can be delivered to Jupiter and Saturn with minimum transfer times of 2.0 and 3.3 years respectively using a solar sail with a characteristic acceleration of $1 \, \mathrm{mm \, s^{-2}}$. After launch to an Earth escape trajectory, the solar sail makes a loop through the inner solar system to accelerate onto a quasi-ballistic arc. Owing to this transfer mode, payloads will arrive with large hyperbolic excess velocities. A solar sail of similar performance can also be used to deliver payloads to Pluto in approximately 10 years. Although solar sailing provides significant performance for planetary outer solar systems missions, the need to transport retro-propulsion capability, or enclose the payload in an aeroshell, perhaps detracts from their effectiveness. It must be remembered, however, that such retro-propulsion capability is also required for conventional missions using chemical propulsion in combination with multiple gravity assists.

Again, owing to the potentially unlimited Δv capability of solar sails, multiple small body rendezvous missions are possible, as are small body sample returns. Small bodies encompass the range of minor solar system objects from asteroids to comets. For example, a sample return from comet Encke can be achieved with a mission duration of order 5 years, again using a solar sail with a characteristic acceleration of $1 \, \mathrm{mm \, s^{-2}}$. An Encke mission is particularly difficult since the comet has a high eccentricity orbit requiring significant energy for both the rendezvous and the return phase. Many comets also have high inclinations which make them even more difficult targets for other forms of propulsion. Indeed, comet Halley has a retrograde orbit which led to the solar sail development effort at Jet Propulsion Laboratory during the 1970s. Interesting orbits are possible in the vicinity of such small bodies since the local gravitational acceleration can be of the same order as the solar radiation pressure acceleration. Also of interest is the possibility of a survey of multiple asteroids. This is a particularly attractive and cost-effective concept since the mission is essentially open ended, allowing repeated science returns using the same suite of instruments. A solar sail with autonomous on-board navigation and decision-making software offers exciting possibilities for such missions.

For fast solar system escape missions solar sailing offers significant performance gains. For these missions, to the heliosphere at 100 au or the focus of the solar gravitational lens at 550 au, retro-propulsion is not required so that inner solar system looping trajectories can be used to accelerate payloads to extremely high cruise speeds. Indeed, due to the rapidly diminishing solar radiation pressure the solar sail can be jettisoned after several astronomical units. Using a high-performance solar sail with a characteristic acceleration of $6 \, \mathrm{mm \, s^{-2}}$ and a close solar pass at

0.2 au, cruise speeds of over 20 au per year can be achieved. This compares well with the 3 au per year cruise speed of the Voyager spacecraft, accelerated using gravity assists. While solar sails can generate extremely high cruise speeds, nuclear electric propulsion offers competition. In particular, the nuclear fission or radio-isotope power source used for propulsion can also be used to provide electrical power to the payload. The solar sail must deliver some power source in addition to the payload itself, since photo-voltaic solar cells are of little use at such vast distances.

1.4.4 Non-Keplerian orbits

Owing to the continuously available solar radiation pressure force, solar sails are capable of exotic non-Keplerian orbits that are impossible for any other type of spacecraft. Although some of these missions require advanced, high-performance solar sails others are possible using relatively modest solar sail designs. The solar sail performance required for these orbits is a function of the local gravitational acceleration; therefore, the displacement of orbits high above the ecliptic plane, for example, requires a characteristic acceleration of order $6 \, \text{mm} \, \text{s}^{-2}$, while the generation of an artificial Lagrange point may only require a characteristic acceleration of order $0.25 \, \text{mm} \, \text{s}^{-2}$. While these orbits are not forbidden for other forms of low-thrust propulsion, they can only be achieved for a limited duration, which is fixed by the propellant mass fraction of the vehicle.

Firstly, using an advanced solar sail it would be possible to choose its total mass per unit area so that the solar radiation pressure force exactly balances the solar gravitational force. This is possible since both of these forces have an inverse square variation with heliocentric distance (see Chapter 2, however). The required characteristic acceleration for such a balance point is approximately $6 \, \text{mm} \, \text{s}^{-2}$, corresponding to a solar sail loading of only $1.5 \, \text{g} \, \text{m}^{-2}$. Such a solar sail would enable solar physics missions which levitate above the solar poles, providing continuous observations, or indeed hovering at any particular location in the solar system. Such a solar sail could also be used to displace circular Sun-centred orbits high above the ecliptic plane, with the orbit period chosen to be synchronous with the Earth or some other solar system body, as shown in Fig. 1.18. When solar sails are available with a characteristic acceleration of order $6 \, \text{mm} \, \text{s}^{-2}$, large families of exotic new orbits will be available for mission applications.

For high-performance solar sails operating in the vicinity of the Earth, it has been demonstrated that they may be used to displace communication satellites above and below the geostationary orbit plane. This concept, proposed by physicist Robert Forward, would allow satellites to be stacked above and below the equatorial plane, greatly increasing the number of available slots. Using more modest solar sails the location of the Earth–Sun Lagrange balance points can be artificially displaced. The Lagrange points are locations where a spacecraft will remain in equilibrium with respect to the Earth and the Sun. For example, the interior L_1 point 1.5 million km Sunward of the Earth is a favoured location for solar physics missions. Since the solar sail adds an extra force to the dynamics of the orbit, the location of the L_1 point can be artificially displaced closer to the Sun or even above the ecliptic plane.

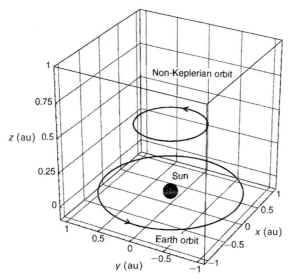

Fig. 1.18. Sun-centred non-Keplerian orbit.

Since the local gravitational acceleration in the vicinity of L_1 is small, only modest solar sails are required. For example, a solar sail with a characteristic acceleration of only $0.25 \, \mathrm{mm \, s^{-2}}$ can double the distance of the L_1 point from Earth. As will be seen, this new equilibrium location appears useful for providing early warning of solar plasma storms before they reach Earth. A solar sail with double the performance can be permanently stationed high above the L_1 point so that it appears above the Arctic regions of the Earth. Even allowing for the long path length involved, such an equilibrium location would enable continuous communications with high-latitude regions or continuous, real-time polar imaging. This is in stark comparison to conventional geostationary satellites which appear low or below the horizon in high-latitude and polar regions. The 'Statite' concept for high-latitude communications, also patented by Robert Forward, exploits these advantages.

1.5   FUTURE DEVELOPMENT

1.5.1   Near term

While solar sailing offers significant advantages for solar system exploration, it is not a general-purpose propulsion system which can meet the requirements of every conceivable mission. Neither is it a technology worth developing for its own sake. While the notion of sailing through the solar system using nothing more than light pressure is a truly wonderful thought, solar sailing must be pulled forward by mission applications at the same time as it is pushed by technology development. Solar sailing proponents must therefore be practical, hard-headed romantics. This rule also holds true for initial flight tests of solar sailing. Unless such flight tests provide confidence in the technology and a clear path towards some enabling capability, they will not perform a useful function. On this note it is perhaps worth considering the use of low-cost sounding rockets to test sail deployment mechanisms during the short period of free-fall provided. Such development campaigns would allow for several tests, perhaps of scaled prototypes, for the same cost as a single launch to orbit. By spreading the risk over several tests the inevitable unforeseen serial point failures of deployment can be identified. In comparison, the consequences of a deployment failure on a single orbital test may be profound. If the first true solar sail fails to deploy it may postpone solar sailing for another twenty years.

Since the JPL studies of the mid-1970s technology has matured considerably. The opportunities provided by micro-spacecraft technologies have already been discussed. However, there are other developments which have yet to be exploited. In particular there is much work to be done to devise new ways of fabricating ultra-thin deployable sail films. A conventional solar sail film requires a plastic substrate to allow handling and folding, typically between 1 and 7 μm thick. This substrate is then coated with a thin layer of aluminium, typically 0.1 μm thick, to provide a highly reflective surface. However, for a well-designed solar sail structure the stresses imposed on the sail film are so low that the plastic substrate is somewhat redundant after deployment. Since the substrate comprises most of the mass of the sail film, removal after deployment would enable truly high-performance solar sails. While concepts exist, such as plastic substrates which will sublimate under the action of solar UV radiation, little experimental work has been undertaken. Recently, however, experiments by Salvatore Scaglione and colleagues in Italy have demonstrated that diamond-like carbon (DLC), a metastable form of carbon, can be used as a UV-sensitive buffer between the substrate and the reflector. This scheme will in principle allow the substrate to detach shortly after deployment. Removable substrates are not the only means of improving solar sail performance. If the sail film is perforated with holes smaller than the mean wavelength of Sunlight, the mass of the sail film is greatly reduced, but without significantly degrading its reflectivity. This is analogous to wire mesh reflectors used in radio-frequency antennae. Again, the technology to fabricate such films is at hand since the semi-conductor industry uses similar techniques on a daily basis. A deployable, all-metal film with perforations would enable quite remarkable solar sails to be manufactured.

The previous section has indicated some missions where solar sailing is used to its optimum advantage. High-energy and/or long-duration missions are the key to solar sailing where it can be used most efficiently. For this reason the possibilities for the long-term development of solar sailing relies on the investigation of those niche missions and applications which are unique to solar sails. There are too many long-term vested interests to expect solar sailing to be developed for non-optimum missions where existing technologies meet the challenge – the exception being cases where solar sails may lead to significant cost reductions, for example, by allowing a smaller, lower cost launch vehicle to be used. In the near term the main application of interest appears to be artificial Lagrange points displaced sunward of

L_1 for solar storm warning. This is the Geostorm mission, to be described in detail in Chapter 6, which is some way towards graduating from a study to a funded mission.

Geostorm is an ideal first mission for solar sailing as it brings together several key elements into a single compelling concept. Firstly, only a modest solar sail is required with a characteristic acceleration of order $0.25 \, \text{mm} \, \text{s}^{-2}$. Then, the mission application has a powerful group of users in the US military and Department of Energy. Advanced warning of solar storms will be used to protect satellite communication links and electricity distribution networks. It is important to note that these users are not interested in solar sailing *per se*, but only in the unique mission application which it enables. There are lessons here for future missions. Lastly, the user group are insured against the use of an untested technology. If the solar sail fails to deploy, the mission can continue at the natural L_1 point sunward of the Earth. Because of this insurance Geostorm can in fact be used as a fast track to a first solar sail mission. If deployment fails, then the mission is only degraded, not lost.

Other than artificial Lagrange points, the near-term applications which appear particularly attractive are the solar polar and Mercury orbiter missions. Both of these missions are confined to the inner solar system where the solar sail will experience enhanced acceleration. As such, the effective Δv and specific impulse delivered by the solar sail are quite immense. Either of these would be a spectacular demonstration of the mission enabling properties of solar sailing. While a limited Mercury orbiter mission is certainly possible in the near term using chemical propulsion with gravity assists, a close solar polar orbit is not. Again, the competition is from the more mature solar-electric propulsion technology, but the electric propulsion mission duration and launch mass appear quite excessive in comparison to solar sailing.

1.5.2 Autonomous explorers

As repeated many times in this chapter, solar sails offer the potential of unlimited Δv. The Δv capacity of solar sailing is limited only by the lifetime of the sail film in the space environment. While this is a parameter which must be characterised by real flight experience, there is no reason to expect sail films to be short lived. Combining this Δv capacity with current developments in autonomous spacecraft systems quickly leads to the prospect of future solar sails as autonomous solar system explorers. In fact these factors were inherent in the solar sail interplanetary shuttle concept devised following the JPL comet Halley studies of the mid-1970s. By using a square solar sail inherited from the comet Halley vehicle along with autonomous onboard systems, it was proposed that multiple payloads could be transported to bodies within the solar system, as illustrated in Fig. 1.19. After payload delivery, the solar sail would then return to Earth orbit for refurbishment and a new set of payloads. While this refurbishing concept may not fit into current thinking for low-cost robotic missions, the possibility of multiple objective missions certainly does.

A small solar sail with a suite of miniaturised sensors and self-directing autonomous software would be a truly powerful combination. The solar sail would provide the unlimited agility required for open-ended missions. As such the

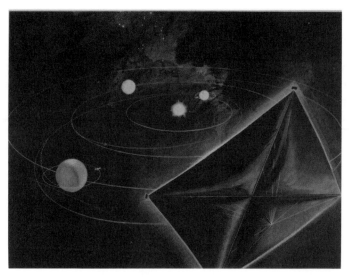

Fig. 1.19. Interplanetary shuttle concept (NASA/JPL).

spacecraft would be a true autonomous explorer. Interacting with the environment through its sensors and actuating itself with the solar sail, the entire integrated system would mimic terrestrial sensory-based mobile robots, but in a much larger arena. Launched on an Earth escape trajectory with only high-level instructions, the solar sail would plan its own actions based on findings obtained from earlier phases of its mission. Indeed, constellations of such autonomous explorers may slowly diffuse through the solar system, returning information which they determine to be of interest to their creators on Earth. Such schemes may in the future allow human awareness to expand to every corner of the solar system. However, it will be a virtual exploration of our neighbourhood, conducted by proxy through machines of our own making.

1.5.3   Speculation

Crystal ball gazing is always a danger as, with hindsight, any predictions appear hopelessly optimistic or pessimistic (witness the early **IBM** chairman who predicted a world market for 'about five computers'). On a pessimistic note, it may be that solar sailing is a concept whose time has already come and gone. Future developments in reaction propulsion technology may lead solar sails the way of airships in aviation history. That said, modern high-technology airships seem set to make a return for long-haul cargo transportation in the twenty-first century. Perhaps solar sails will follow the same course and reappear as the gigaton freighters of future interplanetary commerce.

Taking the more preferable optimistic view there appears to be a bright future for solar sailing. As spacecraft shrink in both mass and volume, solar sailing offers the ideal propulsion mode for autonomous self-directing agents to explore the solar system on our behalf. Nanotechnology may allow the manufacture of immense self-supporting, self-repairing solar sail films with a vanishingly small loading, formed from linked chains of carbon atoms. Molecular level computing may be distributed throughout the film so that the sail becomes both propulsion and payload. These ultra-high-performance solar sails could: allow the delivery of interstellar probes, perhaps using directed laser energy; open up rapid transportation within the solar system at negligible cost; and even save the Earth by deflecting a killer asteroid! Indeed, perhaps this is how we will travel to the stars, with our senses encoded in monomolecular films of carbon and silicon, drifting ghost-like through interstellar space. While these thoughts are pure speculation, it is clear that we have quite a privileged place in the history of space exploration. A true solar sail has yet to fly, so the opportunity is at hand to launch the technology which may lead to the fruition of such ideas in the future. Whether we seize this opportunity which is now at hand remains to be seen.

1.6 FURTHER READING

Historical interest

Maxwell, J. C., *Electricity and Magnetism*, Oxford University Press, 1873.

Lebedew, P., 'The Physical Causes of Deviation from Newton's Law of Gravitation', *Astrophysical Journal*, **10**, 155–161, 1902.

Tsiolkovsky, K.E., *Extension of Man into Outer Space*, 1921 [also, Tsiolkovsky, K.E., *Symposium Jet Propulsion*, No. 2, United Scientific and Technical Presses, 1936].

Tsander, K., *From a Scientific Heritage*, NASA Technical Translation *TTF-541*, 1967 [quoting a 1924 report by the author].

Wiley, C. [pseudonym Sanders, R.], 'Clipper Ships of Space', *Astounding Science Fiction*, p. 135, May 1951.

Garwin, R.L., 'Solar Sailing – A Practical Method of Propulsion Within the Solar System', *Jet Propulsion*, **28**, 188–190, 1958.

Clarke, A.C., 'The Wind from the Sun' (first published 1963), see, for example, *Project Solar Sail*, ed. Clarke, A.C., Penguin Group Books, New York, 1990.

Friedman, L., 'Solar Sailing: The Concept Made Realistic', AIAA-78-82, *16th AIAA Aerospace Sciences Meeting*, Huntsville, January 1978.

Drexler, K.E., 'High Performance Solar Sails and Related Reflecting Devices', AIAA-79-1418, *4th Princeton/AIAA Conference on Space Manufacturing Facilities*, Princeton, May 1979.

Logsdon, J.M., 'Missing Halley's Comet: The Politics of Big Science', *ISIS*, **80**, 302, 254–280, 1989.

Perret, A., LaBombard, E. & Miura, K., 'The Solar Sail Race to the Moon', IAF-89-539, *40th International Astronautical Federation Congress*, Malaga, October 1989.

Selected introductory papers

Tsu, T.C., 'Interplanetary Travel by Solar Sail', *American Rocket Society Journal*, **29**, 422–427, 1959.

London, H.S., 'Some Exact Solutions of the Equations of Motion of a Solar Sail With a Constant Setting', *American Rocket Society Journal*, **30**, 198–200, 1960.

Sands, N., 'Escape from Planetary Gravitational Fields by use of Solar Sails', *American Rocket Society Journal*, **31**, 527–531, 1961.

Fimple, W.R., 'Generalised Three-Dimensional Trajectory Analysis of Planetary Escape by Solar Sail', *American Rocket Society Journal*, **32**, 883–887, 1962.

Wright, J.L. & Warmke, J.M., 'Solar Sail Mission Applications' AIAA-76-808, *AIAA/AAS Astrodynamics Conference*, San Diego, August 1976.

Van der Ha, J.C. and Modi, V.J., 'Long-term Evaluation of Three-Dimensional Heliocentric Solar Sail Trajectories with Arbitrary Fixed Sail Setting', *Celestial Mechanics*, **19**, 113–138, 1979.

Svitek, T. *et al.*, 'Solar Sail Concept Study', IAF-ST-82-12, *33rd International Astronautical Congress*, Paris, October 1982.

Staehle, R.L., 'An Expedition to Mars Employing Shuttle-Era Systems, Solar Sail and Aerocapture', *Journal of the British Interplanetary Society*, **35**, 327–335, 1982.

Forward, R.L., 'Light-Levitated Geostationary Cylindrical Orbits: Correction and Expansion', *Journal of Astronautical Sciences*, **38**, 3, 335–353, 1990.

McInnes, C.R., 'Solar Sail Halo Trajectories: Dynamics and Applications', IAF-91-334, *42nd International Astronautical Congress*, Montreal, October 1991.

Forward, R.L., 'Statite: A Spacecraft That Does Not Orbit', *Journal of Spacecraft and Rockets*, **28**, 5, 606-611, 1991.

McInnes, C.R., McDonald, A.J.C., Simmons, J.F.L. & MacDonald, E.W., 'Solar Sail Parking in Restricted Three-Body Systems', *Journal of Guidance Dynamics and Control*, **17**, 2, 399–406, 1994.

Maccone, C., 'Space Missions Outside the Solar System to Exploit the Gravitational Lens of the Sun', *Journal of the British Interplanetary Society*, **47**, 45–52, 1994.

Leipold, M., Borg, E., Lingner, S., Pabsch, A., Sachs, R. and Seboldt, W., 'Mercury Orbiter with a Solar Sail Spacecraft', *Acta Astronautica*, **35**, Suppl. 635–644, 1995.

Prado, J.Y., Perret, A., Pignolet, G. & Dandouras, I., 'Using a Solar Sail for a Plasma Storm Early Warning System', IAA-96–IAA.3.3.06, *47th International Astronautical Congress*, October 1996.

Leipold, M., 'ODISSEE – A Proposal for Demonstration of a Solar Sail in Earth Orbit', IAA-L98-1005, *3rd International Academy of Astronautics Conference on Low-Cost Planetary Missions*, Pasadena, April 1998.

McInnes, C.R., 'Mission Applications for High Performance Solar Sails', IAA-L.98-1006, *3rd IAA International Conference on Low-Cost Planetary Missions*, Pasadena, April 1998.

Gershman, R. & Seybold, C., 'Propulsion Trades for Space Science Missions', IAA-L.98-1001, *3rd IAA International Conference on Low-Cost Planetary Missions*, Pasadena, April 1998.

Carroll, K.A., 'Economical Planetary Space Science Using Small Solar Sail Propelled Spacecraft', *Proceedings of the 10th Conference on Astronautics*, CASI (Canadian Aeronautics and Space Institute), Ottawa, October 1998.

Solar sailing books

Polyakhova, E., *Space Flight Using a Solar Sail – The Problems and the Prospects*, Kosmicheskiy Polet Solnechnym Parusom, Moscow, 1986 [in Russian].
Friedman, L., *Star Sailing: Solar Sails and Interstellar Travel*, Wiley Science Publications, New York, 1988.
Clark, A.C. [ed.], *Project Solar Sail*, Penguin Group Books, New York, 1990.
Wright, J.L., *Space Sailing*, Gordon & Breach Science Publishers, Philadelphia, 1992.
Souza, D.M., *Space Sailing*, Lerner Publications Company, Minneapolis, 1994.

Solar sail internet sites

http://www.ugcs.caltech.edu/~diedrich/solarsails/
http://www.ec-lille.fr/~u3p/index.html
http://www.kp.dlr.de/solarsail/

2

Solar radiation pressure

2.1 INTRODUCTION

The source of motive force for solar sail spacecraft is the momentum transported to the sail by radiative energy from the Sun. While the observation that light can push matter is quite contrary to everyday experience, it is a commonplace mechanism in the solar system. Perhaps the most striking example is the elegant beauty of comet tails. Comets have in fact two distinct tails: an ion tail swept out by the solar wind and a dust tail swept out by solar radiation pressure. As will be seen, however, light pressure is by far the dominant effect on solar sails.

Interplanetary dust is also effected by solar radiation pressure. The Poynting–Robertson effect is a mechanism whereby dust grains experience a transverse drag as well as radial light pressure. This is due to the relativistic aberration of light as the dust grains orbit the Sun. The resulting drag then causes dust to spiral very slowly inwards towards the Sun. More remarkably, under certain conditions when the grains spiral close to the Sun, they begin to evaporate, thus reducing the ratio of their mass to cross-section. The effect of solar radiation pressure then greatly increases, sometimes to the extent that light pressure can exceed solar gravity, thereby ejecting dust into interstellar space. How appropriate, then, that the same mechanism which sweeps out comet tails and accelerates interplanetary material to the stars can also be harnessed to propel spacecraft through the solar system.

In this chapter the physics of solar radiation pressure will be explored by considering the two physical descriptions of the momentum transfer process. Firstly, using the quantum description of radiation as packets of energy, photons can be visualised as travelling radially outwards from the Sun and scattering off the sail thus imparting momentum. Secondly, using the electromagnetic description of radiation, momentum is transported from the Sun through the vacuum of space to the sail via electromagnetic waves. Both of these modern descriptions are, however, predated by many other theories of the mechanism by which light can push matter, as will be discussed.

Following the discussion of the physics of radiation pressure, the standard inverse square law for solar radiation pressure will be derived by calculating the momentum flux across an oriented solar sail. The solar sail performance is then parameterised by three metrics: the solar sail loading, characteristic acceleration and lightness number. Radiative transfer methods are then briefly introduced as a rigorous means of calculating the properties of arbitrary radiation fields. The methods presented are used to derive an expression for solar radiation pressure taking account of the finite angular size of the solar disc. It is shown that this exact form of the solar radiation pressure law is not in fact inverse square, but has a more complex functional form. The consequences of this deviation from an inverse square law are explored in Chapter 5. A realistic solar sail force model is also presented which incorporates the non-perfect optical properties of the sail film. A comparison is then made between the realistic force model and an ideal solar sail. Lastly, other forms of momentum transport, such as the solar wind, are also briefly considered.

2.2 HISTORICAL VIEW OF SOLAR RADIATION PRESSURE

In 1619 Johannes Kepler proposed that comet tails were pushed outwards from the Sun due to pressure from sunlight. At this time the corpuscular theory of light was the favoured view of optics and the outward pressure due to sunlight was a natural consequence of this theory. With hindsight it is somewhat ironic that this early explanation is qualitatively the same as the current view of solar radiation pressure, namely momentum transport by photons. Newton, a strong proponent of the corpuscular theory, accepted Kepler's view as a possible explanation. However, in 1687 he attempted to explain the phenomenon solely within his theory of universal gravitation. He advanced the hypothesis that there was an ambient ether denser than the material of comet tails; therefore, the observed repulsion was merely due to buoyancy forces and the Sun had only an attractive, gravitational force. Some time later, in 1744, Euler returned to Kepler's original view. However, Euler adopted the longitudinal wave theory of light due to Huygens. With this theory Euler was able to show that a longitudinal wave would exert a repulsive force on a body in its path. Aside from these theoretical developments, the first experimental attempts to verify and measure radiation pressure were made by de Marian and du Fay in 1754. However, owing to the effect of residual air currents in their apparatus, these early, but ambitious, investigations proved inconclusive. It was not until the beginning of the twentieth century, with the refined experimental apparatus of the day, that radiation pressure was finally characterised in the laboratory.

In 1785, some time after de Marian and du Fay, Coulomb performed new experiments with electrostatics. Knowing of these experiments, in 1812 Olbers rejected all previous explanations of comet tail repulsion and proposed that the Sun had a net electrical charge. Particles leaving a comet nucleus then had to be charged with the same sign as that of the Sun. The fact that electrostatic forces have an inverse square variation (as does solar radiation pressure) supported the theory of Olbers which then became prevalent. Although initially successful, his theory was eventually seen as flawed when no possible mechanism of charging could be either proposed or observed.

Ultimately, the correct theoretical basis for the existence of radiation pressure came independently of the astronomical theories. The Scottish physicist James Clerk Maxwell predicted the existence of radiation pressure in 1873 as a consequence of his unified theory of electromagnetic radiation. Apparently independent of Maxwell, Bartoli, in 1876, demonstrated the existence of radiation pressure as a consequence of the second law of thermodynamics. Also in 1873, Crookes mistakenly believed that he had demonstrated the existence of radiation pressure using his newly devised radiometer. Even today this device is occasionally used as a flawed demonstration of radiation pressure. The forces exerted on the radiometer, which are in fact due to molecular scattering, are orders of magnitude greater than radiation pressure forces. The true experimental verification of the existence of radiation pressure and the verification of Maxwell's quantitative results came at last in 1900. At the University of Moscow, Peter Lebedew finally succeeded in isolating radiation pressure using a series of elegant torsion balance experiments. Independent verification was also obtained in 1901 by Nichols and Hull at Dartmouth College, New Hampshire.

2.3   THE PHYSICS OF RADIATION PRESSURE

2.3.1   Quantum description

Using quantum mechanics, radiation pressure can be envisaged as being due to momentum transported by photons, the quantum packets of energy of which light is composed. The concept of photons has its origins in the investigation of thermal radiation by Planck and others at the beginning of the twentieth century. While attempting to provide a theoretical explanation for the distribution of energy as a function of wavelength in a black body thermal cavity, Planck was led to the conclusion that the energy of the cavity must be quantised at discrete levels. While the energy levels were to be discrete, Planck still considered that radiation propagated as a wave after emission. In 1905 Einstein, unhappy with Planck's view, proposed that the radiative energy was itself emitted in discrete packets, preserving its identity as a photon as it propagated. It was clear that Maxwell's wave theory of light could describe optical phenomena such as diffraction and inter- ference; however, such phenomena are time averaged so that the wave theory does not describe events of emission and absorption. Einstein proposed that not only were the energy levels in sources of thermal radiation discrete, but also that the energy was emitted and absorbed in discrete packets. This view at once explained the puzzling observations of the photoelectric effect. Einstein's Nobel prize in 1921 was noted for the explanation of the photoelectric effect and 'contributions to mathematical physics'.

From Planck's law, a photon of frequency ν will transport energy E given by the famous equation

$$E = h\nu \tag{2.1}$$

where h is Planck's constant. Although this equation is now termed Planck's law, Einstein derived an equivalent equation from statistical thermodynamics. Only later did he realise that his constant of proportionality between energy and frequency was in fact the Planck constant used to describe the discrete energy levels of sources of thermal radiation. The term 'photon' was later coined by G.N. Lewis in 1925.

In addition to this quantum view, the mass–energy equivalence of special relativity allows the total energy E of a moving body to be written as

$$E^2 = m_0^2 c^4 + p^2 c^2 \tag{2.2}$$

where m_0 is the rest mass of the body, p is its momentum and c is the speed of light. The first term of the equation represents the rest energy of the body while the second term represents the energy due to its motion. Since a photon has zero rest mass, its energy may be written as

$$E = pc \tag{2.3}$$

Therefore, using the photon energy defined by Eqs (2.1) and (2.3), the momentum transported by a single photon is given by

$$p = \frac{h\nu}{c} \tag{2.4}$$

It should be noted that this equation has been derived by combining quantum mechanics with special relativity, a theory of classical physics.

In order to calculate the pressure exerted on a body, the momentum transported by a flux of photons must now be considered. The energy flux W (the energy crossing unit area in unit time) at a distance r from the Sun may be written in terms of the solar luminosity L_S and scaled by the Sun–Earth distance R_E as

$$W = W_E \left(\frac{R_E}{r} \right)^2 \tag{2.5a}$$

$$W_E = \frac{L_S}{4\pi R_E^2} \tag{2.5b}$$

where W_E is the energy flux measured at the Earth's distance from the Sun. As noted in Chapter 1, the Sun–Earth distance is also termed the astronomical unit (au). Using Eqs (2.5) the energy ΔE transported across a surface of area A normal to the incident radiation in time Δt is given by

$$\Delta E = WA\,\Delta t \tag{2.6}$$

From Eq. (2.3) this energy then transports a momentum Δp, given by

$$\Delta p = \frac{\Delta E}{c} \tag{2.7}$$

The pressure P exerted on the surface is then defined as the momentum transported per unit time, per unit area, so that

$$P = \frac{1}{A}\left(\frac{\Delta p}{\Delta t}\right)$$

(2.8)

Therefore, using Eq. (2.6), the pressure exerted on the surface due to momentum transport by photons is given by

$$P = \frac{W}{c}$$

(2.9)

For a perfectly reflecting surface the observed pressure is twice the value provided by Eq. (2.9) due to the momentum transferred to the surface by the incident photons and the reaction provided by reflected photons.

Using Eq. (2.9) the solar radiation pressure exerted on a solar sail at the Earth's distance from the Sun (1 au) may now be calculated. Since the orbit of the Earth about the Sun is slightly elliptical, the energy flux received at the Earth varies by approximately 3.5% during the year. However, an accepted mean value is the solar constant W_E of $1368\,\mathrm{J\,s^{-1}m^{-2}}$. Therefore, the pressure exerted on a perfectly reflecting solar sail at 1 au is taken to be $9.12 \times 10^{-6}\,\mathrm{N\,m^{-2}}$.

2.3.2   Electromagnetic description

Using the electromagnetic description of light, momentum is transported to the solar sail by electromagnetic waves. Physically, the electric field component of the wave $\mathbf{E}$ induces a current $\mathbf{j}$ in the sail, as shown in Fig. 2.1. The magnetic component of the incident wave $\mathbf{B}$ then generates a Lorentz force $\mathbf{j} \times \mathbf{B}$ in the direction of propagation of the wave. The induced current generates another electromagnetic wave which is observed as the reflection of the incident wave. For a wave propagating along the x axis the force exerted on a current element is then given by

$$\mathrm{d}f = j_z B_y\, \mathrm{d}x\, \mathrm{d}y\, \mathrm{d}z$$

(2.10)

where j_z is the current density induced in the surface of the reflector. The resulting pressure on the current element, defined as the force per unit area, can then be written as

$$\mathrm{d}P = j_z B_y\, \mathrm{d}x$$

(2.11)

From Maxwell's equations of electrodynamics the current term in Eq. (2.11) can be replaced by field terms. Following this substitution it can be demonstrated that the time average pressure is then given by

$$\langle \mathrm{d}P \rangle = -\frac{\partial}{\partial x}\left(\frac{1}{2}\varepsilon_0 E_z^2 + \frac{1}{2\mu_0}B_y^2\right)\mathrm{d}x$$

(2.12)

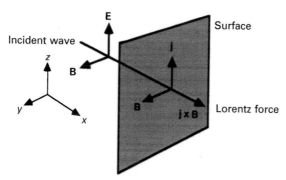

Fig. 2.1. Electromagnetic description of radiation pressure.

The term in parentheses is identified as the energy density U for the electric component E and magnetic component B of the incident wave, defined as

$$U = \frac{1}{2}\varepsilon_0 E^2 + \frac{1}{2\mu_0}B^2 \tag{2.13}$$

where ε_0 is the permittivity of free space and μ_0 is the permeability of free space. The pressure exerted on a surface of thickness Δl is now obtained, by integrating Eq. (2.12), as

$$\langle P \rangle = -\int_0^{\Delta l} \frac{\partial U}{\partial x} \, dx \tag{2.14}$$

For a perfectly absorbing medium the pressure exerted on the surface is then given by the total energy density of the electromagnetic wave, viz.

$$\langle P \rangle = \langle U \rangle \tag{2.15}$$

Consider now two plane waves which are separated by a distance Δx and are incident on a surface of area A, as shown in Fig. 2.2. The volume of space between the two waves which will impinge on the surface is then $A \, \Delta x$. In addition, the spacing Δx between the waves is equivalent to $c \, \Delta t$, where Δt is the travel time between the wave fronts. The energy density of the electromagnetic wave is therefore given by

$$U = \frac{\Delta E}{A(c \, \Delta t)} \tag{2.16}$$

where ΔE is the energy contained within the volume element. In addition, the energy flux W across the surface may be written as

$$W = \frac{1}{A}\left(\frac{\Delta E}{\Delta t}\right) \tag{2.17}$$

Therefore, using Eq. (2.16) it can be seen that

$$U = \frac{W}{c} \tag{2.18}$$

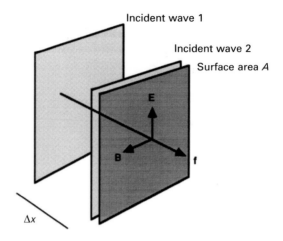

Fig. 2.2. Energy density of an electromagnetic wave.

However, this is the same expression as Eq. (2.9), the radiation pressure derived from the quantum description of light in section 2.3.1. It is therefore concluded that the quantum and electromagnetic description of radiation pressure are equivalent, as expected. For the electromagnetic description of light, the radiation pressure is just the energy density of the electromagnetic wave, as stated by Maxwell in his 1873 treatise on electricity and magnetism: '*Hence in a medium in which waves are propagated there is a pressure in the direction normal to the waves and numerically equal to the energy in unit volume.*'

2.3.3   Force on a perfectly reflecting solar sail

Now that the physics of solar radiation pressure has been discussed, the force exerted on an ideal, perfectly reflecting solar sail will be calculated. This calculation will be used to form the basis of a realistic solar sail force model provided later in this chapter. In addition, several parameters which characterise the solar sail performance are introduced. These key performance metrics will be used frequently in later chapters.

A solar sail is an oriented surface so that the acceleration experienced by the solar sail is a function of the sail attitude. For a solar sail of area A with a unit vector $\mathbf{n}$ directed normal to the sail surface, the force exerted on the sail due to photons incident from the $\mathbf{u}_i$ direction is given by

$$f_i = PA(\mathbf{u}_i \cdot \mathbf{n})\mathbf{u}_i \tag{2.19a}$$

where $A(\mathbf{u}_i \cdot \mathbf{n})$ is the projected sail area in the $\mathbf{u}_i$ direction, as shown in Fig. 2.3. Similarly, the reflected photons will exert a force of equal magnitude on the solar sail, but in the specular reflected direction $-\mathbf{u}_r$, viz.

$$f_r = -PA(\mathbf{u}_i \cdot \mathbf{n})\mathbf{u}_r \tag{2.19b}$$

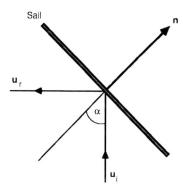

Fig. 2.3. Perfectly reflecting flat solar sail.

Using the vector identity $\mathbf{u_i} - \mathbf{u_r} = 2(\mathbf{u_i} \cdot \mathbf{n})\mathbf{n}$, the total force f exerted on the solar sail is therefore given by

$$f = 2PA(\mathbf{u_i} \cdot \mathbf{n})^2 \mathbf{n} \tag{2.20}$$

Then, using Eqs (2.5) and Eq. (2.9) this total force may be written as

$$f = \frac{2AW_E}{c}\left(\frac{R_E}{r}\right)^2 (\mathbf{u_i} \cdot \mathbf{n})^2 \mathbf{n} \tag{2.21}$$

where again W_E is the solar constant defined in section 2.3.1.

The solar sail performance may now be parameterised by the total spacecraft mass per unit area m/A. This constant will be termed the sail loading σ and is a key design parameter, as discussed in section 1.3.2. In addition, the sail pitch angle α will be defined as the angle between the sail normal and the incident radiation, as shown in Fig. 2.3. Using these definitions, the solar sail acceleration may now be written as

$$\mathbf{a} = \frac{2W_E}{c}\frac{1}{\sigma}\left(\frac{R_E}{r}\right)^2 \cos^2\alpha\,\mathbf{n} \tag{2.22}$$

Following section 1.3.2 the characteristic acceleration of the solar sail a_0 will again be defined as the acceleration experienced at 1 au with the sail normal to the Sun such that $\alpha = 0$. The characteristic acceleration is an equivalent design parameter to the solar sail loading and may be conveniently written as

$$a_0 = \frac{9.12\eta}{\sigma\,[\mathrm{g\,m^{-2}}]}\,[\mathrm{mm\,s^{-2}}] \tag{2.23}$$

where η is again some overall efficiency of the solar sail used to account for the finite reflectivity of the sail film and sail billowing. Typically the total solar sail efficiency is of order 0.9, although a conservative value of 0.85 will frequently be assumed in later chapters.

For a solar sail in a heliocentric orbit the direction of incidence of the radiation $\mathbf{u_i}$ is defined by the unit radial vector $\hat{\mathbf{r}}$ from the Sun to the solar sail. Furthermore, the solar sail acceleration may also be written in terms of the solar gravitational acceleration as

$$\mathbf{a} = \beta \frac{GM_S}{r^2} (\hat{\mathbf{r}} \cdot \mathbf{n})^2 \mathbf{n} \tag{2.24}$$

where M_S is the solar mass and G is the universal gravitational constant. The dimensionless sail loading parameter β will now be defined as the ratio of the solar radiation pressure acceleration to the solar gravitational acceleration. This parameter is also referred to as the lightness number of the solar sail. Since both the solar radiation pressure acceleration and the solar gravitational acceleration are assumed to have an inverse square variation, the lightness number is independent of the Sun–sail distance. Using Eq. (2.22), Eq. (2.24) and Eq. (2.5b) the solar sail lightness number may be written as

$$\beta = \frac{\sigma^*}{\sigma} \tag{2.25a}$$

$$\sigma^* = \frac{L_S}{2\pi G M_S c} \tag{2.25b}$$

The critical solar sail loading parameter σ^* is found to be $1.53 \, \mathrm{g\,m^{-2}}$. This is a unique constant which is a function of the solar mass and the solar luminosity. With this mass per unit area the solar sail lightness number is unity, so that the solar radiation pressure acceleration is exactly equal to the solar gravitational acceleration. As will be seen in later chapters, such a sail loading is an extremely challenging requirement, but can enable unique orbits and mission applications for high-performance solar sails.

2.4   RADIATIVE TRANSFER METHODS

2.4.1   Specific intensity

A more rigorous means of calculating the effect of radiation pressure on a solar sail is through the use of radiative transfer methods. Radiative transfer provides a consistent methodology for the analysis of both energy and momentum transport by a radiation field. For an arbitrary radiation field, the properties of the field are a function of both position and time. Furthermore, at a particular position within the radiation field its properties may have a distribution in direction and frequency. These properties of the radiation field may be completely, and compactly, described by the specific intensity of the field. The specific intensity $I_\nu(\mathbf{r}, \mathbf{u}; t)$ of radiation at position $\mathbf{r}$ and time t propagating in direction $\mathbf{u}$ with frequency ν is defined to be the energy dE transported across a directed surface element $d\mathbf{A}$ in time dt into a solid angle $d\Omega$ about direction $\mathbf{u}$ in the frequency range $(\nu, \nu + d\nu)$, viz.

$$dE = I_\nu(\mathbf{r}, \mathbf{u}; t)(\mathbf{u} \cdot d\mathbf{A}) \, d\Omega \, dt \, d\nu \tag{2.26}$$

where $\mathbf{u} \cdot d\mathbf{A}$ is the projected surface area normal to the direction of propagation $\mathbf{u}$, as shown in Fig. 2.4.

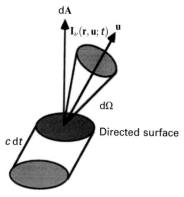

dA

$I_\nu(\mathbf{r}, \mathbf{u}; t)$ **u**

$d\Omega$

Directed surface

$c\,dt$

Fig. 2.4. Radiation field specific intensity.

From a microscopic perspective the radiation field is composed of individual photons which may be described by the photon number density function $\psi_\nu(\mathbf{r}, \mathbf{u}; t)$. This function is defined such that $\psi_\nu(\mathbf{r}, \mathbf{u}; t)\,d\Omega\,d\nu$ is the number of photons per unit volume at position $\mathbf{r}$ at time t and in a frequency range $(\nu, \nu + d\nu)$ propagating with speed c into a solid angle $d\Omega$ about direction $\mathbf{u}$. Therefore, the number of photons dN crossing a surface element dA in time dt is given by

$$dN = \psi_\nu(\mathbf{r}, \mathbf{u}; t)(\mathbf{u}\cdot d\mathbf{A})(c\,dt)\,d\Omega\,d\nu \qquad (2.27)$$

where $(\mathbf{u}\cdot d\mathbf{A})(c\,dt)$ is the volume element containing photons emerging through $d\mathbf{A}$ in time dt in direction $\mathbf{u}$, as shown in Fig. 2.4. Since each photon has energy $h\nu$, the energy transported is also given by

$$dE = ch\nu\,\psi_\nu(\mathbf{r}, \mathbf{u}; t)(\mathbf{u}\cdot d\mathbf{A})\,d\Omega\,dt\,d\nu \qquad (2.28)$$

Comparing Eqs (2.26) and (2.28) it is clear that the specific intensity is related to the photon number density function by

$$I_\nu(\mathbf{r}, \mathbf{u}; t) = ch\nu\,\psi_\nu(\mathbf{r}, \mathbf{u}; t) \qquad (2.29)$$

The photon number density function may now be related to the photon distribution function $\Psi_\nu(\mathbf{r}, \mathbf{u}; t)$. This function is defined such that the number of photons per unit volume in frequency range $(\nu, \nu + d\nu)$ and with momenta in the range $(\mathbf{p}, \mathbf{p} + d\mathbf{p})$ is given by $\Psi_\nu(\mathbf{r}, \mathbf{u}; t)\,d^3p$, where $\mathbf{p} = (h\nu/c)\mathbf{u}$ is the photon momentum. Using the relation $d^3p = p^2 dp\,d\Omega$ and equating photon numbers in a unit volume of space using the functions $\psi_\nu(\mathbf{r}, \mathbf{u}; t)$ and $\Psi_\nu(\mathbf{r}, \mathbf{u}; t)$ it is found that

$$\left(\frac{h^3\nu^2}{c^3}\right)\Psi_\nu(\mathbf{r}, \mathbf{u}; t)\,d\nu\,d\Omega = \psi_\nu(\mathbf{r}, \mathbf{u}; t)\,d\nu\,d\Omega \qquad (2.30)$$

Therefore, from Eq. (2.29) the specific intensity of the radiation field may also be written in terms of the photon distribution function as

$$I_\nu(\mathbf{r}, \mathbf{n}; t) = \left(\frac{h^4 \nu^3}{c^2}\right) \Psi_\nu(\mathbf{r}, \mathbf{n}; t) \tag{2.31}$$

For a black-body source of radiation (one that is in thermal equilibrium), as the Sun is assumed to be, it can be shown that the photon distribution function is defined as

$$\Psi_\nu = \frac{2}{h^3}(e^{h\nu/kT} - 1) \tag{2.32}$$

where T is the black-body temperature of the Sun and k is Boltzmann's constant. With this choice of distribution function the specific intensity is then defined by the black-body Planck function. Therefore, by knowing the photon distribution function of the radiation field Ψ_ν, and so the specific intensity, the properties of the entire black-body radiation field are known.

2.4.2   Angular moments of specific intensity

Now that the concept of specific intensity has been defined, derived quantities such as flux and radiation pressure may be calculated by taking angular moments of the specific intensity. Firstly, the flux $\mathbf{F}(\mathbf{r}; t)$ will be defined as the vector such that $\mathbf{F}(\mathbf{r}; t) \cdot d\mathbf{A}$ provides the net rate of flow of radiative energy across the directed surface element $d\mathbf{A}$ in all frequencies. The flux can be obtained by firstly calculating the net number of photons crossing $d\mathbf{A}$ from all solid angles in time dt and in the frequency range $(\nu, \nu + d\nu)$. From Eq. (2.27) the net number of photons in this frequency range is given by

$$N = \left[\oint_{4\pi} c\psi_\nu(\mathbf{r}, \mathbf{u}; t)\mathbf{u}\, d\Omega\, d\nu\right] \cdot d\mathbf{A} \tag{2.33}$$

Therefore, multiplying N by $h\nu$, the radiative energy flux for photons in the range $(\nu, \nu + d\nu)$ is obtained. Using Eq. (2.29) and integrating over the entire frequency range the flux may then be written as

$$\mathbf{F}(\mathbf{r}; t) = \int_0^\infty \oint_{4\pi} I_\nu(\mathbf{r}, \mathbf{u}; t)\mathbf{u}\, d\Omega\, d\nu \tag{2.34}$$

The radiation pressure tensor $\mathbf{P}(\mathbf{r}; t)$ will now be defined as the second angular moment of specific intensity. The radiation pressure tensor is a matrix quantity defined such that the element P^{ij} is the net rate of transport of the ith component of momentum through area dA oriented normal to jth co-ordinate axis. From Eq. (2.27), the number of photons with frequency ν propagating in solid angle $d\Omega$ about direction u^j and crossing surface area dA in time dt is $c\psi_\nu(\mathbf{r}, \mathbf{u}; t)u^j$ with each photon transporting momentum $(h\nu/c)u^i$ in the ith direction. Therefore, integrating over all solid angles, the radiation pressure tensor due to photons in the frequency range $(\nu, \nu + d\nu)$ is obtained as

$$P^{ij}(\mathbf{r}; t) = \oint_{4\pi} c\psi_\nu(\mathbf{r}, \mathbf{u}; t)\left(\frac{h\nu}{c}u^i\right)u^j\, d\Omega \tag{2.35}$$

However, it can be seen from Eq. (2.29) that specific intensity is defined as $I_\nu(\mathbf{r}, \mathbf{u}; t) = ch\nu\psi_\nu(\mathbf{r}, \mathbf{u}; t)$ so that by integrating over the entire frequency range Eq. (2.35) may be written as

$$P(\mathbf{r}; t) = \frac{1}{c} \int_0^\infty \oint_{4\pi} I_\nu(\mathbf{r}, \mathbf{u}; t) \mathbf{u}\mathbf{u} \, d\Omega \, d\nu \qquad (2.36)$$

where $\mathbf{u}\mathbf{u}$ is the tensor quantity $[u^i u^j]$. Therefore, for a given radiation field specific intensity, Eq. (2.36) may be used to obtain the radiation pressure at any point and in any direction within the field. Furthermore, by knowing the surface geometry and optical properties of an arbitrary body placed within the radiation field, the radiation pressure experienced by the body may be precisely obtained.

2.5   RADIATION PRESSURE WITH A FINITE SOLAR DISC

2.5.1   Why the inverse square law is inadequate

It has long been assumed that solar radiation pressure has an inverse square variation with heliocentric distance, as derived in section 2.3. In this section it will be shown that for a planar solar sail the assumption of an inverse square variation is in fact not valid when account is taken of the finite angular size of the solar disc. This modelling of the source of solar radiation pressure is quite distinct from modelling of the solar radiation pressure force, which is dependent on the optical properties of the sail film, discussed later in this chapter. In section 2.3 the solar radiation incident on the sail surface was implicitly assumed to be incident along parallel rays. The varying direction of incidence of radiation from different parts of the solar disc will now be included through the use and integration of the radiation pressure tensor. Therefore, Eq. (2.36) will provide the exact radiation pressure obtained from the finite angular sized solar disc. Although the deviation from an inverse square law is small, long-duration orbits close to the Sun will experience a cumulative perturbation, as will future high-performance solar sails using grazing solar passes *en route* to the outer solar system. In addition, the long-term stability of certain classes of orbits will also be affected, as will be seen in Chapter 5.

2.5.2   Uniformly bright solar disc

Firstly, it will be assumed that the solar disc has uniform brightness; that is, an element of the solar disc will appear equally bright when viewed from any aspect angle. In this case the specific intensity is time independent and isotropic across the solar disc. Therefore, using Eq. (2.36) the solar radiation pressure exerted on a radially oriented, perfectly reflecting sail at a heliocentric distance r may be written as

$$P(r) = \frac{2}{c} \int_0^\infty \int_0^{2\pi} \int_0^{\theta_0} I_\nu \cos^2\theta \, d\Omega \, d\nu, \qquad d\Omega = \sin\theta \, d\theta \, d\phi \qquad (2.37)$$

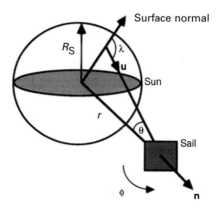

Fig. 2.5. Solar radiation pressure with a finite angular sized solar disc.

where the angular radius of the solar disc θ_0 is given by $\sin^{-1}(R_S/r)$, as shown in Fig. 2.5. Noting the azimuthal symmetry of the geometry and that the specific intensity is independent of r, Eq. (2.37) then reduces to the integral

$$P(r) = \frac{4\pi}{c} I_0 \int_{\xi_0}^1 \xi^2 d\xi, \qquad \xi = \cos\theta, \qquad \xi_0 = \cos\theta_0 \qquad (2.38)$$

where I_0 is the frequency integrated specific intensity. Performing this integration and substituting for ξ_0 it is found that

$$P(r) = \frac{4\pi}{3c} I_0 \left\{ 1 - \left[1 - \left(\frac{R_S}{r} \right)^2 \right]^{3/2} \right\} \qquad (2.39)$$

In order to proceed, Eq. (2.39) may now be expanded in powers of $(R_S/r)^2$ and for $r \gg R_S$ may be written to first order as

$$P(r) = \frac{2\pi}{c} I_0 \left(\frac{R_S}{r} \right)^2 + O\left(\frac{R_S}{r} \right)^4 \qquad (2.40)$$

However, at large values of r this expansion must match asymptotically with the expression for the radiation pressure from a distant point source, viz.

$$P^*(r) = \frac{2}{c} \left(\frac{L_S}{4\pi R_S^2} \right) \left(\frac{R_S}{r} \right)^2 \qquad (2.41)$$

Hence, by comparing Eqs (2.40) and (2.41) the frequency integrated specific intensity I_0 is identified as

$$I_0 = \frac{L_S}{4\pi^2 R_S^2} \qquad (2.42)$$

Substituting for I_0 in Eq. (2.39) an expression for the solar radiation pressure exerted

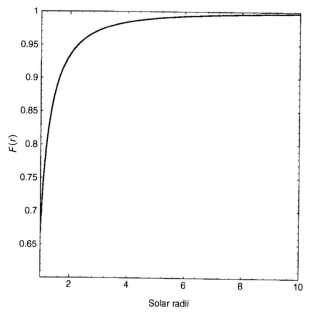

Fig. 2.6. Deviation of the uniformly bright finite disc model from an inverse square law.

on a radially oriented solar sail from a uniformly bright, finite angular sized solar disc is obtained as

$$P(r) = \frac{L_S}{3\pi c R_S^2}\left\{1 - \left[1 - \left(\frac{R_S}{r}\right)^2\right]^{3/2}\right\} \qquad (2.43)$$

A more useful way of representing Eq. (2.43) is to express it in terms of the usual inverse square law $P^*(r)$, viz.

$$P(r) = P(r)^* F(r) \qquad (2.44a)$$

$$F(r) = \frac{2}{3}\left(\frac{r}{R_S}\right)^2\left\{1 - \left[1 - \left(\frac{R_S}{r}\right)^2\right]^{3/2}\right\} \qquad (2.44b)$$

The function $F(r)$ now describes the deviation of the true solar radiation pressure from an inverse square law. It can be seen from Eq. (2.44b) that $F(r)$ attains its minimum value at $r = R_S$ where $F(R_S) = \frac{2}{3}$, giving the greatest deviation of the solar radiation pressure from an inverse square law. Furthermore, as $r \to \infty$, $F(r) \to 1$ since the solar disc becomes more point-like. From Fig. 2.6 it can also be seen that $F(r)$ approaches unity over a scale of order 10 solar radii (0.047 au) so that the magnitude of the deviation from an inverse square form is small at large heliocentric distances. Physically, the deviation from an inverse square law is due to photons from the solar limb intercepting the sail at an oblique angle to the sail surface, whereas photons from the centre of the disc are incident along the normal to the sail.

The photons from the solar limb therefore transfer a smaller amount of momentum to the solar sail than those from the centre of the disc. At large heliocentric distances, however, photons from all parts of the solar disc are incident along near-parallel rays.

2.5.3   Limb-darkened solar disc

A more accurate model of the solar radiation pressure may be obtained by including solar limb darkening in the functional form of the specific intensity. Limb darkening is an effect due to the specific intensity of the solar radiation field having a directional dependence; that is, as the radiation from a point on the solar surface is viewed from an oblique angle, the associated specific intensity falls so that the limb of the solar disc appears darker than the disc centre. Empirically, solar limb darkening has a complex functional form. However, using an approximate model of the solar atmosphere an analytic expression for the limb-darkening effect can be obtained. The grey solar atmosphere model, which assumes that the solar atmosphere is in both radiative and local thermodynamic equilibrium, allows such an analytic expression for the limb-darkened specific intensity. The specific intensity of the solar radiation field may then be written as

$$I = \frac{I_0}{4}(2 + 3\cos\lambda) \tag{2.45}$$

where again I_0 is the frequency integrated specific intensity and the aspect angle λ is defined in Fig. 2.5. It is seen from Eq. (2.45) that the solar limb will appear darker than the centre of the solar disc by a factor of 0.4 using the grey atmosphere approximation. For the limb-darkened specific intensity it is found that $F(R_S) = 0.708$ so that the limb-darkened solar radiation pressure deviates less from an inverse square form than the non-limb-darkened pressure. This is due to the reduced momentum transfer from the solar limb. Again, at large heliocentric distances it is found that $F(r) \rightarrow 1$ as expected.

2.6   SOLAR SAIL FORCE MODELS

In previous sections of this chapter it has been implicitly assumed that the solar sail is a perfect reflector. Clearly such an idealised model is not suitable for realistic trajectory design and mission analysis. Two solar sail force models are now presented. Firstly, the force exerted on a non-perfect solar sail is obtained by considering reflection, absorption and re-radiation by the sail. This theoretical model is parameterised by a number of coefficients which represent the optical properties of the sail film. However, there are still several implicit assumptions made in this model, the most important of these being the assumption that the sail is flat. In practice, however, a real solar sail will billow under load. To account for these additional effects a numerical parametric force model developed at JPL for the Halley rendezvous mission is presented. Future solar sail design studies will require

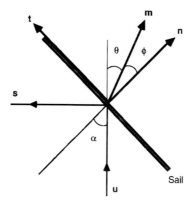

Fig. 2.7. Non-perfect flat solar sail.

detailed force models which account for the sail film optical properties, sail shape and changing sail temperature. In addition to such pre-flight modelling, on-orbit parameter estimation will also be required for orbit determination and orbit propagation. Such estimation will be required at regular intervals if the optical properties are expected to change with time due to sail film degradation.

2.6.1 Optical force model

If the assumption of perfect reflectivity is relaxed, a more exact model of the solar radiation pressure force may be developed. Taking into account the reflectance, absorption and emissivity of the sail film, the total force exerted on the solar sail due to solar radiation pressure may be written as

$$f = f_r + f_a + f_e \tag{2.46}$$

where f_r is the force due to reflection, f_a is the force due to absorption and f_e is the force due to emission by re-radiation. The main optical properties of the sail film can now be defined by the reflection coefficient $\tilde{r}$, absorption coefficient a and transmission coefficient τ with the constraint

$$\tilde{r} + a + \tau = 1 \tag{2.47}$$

where the tilde notation $\tilde{r}$ is used to avoid confusion with the symbol for orbit radius r. However, since $\tau = 0$ on the reflecting side of the solar sail it is clear that

$$a = 1 - \tilde{r} \tag{2.48}$$

The solar sail orientation is again defined by a vector **n** normal to the sail surface with a transverse unit vector **t** normal to **n**, as shown in Fig. 2.7. In addition, the direction of incidence of photons will be defined by a unit vector **u** and the direction of specularly reflected photons by a unit vector **s**. The following relationships are found to hold between these unit vectors:

$$\mathbf{u} = \cos\alpha\,\mathbf{n} + \sin\alpha\,\mathbf{t} \tag{2.49a}$$

$$\mathbf{s} = -\cos\alpha\,\mathbf{n} + \sin\alpha\,\mathbf{t} \tag{2.49b}$$

where α is again the pitch angle of the solar sail relative to the Sun-line. In addition, by combining Eqs (2.49a) and (2.49b) it can be seen that

$$\mathbf{s} = \mathbf{u} - 2\cos\alpha\,\mathbf{n} \tag{2.49c}$$

Initially it will be assumed that all incident photons are absorbed by the sail. Then the fraction of these incident photons which are re-emitted by reflection and re-radiation will be considered.

The force exerted on the solar sail due to absorbed photons is given by $PA\cos\alpha\,\mathbf{u}$, where $A\cos\alpha$ is the projected sail area in direction $\mathbf{u}$. Resolving this force into normal and transverse components using Eq. (2.49a), it is found that

$$f_{\mathrm{a}} = PA(\cos^2\alpha\,\mathbf{n} + \cos\alpha\sin\alpha\,\mathbf{t}) \tag{2.50}$$

A fraction $\tilde{r}$ of the incident photons will now be reflected. Of that fraction of photons another fraction s will be specularly reflected in direction $\mathbf{s}$, so providing a force f_{rs} in $-\mathbf{s}$ direction given by

$$f_{rs} = -(\tilde{r}s)PA\cos\alpha\,\mathbf{s} \tag{2.51}$$

In addition, another fraction of photons will be uniformly scattered from the reflecting surface of the sail due to non-specular reflection. This component will generate a force f_{ru} in direction $\mathbf{n}$ given by

$$f_{ru} = B_{\mathrm{f}}\tilde{r}(1-s)PA\cos\alpha\,\mathbf{n} \tag{2.52}$$

where the coefficient B_{f} indicates that the front surface of the sail will be non-Lambertian. A Lambertian surface is one which appears equally bright when viewed from any aspect angle. The non-Lambertian coefficient then describes the deviation from this condition. The total force due to reflected photons is now given by $f_{rs} + f_{ru}$. Using Eq. (2.49b) to again write the total force in terms of the normal and transverse directions yields

$$f_r = PA[(\tilde{r}s\cos^2\alpha + B_{\mathrm{f}}(1-s)\tilde{r}\cos\alpha)\mathbf{n} - \tilde{r}s\cos\alpha\sin\alpha\,\mathbf{t}] \tag{2.53}$$

The final force component is that due to photons which have been absorbed and then re-emitted as thermal radiation from both the front and back surfaces of the sail. The power emitted from a unit area of the sail at temperature T is $\varepsilon\tilde{\sigma}T^4$, where $\tilde{\sigma}$ is the Stefan–Boltzmann constant and ε is the surface emissivity. Again, the tilde notation $\tilde{\sigma}$ is used to avoid confusion with the sail loading σ. Therefore, allowing for the non-Lambertian nature of the front and back sail surfaces, and assuming the sail has uniform temperature, the force due to emission by re-radiation is given by

$$f_{\mathrm{e}} = \frac{\tilde{\sigma}T^4}{c}(\varepsilon_{\mathrm{f}}B_{\mathrm{f}} - \varepsilon_{\mathrm{b}}B_{\mathrm{b}})\mathbf{n} \tag{2.54}$$

where ε_{f} and ε_{b} are the front and back emissivities. The sail temperature may now be

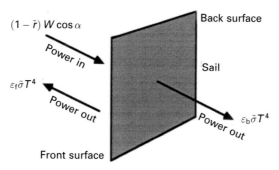

Fig. 2.8. Solar sail thermal balance.

obtained from the balance between the thermal input and the thermal output of $(\varepsilon_f + \varepsilon_b)\tilde{\sigma}T^4$, as shown in Fig. 2.8. The thermal input to the sail is given by $(1 - \tilde{r})W\cos\alpha$, where W is the solar flux incident on the sail. However, since the radiation pressure is just W/c the sail temperature may be written as

$$T = \left[\frac{(1 - \tilde{r})cP\cos\alpha}{\tilde{\sigma}(\varepsilon_f + \varepsilon_b)}\right]^{1/4} \tag{2.55}$$

Therefore, substituting for the sail equilibrium temperature, the force exerted on the solar sail due to emission by re-radiation is given by

$$f_e = PA(1 - \tilde{r})\frac{\varepsilon_f B_f - \varepsilon_b B_b}{\varepsilon_f + \varepsilon_b}\cos\alpha\,\mathbf{n} \tag{2.56}$$

It is clear from Eq. (2.56) that a sail backing of low emissivity is required to maximise the normal force. However, the emissivity of the backing is the only coefficient which can be used to control the sail temperature.

Collecting all of these contributions, the total force exerted on the solar sail may be written in terms of normal and transverse components as

$$f_n = PA\left\{(1 + \tilde{r}s)\cos^2\alpha + B_f(1 - s)\tilde{r}\cos\alpha + (1 - \tilde{r})\frac{\varepsilon_f B_f - \varepsilon_b B_b}{\varepsilon_f + \varepsilon_b}\cos\alpha\right\}\mathbf{n} \tag{2.57a}$$

$$f_t = PA(1 - \tilde{r}s)\cos\alpha\sin\alpha\,\mathbf{t} \tag{2.57b}$$

The total force vector and force magnitude may then be written as

$$f = f\mathbf{m} \tag{2.58a}$$

$$f = (f_n^2 + f_t^2)^{1/2} \tag{2.58b}$$

where $\mathbf{m}$ is the unit vector in the direction of the total force, as shown in Fig. 2.7. The angle of the force vector relative to the incident radiation will be defined by the cone angle θ. Since a real sail is not a perfect reflector, the direction of the force vector will not be normal to the sail surface as the force due to absorbed photons is somewhat greater than that due to reflected photons. Therefore, the force vector direction is

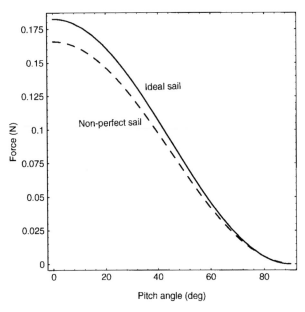

Fig. 2.9. Force exerted on a 100×100 m ideal square solar sail and non-perfect square solar sail at 1 au.

Table 2.1. Optical coefficients for an ideal solar sail, JPL square sail and JPL heliogyro.

	$\tilde{r}$	s	ε_f	ε_b	B_f	B_b
Ideal sail	1	1	0	0	2/3	2/3
Square sail	0.88	0.94	0.05	0.55	0.79	0.55
Heliogyro	0.88	0.94	0.05	0.55	0.79	0.55

biased towards the direction of incident radiation. The angle between the force vector and the sail normal is the centre-line angle ϕ given by

$$\tan \phi = \frac{f_t}{f_n} \tag{2.59}$$

again shown in Fig. 2.7. As will be seen, the centre-line angle constrains the direction of the force vector cone angle. For a perfectly reflecting solar sail, of course, the centre-line angle vanishes since the force vector is always directed normal to the sail surface.

Sail optical parameters derived from the JPL comet Halley rendezvous study are listed in Table 2.1 for a square sail and heliogyro. These parameters will now be used to compare the optical solar sail model to an idea solar sail, as shown in Fig. 2.9, where the total force exerted on a 100×100 m square solar sail located at 1 au has been calculated. It can be seen that there is a significant difference in force magnitude

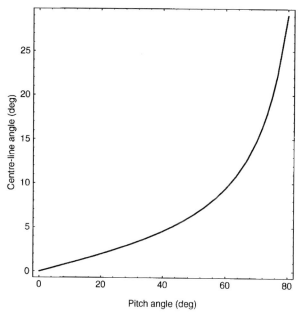

Fig. 2.10. Centre-line angle for a non-perfect solar sail model.

between the two models. Firstly, the force exerted on the realistic solar sail is less than that on the ideal solar sail when oriented along the Sun-line, as expected. In addition, due to optical absorption the realistic solar sail has a centre-line angle as discussed above. The variation of the centre-line angle is shown in Fig. 2.10. An important effect of the centre-line angle for the realistic square solar sail model is that it can only direct its force vector to a maximum cone angle of 55.5°, corresponding to a sail pitch angle of 72.6°, as shown in Fig. 2.11. This is in contrast to the idea solar sail where the force vector is always directed normal to the sail surface and can in principle be directed up to 90° from the Sun-line. This limitation on the direction of the force vector for real solar sails poses constraints for some mission applications, as will be seen in later chapters. Lastly, the relative magnitude of the force due to absorption, reflection and emission are shown in Fig. 2.12. It can be seen that the effect of thermal emission from the sail film is to generate a small sunward force due to the high emissivity of the sail coating required for thermal control.

2.6.2   Parametric force model

The analysis provided in section 2.6.1 has assumed implicitly that the solar sail is flat. However, any real solar sail will billow when under load. The exact shape of the sail then couples in a complex manner to the radiation pressure force. During the Jet Propulsion Laboratory comet Halley rendezvous study an exact force model was derived by calculating the sail shape and numerically integrating the radiation

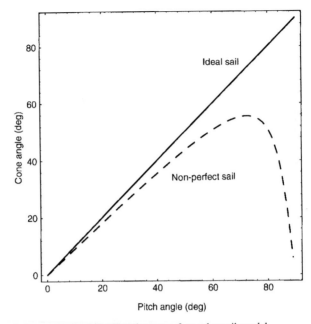

Fig. 2.11. Cone angle for an ideal solar sail and non-perfect solar sail model.

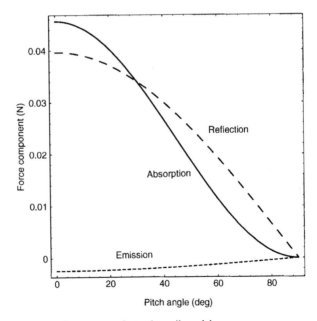

Fig. 2.12. Force components for a non-perfect solar sail model.

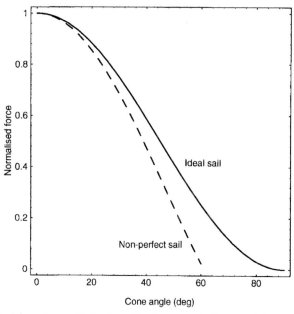

Fig. 2.13. Normalised force for an ideal solar sail and parametric force model.

Table 2.2. Force coefficients for an ideal solar sail, JPL square sail and JPL heliogyro.

	C_1	C_2	C_3
Ideal sail	0.5	0.5	0
Square sail	0.349	0.662	−0.011
Heliogyro	0.367	0.643	−0.010

pressure force across the curved sail film. This numerical force model is para-meterised as a function of the cone angle between the incoming radiation and the direction of the force vector, viz.

$$\boldsymbol{f} = f_0(c_1 + c_2\cos 2\theta + c_3\cos 4\theta)\mathbf{m} \qquad (2.60)$$

where f_0 is the force exerted when the solar sail is oriented normal to the Sun-line. Sets of parameters for the JPL comet Halley square sail and heliogyro are listed in Table 2.2. Again, the realistic force model can be compared with an ideal solar sail, as shown in Fig. 2.13, where the forces have been normalised to the force exerted when the solar sail is oriented normal to the Sun-line. It can be seen that again there is a significant deviation from the ideal solar sail model. Most importantly, using this model the nett force exerted on the realistic solar sail vanishes when the cone angle reaches $61.1°$ due to the centre-line effect, as discussed in section 2.6.1. This is the cut-off cone angle which constrains the direction of action of the solar radiation pressure force vector.

2.7   OTHER FORCES

Although solar radiation pressure generates the largest force on a solar sail, other secondary forces are also present. For example, the solar wind will exert a small pressure on the sail due to the momentum transported by solar wind protons. During periods of high solar wind speed the mean proton number density ρ at the Earth rises to approximately $4 \times 10^6 \, \text{m}^{-3}$ with a wind speed v_w of order $700 \, \text{km} \, \text{s}^{-1}$. The solar wind pressure P_w exerted on the solar sail can then be estimated from the momentum transported as

$$P_\text{w} \sim m_\text{p} \rho v_\text{w}^2 \tag{2.61}$$

where m_p is the mass of a proton. Using the above parameters a solar wind pressure of order $3 \times 10^{-9} \, \text{N} \, \text{m}^{-2}$ is obtained, which is nearly 10^{-4} less than the direct solar radiation pressure exerted on the sail at 1 au. In addition, first-order relativistic effects are proportional to the ratio of the solar sail speed to the speed of light, typically of order 10^{-4}. For solar sails in Earth orbit the secondary pressure due to radiation scattered from the Earth is also small, being at least three orders of magnitude less than that due to the direct solar radiation pressure. The main perturbation in Earth orbit is residual atmospheric drag experienced at low altitudes. This perturbation is a strong function of solar activity which is modulated in an 11-year cycle. In general, air drag begins to exceed direct solar radiation pressure below altitudes from approximately 600 to 900 km.

2.8   SUMMARY

In this chapter the physics of solar radiation pressure has been investigated. It has been shown that the quantum and wave descriptions of light yield the same expression for solar radiation pressure, as expected. The pressure exerted on an ideal solar sail has then been calculated and the sail performance characterised by the solar sail loading, characteristic acceleration and lightness number. It has also been shown that the finite angular size of the solar disc means that the solar radiation pressure exerted on a flat, Sun-facing solar sail does not in fact follow an inverse square law. In general, the effect of this deviation will be to introduce small, cumulative errors in trajectory propagation. This is important for long-duration trajectories in the inner solar system and for future high-performance solar sail missions making extremely close passes to the Sun. Lastly, two solar sail force models have been presented. Firstly, a flat solar sail model which accounts for the real optical properties of the sail film, and then a numerical force model which also accounts for the sail shape under load. Comparison with an ideal solar sail shows that the direction of the force vector is constrained due to sail absorption effects. This constraint has an important bearing on trajectory design.

2.9   FURTHER READING

Historical interest

Maxwell, J. C., *Electricity and Magnetism*, Oxford University Press, 1873.
Lebedew, P., 'The Physical Causes of Deviation from Newton's Law of Gravitation', *Astrophysical Journal*, **10**, 155–161, 1902.
Nichols, E.F. & Hull, G.F., 'The Experimental Investigation of the Pressure of Light', *Astrophysical Journal*, **15**, 60–65, 1902.
Nichols, E.F. & Hull, G.F., 'The Pressure due to Radiation', *Physical Review*, **17**, 26–50, 1903.

Radiative transfer

Mihalas, D. & Mihalas, B.W., *Foundations of Radiation Hydrodynamics*, Oxford University Press, 1984.
McInnes, C.R. & Brown, J.C., 'Solar Sail Dynamics with an Extended Source of Radiation Pressure', IAF-89-350, *40th International Astronautical Federation Congress*, Malaga, October 1989.
McInnes, C.R. & Brown, J.C., 'The Dynamics of Solar Sails with a Non-point Source of Radiation Pressure', *Celestial Mechanics and Dynamical Astronomy*, **9**, 249–264, 1990.

Solar sail force model

Georgevic, R.M., 'The Solar Radiation Pressure Force and Torques Model', *Journal of Astronautical Sciences*, **20**, 257–274, 1973.
Wright, J.L., *Space Sailing*, Gordon & Breach Science Publishers, 1992 (Appendix A, B).

3

Solar sail design

3.1   INTRODUCTION

The romantic notion of solar sailing portrayed in science fiction is of a huge billowing hemispherical film joined to a payload through shroud lines attached at the rim of the sail. Indeed, this picture of solar sailing is described in some early technical publications on the subject. While a hemispherical sail is an obvious initial choice, some thought quickly leads us away from this view – principally, since a hemispherical sail is much less efficient than a flat sail of the same area, and light pressure does not inflate the sail in the same manner as kinetic gas pressure. The pursuit of practical solar sail designs which can be packed, deployed and controlled has led to many varied concepts. The diverse range of solar sail design concepts is somewhat reminiscent of early developments in aviation, when a large number of competing designs were considered. From those initial concepts practical experience led to a limited selection and ultimate evolution towards the standard aircraft configurations seen today, and perhaps the long-term development of solar sailing will follow a similar route. When practical experience of solar sails is at hand it may become apparent that there only a few design concepts that can be truly optimised for certain common classes of mission.

One of the key problems in solar sail design is the packing and subsequent deployment of large areas of thin sail film. Since the dimensional expansion ratio between a deployed and stowed solar sail can be over 100, innovative structural engineering solutions are required. Indeed, the packing and deployment problem has perhaps been the greatest impediment to practical solar sail utilisation. In addition, since the sail is folded for packing, the reflecting medium of the sail must be mounted on a thin substrate, and the presence of a substrate leads to a fundamental limitation on solar sail performance if conventional materials are used. In order to deploy the sail and maintain a flat surface, tension must also be applied, either by a deployable structure or in part by spin-induced tension. Once deployed this tensioned sail film must then be oriented to direct the solar radiation pressure force for orbit manoeuvring. Owing to the large moments of inertia of solar sails, innovative engineering solutions are again required.

In this chapter these design problems will be addressed along with other key issues such as sail film manufacture, structures and thermal control. Firstly, the basic design parameters for solar sails will be defined and sensitivity functions derived. These functions provide a measure of the sensitivity of the solar sail performance to variations in design parameters, such as the sail area and payload mass. The key technologies of solar sail manufacture are then reviewed. Candidate materials for sail film substrates and coatings are investigated and their relative merits discussed. These coatings provide both a reflective front surface and a high-emissivity rear surface for passive thermal control. The interaction of the sail film with the space environment is also discussed and potential problems due to electrical charging identified. The logistics of manufacturing and packing a large sail film are also investigated and candidate folding schemes presented. The subsequent deployment problem is considered and various structural concepts explored. Of particular interest are current developments in inflatable structures which may provide simpler and more reliable schemes than conventional approaches.

Following the discussion of the key technologies required for solar sailing the three main solar sail configurations will be considered in detail. Design issues for the square solar sail, heliogyro and disc solar sail will be discussed along with control of the sail orientation. In the case of the heliogyro and disc solar sail, the equilibrium shape of the sail film will also be determined. As an alternative to these conventional approaches, a novel concept is presented which offers potentially superior performance and less complex control. The solar photon thruster was initially devised in the Soviet Union in the early 1970s and was later re-invented by physicist Robert Forward. This solar sail concept separates the functions of collecting and reflecting incident photons and so does not required the entire solar sail to be rotated.

Looking to the future, one of the great promises of solar sailing lies with high-performance sail films with a lightness number of order unity. Developments in this area will be described and possible future directions of investigation for the production of deployable thin-film metallic sails presented. A similar promise for future development and utilisation is held by micro-solar sails with a mass of under $1\,\mathrm{kg}$ and a surface area of $100\,\mathrm{m}^2$ or less. Using recent advances in micro-electromechanical systems (MEMS) technology, ultra-low-mass but highly capable spacecraft appear possible in the future. These low-mass spacecraft will only require small solar sails and offer new, innovative means of sail deployment and control. Lastly, a review of a representative sample of recent solar sail design concepts is presented and evaluated. These designs draw on the main solar sail configurations and illustrate some new and innovative engineering solutions to the unique and challenging problems posed by solar sailing.

3.2   DESIGN PARAMETERS

As discussed in Chapters 1 and 2, the fundamental design parameter for a solar sail is its characteristic acceleration. This parameter determines the transfer time to a particular target object or even whether a particular class of orbits is possible. The characteristic acceleration is, however, a function of both the efficiency of the solar sail design and the payload mass. As discussed in section 2.3.3, the characteristic acceleration is defined as the solar radiation pressure acceleration experienced by a solar sail oriented normal to the Sun-line at a heliocentric distance of one astronomical unit. At this distance from the Sun the magnitude of the solar radiation pressure P exerted on a perfectly absorbing surface is $4.56 \times 10^{-6}\,\mathrm{N\,m^{-2}}$. Therefore, again allowing for a finite sail efficiency η, the characteristic acceleration is given by

$$a_0 = \frac{2\eta P}{\sigma}, \qquad \sigma = \frac{m}{A} \tag{3.1}$$

where σ is the total solar sail loading, with m the total solar sail mass and A the sail area. The sail efficiency is a function of the optical properties of the sail film and the sail shape, as discussed in section 2.6. The total mass of the solar sail will now be partitioned into two components, m_S due to the sail film and structure and m_P the payload mass. Therefore, the characteristic acceleration of the solar sail may now be written as

$$a_0 = \frac{2\eta P}{\sigma_S + (m_P/A)}, \qquad \sigma_S = \frac{m_S}{A} \tag{3.2}$$

where σ_S is the mass per unit area of the sail assembly. The so-called sail assembly loading is then a key parameter which is a measure of the performance of the sail film and the efficiency of the solar sail structural design.

Using Eq. (3.2) the influence which various design parameters have on the solar sail characteristic acceleration will now be investigated. For a fixed sail area and efficiency, Eq. (3.2) becomes a function of two variables: the sail assembly loading σ_S and the payload mass m_P. Therefore, the functional form of the characteristic acceleration may be illustrated as a surface, shown in Fig. 3.1 for a $100 \times 100\,\mathrm{m}$ solar sail with an efficiency of 0.85. It can be seen that for large values of σ_S the solar sail characteristic acceleration is relatively insensitive to variations in the payload mass due to the sail film and structural mass dominating the total mass of the solar sail. Conversely, for a large payload mass, the characteristic acceleration becomes insensitive to variations in the sail assembly loading. It is clear from Fig. 3.1 that to obtain a high characteristic acceleration not only must the sail film and structure be light, but the payload mass must also be small. As will be seen, there is a trade-off between investing effort in designing an efficient, high-performance solar sail and investing effort in developing a low-mass miniaturised payload.

During the solar sail design process a mass margin must be allocated to allow for growth in the mass of the solar sail and payload as the design matures. For solar sails a mass margin in the order of 20% or more is required due to the lack of experience in the manufacture of a real solar sail spacecraft. For significant increases in payload mass (due to additional instruments, for example) the entire design may be re-scaled by increasing the sail area. For example, inverting Eq. (3.2) an expression for the payload mass is obtained as

$$m_P = \left(\frac{2\eta P}{a_0} - \sigma_S \right) A \tag{3.3}$$

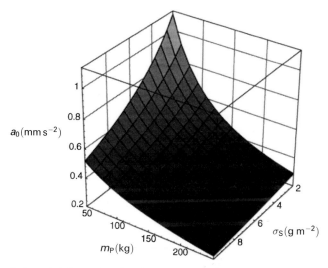

Fig. 3.1. Surface of characteristic acceleration for a 100 × 100 m solar sail with an efficiency of 0.85.

so that for a fixed characteristic acceleration and sail assembly loading, any increase in payload mass can be offset with a corresponding proportional increase in sail area. Any increase in sail area will, of course, increase the total mass of the solar sail, which is ultimately constrained by the choice of launch vehicle. Similarly, growth in the sail assembly loading (due to additional sail coatings, for example) can also be compensated for by an increase in sail area. However, the required increase in sail area is a strong function of the solar sail design. Such design sensitivities can now be determined quantitatively using sensitivity functions.

Sensitivity functions provide an indication of the relative importance of each design parameter for a given point in the solar sail design space. In general, the solar sail design space comprises the sail area, sail assembly loading and payload mass. For a fixed sail area the design space is a function of only two variables and so can be represented as a surface, as discussed earlier. The sensitivity functions then provide information on the gradient of this surface at a given design point. For example, the variation in characteristic acceleration due to variation in the sail assembly loading can be obtained from the relation

$$\Delta a_0 = \left(\frac{\partial a_0}{\partial \sigma_S} \right) \Delta \sigma_S \tag{3.4}$$

Therefore, using Eq. (3.2) to calculate the gradient due to variations in the sail assembly loading it is found that

$$\frac{\Delta a_0}{a_0} = \Lambda_1 \left(\frac{\Delta \sigma_S}{\sigma_S} \right), \qquad \Lambda_1 = \frac{-1}{1 + (m_P / m_S)} \tag{3.5a}$$

This relation then provides a measure of the sensitivity of the solar sail characteristic acceleration to growth in the sail assembly loading. It is clear that the solar sail performance is insensitive to the sail assembly design if the ratio of payload mass to

Table 3.1. Design sensitivity functions for a $100 \times 100\,\text{m}$ solar sail with a characteristic acceleration of $0.5\,\text{mm s}^{-2}$.

Assembly loading (g m^{-2})	Square sail dimensions (m)	Payload mass (kg)	Λ_1	Λ_2	Λ_3
6	100×100	95	-0.39	-0.61	0.39
4	100×100	115	-0.26	-0.74	0.26
2	100×100	135	-0.13	-0.87	0.13

sail mass is large, as discussed earlier. Other design sensitivity functions can also be obtained using a similar analysis, viz.

$$\frac{\Delta a_0}{a_0} = \Lambda_2 \left(\frac{\Delta m_P}{m_P} \right), \qquad \Lambda_2 = \frac{-1}{1 + (m_S/m_P)} \qquad (3.5b)$$

$$\frac{\Delta a_0}{a_0} = \Lambda_3 \left(\frac{\Delta A}{A} \right), \qquad \Lambda_3 = \frac{1}{1 + (m_P/m_S)} \qquad (3.5c)$$

It can be seen that sensitivity to payload mass growth increases with the ratio of payload mass to sail mass, and that increases in sail area are most effective for a low ratio of payload mass to sail mass, as expected. For illustration, example solar sail design points are listed in Table 3.1. Each solar sail design point generates a characteristic acceleration of $0.5\,\text{mm s}^{-2}$ with an assumed efficiency of 0.85. The sail design sensitivities described above are also calculated from Eqs (3.5). It can be seen that as the efficiency of the sail design increases, the design becomes more sensitive to variations in the payload mass and becomes less sensitive to variations in the total sail area. As is clear from Eqs (3.5a) and (3.5c), the sensitivities to sail assembly loading and sail area variations are equal in magnitude, but opposite in sign.

3.3 SAIL FILMS

3.3.1 Design considerations

Aside from the basic solar sail design parameters, there are many secondary design considerations which must be addressed. Firstly, a suitable material must be chosen for the sail substrate. The substrate is required principally to allow handling, folding, packing and deployment of the sail film. The chosen substrate must then be coated with a suitable reflecting material for efficient photon reflection. The sail substrate must have sufficient tensile strength to ensure that, when fully deployed and under tension, the sail film does not fail and create tears which may propagate further through the sail. In addition, since the reflective coating on the sail film will not have perfect reflectivity, a small fraction of the incident solar radiation will be absorbed by the substrate. This absorbed energy may then have to be dissipated through a thermally emitting rear surface. The choice of a suitable high-emissivity coating is yet another design decision. Since the absorbed energy will increase the substrate temperature, there must be dimensional stability and the sail should have low thermal expansion and shrinkage. As the substrate is so thin it has almost no thermal capacity and temperature changes are essentially instantaneous. Once deployed in orbit the sail film must also be free of wrinkles which may cause multiple reflections and intense hot spots. The sail shape can be simulated using finite element methods, in combination with a knowledge of the boundary conditions imposed by the sail structure, to calculate the actual shape due to billowing. In this way multiple reflections can be traced and eliminated. In all of these considerations the fabrication of the sail is an interdisciplinary task requiring the careful integration of materials, structures and thermal control.

3.3.2   Substrates

The sail substrate makes a major contribution to the total mass of the solar sail and hence has a significant impact on the solar sail performance. For example, Kapton®, one of the best potential sail substrates, has an areal density of $1.42\,\mathrm{g\,m^{-2}}$ per micron thickness even without coatings.* Therefore, the production of very thin films with good mechanical and thermal properties is central to solar sail film design and manufacture. Fortunately, there is extensive industrial experience of the manufacture, coating and handling of thin films for a number of applications. At present the main space application of metalised thin films is thermal blankets for conventional spacecraft. However, the thickness of commercially available thin films is at present somewhat coarse for a moderate performance solar sail. Therefore, the production of large areas of suitable thin film is likely to require specialised industrial processes rather than coating off-the-shelf products.

Several thin film materials have been considered as potential sail substrates. The main candidate films and their basic physical properties are listed in Table. 3.2. Although films such as Lexan® can provide low areal density, the optimal sail film is generally considered to be Kapton. Kapton is essentially chemically inert, has a high resistance to radiation and maintains its physical and mechanical properties over a

Table 3.2. Properties of candidate sail film substrates.

	Bulk density $(\mathrm{g\,cm^{-3}})$	Tensile strength $(\mathrm{N\,m^{-2}})$	Tensile modulus $(\mathrm{N\,m^{-2}})$	Surface density $(\mathrm{g\,m^{-2}})$			
				$1\,\mu m$	$3\,\mu m$	$5\,\mu m$	UV life
Kapton	1.42	1.72×10^8	2.96×10^9	1.42	4.26	7.10	Good
Mylar	1.38	1.72×10^8	3.79×10^9	1.38	4.14	6.90	Low
Lexan	1.20	6.89×10^7	2.07×10^9	1.20	3.60	6.00	Poor

*Kapton®, Mylar® and Lexan® are trade names of E.I. duPont de Nemours & Co.

wide range of temperatures. In addition, it has excellent adhesion to metal films, tapes and adhesives. Although Kapton does not have a melting point it does suffer a phase transition (glass transition temperature) above approximately 680 K. For some applications Kapton has been used to a temperature of 670 K, although its mechanical properties are degraded. A safe, long-term maximum operating temperature for solar sail applications is generally considered to be between 520 K and 570 K. Another good candidate as a sail film is Mylar®, although it has low resistance to solar UV radiation and is therefore unsuitable for long-duration exposure without double-sided coatings.

Kapton film is manufactured by extrusion of a polymer solution through dies onto a smooth casting surface. The film is lifted from the casting surface and cured at 670 K to produce its desirable physical properties. The cured film is then fed through tensioners and wound onto rollers for packing. Coating the film requires a similar tensioning and handling process. At present Kapton is commercially available in rolls of up to 1.32 m wide with a thickness of 7.6 µm, although specialised Mylar-like films are commercially available down to a thickness of 0.9 µm. For a moderate performance solar sail, a thickness of order 2 µm is required. However, such thin films are not routinely used for large volume commercial purposes, mainly due to the difficulty in handing during manufacture. Small-scale experimental trials have demonstrated that Kapton can be plasma etched to provide an areal density as small as $1 \, \mathrm{g \, m^{-2}}$. Patterns of unetched Kapton can then be used to carry most of the loads in the sail film. Other techniques for producing thin films include water casting, where a polymer solution forms a thin film floating on a layer of water, and chemical etching using a sodium hydroxide bath. Experiments have used chemical baths to etch standard Kapton film to 2.5 µm, although small-scale tests have demonstrated thicknesses of only 0.4 µm. In addition to solar sails, other space applications such as solar concentrators and space telescope Sun shades also require films that are thinner than commercially available Kapton, which may allow some synergy in technology development. Even if such thin films can be produced in volume, the film must then be lifted and handled with the associated difficulties of avoiding tears in such ultra-thin materials. Of course, the ultimate sail film will be all metal with no plastic substrate, as will be discussed later in this section.

3.3.3   Coatings

In order to provide a reflecting surface, the sail substrate must be coated with a suitable metallic reflector, typically using a vapour deposition process. The coating must provide good reflectivity across the solar spectrum and have a suitably high melting point. The required melting point will be dependent on the mission trajectory and the subsequent closest solar approach distance. The physical properties of several candidate reflective coatings are listed in Table 3.3. In terms of reflectivity it is found that silver has excellent performance at optical wavelengths, but has a sharp window of transparency at UV wavelengths. Therefore, solar UV radiation could penetrate the coating and degrade the plastic substrate. In addition, silver has a rather high density when compared to other candidate reflective coatings. It can also

Table 3.3. Properties of candidate sail film coatings.

		Surface density (g m^{-2})			
	Bulk density (g cm^{-3})	0.1 μm	0.3 μm	0.5 μm	Melting point (K)
Aluminium	2.70	0.27	0.81	1.35	933
Lithium	0.53	0.05	0.16	0.27	453
Silver	10.5	1.05	3.15	5.25	1234

be seen that the lowest density coating is obtained using lithium, but has a rather low melting point. The optimal choice of coating is generally considered to be aluminium, providing a reflectivity between 0.88 and 0.9. A further front coating, such as silicon oxide, may also be required to reduce pre-launch oxidation of the reflecting surface with a resultant loss of reflectivity.

In addition to a reflective coating, the sail may require a high-emissivity back coating for passive thermal control of the sail film by re-radiation. The overall reflectivity and emissivity of the sail film has a significant effect on the film equilibrium temperature. Since the sail film is extremely thin it has almost no thermal capacity, and temperature changes are therefore essentially instantaneous. The sail equilibrium temperature calculated in Chapter 2 can thus be used to investigate the sail temperature as a function of the sail orientation, the sail optical properties and heliocentric distance. From Eq. (2.55) in section 2.6.1 it can be seen that the equilibrium temperature T is given by

$$T = \left[\frac{1 - \tilde{r}}{\varepsilon_f + \varepsilon_b} \frac{W_E}{\tilde{\sigma}} \left(\frac{R_E}{r} \right)^2 \cos \alpha \right]^{1/4} \qquad (3.6)$$

Using Eq. (3.6), the sail equilibrium temperature can be obtained as a function of heliocentric distance for a range of back emissivities, assuming a reflectivity $\tilde{r}$ of 0.85 and zero front emissivity, as shown in Fig. 3.2. It is clear that for orbits in the inner solar system a high emissivity coating is essential. Uncoated aluminised Kapton has a rear emissivity of order 0.34, which is too low to provide passive thermal control for inner solar system missions. However chromium, with an emissivity of order 0.64, appears to be a suitable candidate for a rear surface coating. For an allowable substrate temperature of order 520 K, solar approaches down to 0.2 au appear possible for a well-designed sail film. Advanced thermal control methods which utilise thin film interference effects and micro-machined quarter-wave radiators also appear possible. These methods would allow significantly higher rear surface emissivities and hence much closer solar approaches. Such close approaches would be of great benefit for rapid inclination cranking manoeuvres and for high acceleration solar passes *en route* to the outer solar system.

Finally, a cross-section of a typical Kapton sail film is shown in Fig. 3.3. The sail has a 2 μm Kapton plastic substrate upon which a thin 0.1 μm aluminium layer is deposited. The rear surface of the sail film has a 0.0125 μm chromium coating for

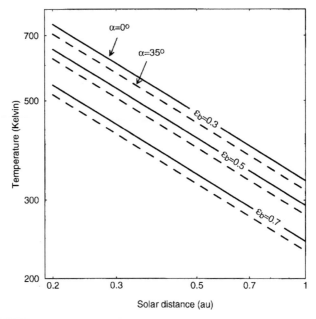

Fig. 3.2. Sail equilibrium temperature as a function of heliocentric distance with a reflectivity of 0.85 and zero front emissivity.

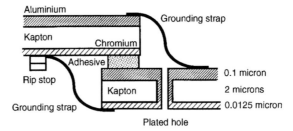

Fig. 3.3. Cross-section of bonded solar sail film panels.

passive thermal control of the sail, as described above. In addition, the sail has grounding straps, which provide electrical paths to prevent electrical discharges, and rip stops, which will be discussed in section 3.3.5.

3.3.4   Metallic sail films

The performance of a conventional solar sail is inevitably limited by the need for a plastic substrate. As noted above, Kapton has an areal density of $1.42\,\mathrm{g\,m^{-2}}$ per micron thickness. Therefore, a sail consisting solely of 1 micron of Kapton is already approaching the limit of areal density required for a solar sail of unit lightness number. In practice, however, the substrate is only required to allow handling and packing during manufacture and to ensure safe deployment without tearing. Once deployed, the loads on the sail film need not be severe for a well-designed structure.

The practical limitations to metalised plastic film comes from the difficulties in handling ultra-thin films in an environment with air currents, electrostatics and particulate contaminants. If such contaminants adhere to the surface of the film, winding under tension onto rollers will only further degrade the integrity of the sail. The question therefore arises whether the substrate could be removed on-orbit, therefore greatly enhancing the performance of the solar sail. While plasma etching of plastic substrates has been demonstrated, it is not a process that could easily be applied on-orbit. A more appropriate solution would be to use a substrate which quickly vaporised under the action of solar UV radiation or hard vacuum, thus leaving a thin reflective metal film. Seams of non-vaporising substrate would be left to provide load paths and rip stops to avoid tear propagation in the thin metal film.

Although such approaches appear attractive and can provide significant performance gains, little work has been performed to date to fabricate such materials. However, recent experiments by Salvatore Scaglione and others in Italy have identified a promising new approach to high-performance sail film design. In this new scheme a layer of diamond-like carbon (DLC) is deposited on a UV transparent plastic substrate. The coated substrate is then aluminised by conventional vapour deposition. Once the sail film is deployed in orbit, the DLC buffer between the substrate and the aluminium layer degrades under the action of penetrating UV radiation, leading to separation of the substrate and an all-metal sail film. Other concepts proposed include etching the substrate by atomic oxygen in Earth orbit.

An alternative to terrestrial manufacturing of the sail is the possibility of manufacture on-orbit. This need not be a large-scale process but, in principle, can be automated using a small vapour deposition unit. The hard vacuum required to allow such vapour deposition is already provided by the space environment. Vapour deposition involves, firstly, evaporating a suitable metal, such as aluminium, from a stock provided in powder form. After evaporation by direct electrical heating the metal vapour is then condensed onto a substrate and finally separated once cooled. As early as the late 1960s terrestrial laboratory experiments were performed to manufacture thin metal films for solar sails. These initial studies used either a rotating, rigid cylindrical substrate or a flexible rotating band, as shown in Fig. 3.4. By altering the speed of rotation of the substrate device, and by using masks, the thickness of the metal film could be varied to provide, for example, additional

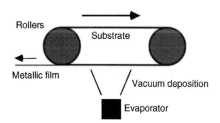

Fig. 3.4. Production of an all-metal sail film using vapour deposition.

strength at the edges of the film. Small-scale laboratory processes fabricated coarse films with a thickness of 12.7 μm and a 25.4 μm edge. On a larger scale the process could directly manufacture heliogyro blades with the thicker blade edges carrying tension loads. After manufacture the vapour deposition unit would be discarded by separation from the solar sail hub.

Much thinner metal films, only a few hundred atoms thick, have also been manufactured in small-scale laboratory experiments. In this process a glass substrate is firstly coated with detergent and a layer of aluminium is added through vapour deposition in a vacuum chamber. The glass substrate is then immersed in water so that the thin metal film separates from the substrate and floats on the layer of water. Experiments in the late 1970s by Eric Drexler have demonstrated this method for producing films as thin as 0.05 μm. Although aluminium has principally been considered, there are other possibilities for metallic films, as listed in Table 3.3. In particular, lithium has a much lower density than aluminium, but also has a lower melting point. For any thin metallic film there will be a minimum thickness at which the optical transmittance of the film becomes significant. For example, aluminium has a transmittance of order 0.01 at a thickness of 0.03 μm, rising to 0.4 at a thickness of 0.005 μm. Therefore, as the thickness of the film is decreased the benefit of lower areal density is rapidly offset by the increased optical transmittance of the film.

For a metallic film thick enough to have a low optical transmittance, the film can in principle be made lighter without greatly degrading its optical properties. This can be achieved by perforating the film with holes smaller than the mean wavelength of sunlight, as shown in Fig 3.5. Such techniques already exist for other applications using focused ion beams and holographic gratings. These techniques can routinely produce structures of order 0.2 μm, albeit on a small scale. For application to solar sails, the perforated metal film could be laid down on a UV-sensitive plastic substrate using techniques adapted from the semiconductor industry. As described above, the substrate would allow easy handling, packing and deployment, but would vaporise or separate after deployment to leave a thin, perforated metallic film. The properties of the film could be further enhanced by configuring short whiskers on the rear surface of the sail to radiate away heat for passive thermal control. The whiskers would act as quarter-wave antennae at infra-red wavelengths. Such micro-fabrication methods hold great promise for the future.

3.3.5   Environmental effects

Any large surface which is exposed in space will be subject to continual degradation from a number of environmental sources. For spar-supported solar sails, micro-meteorite impacts may cause only local damage. However, for spinning solar sails the centrifugal tension in the sail film will certainly require the use of regular rip stops to prevent catastrophic tear propagation. The integration of rip stops during the manufacturing process will be discussed in section 3.3.6. In addition to particulate damage, solar UV radiation can degrade some plastic substrates, such as Mylar, through weakening and reduction in tensile strength. Even for a fully

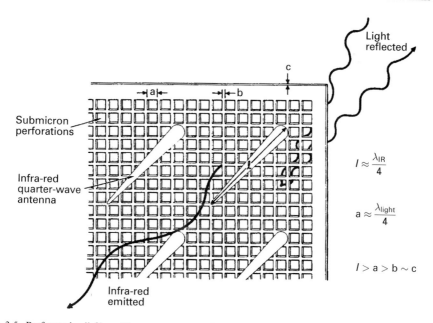

Fig. 3.5. Perforated sail film with quarter-wave radiators for passive thermal control (Robert Forward).

coated sail film, UV radiation may still reach the substrate by penetration through the coating or through local cracks.

Since the sail film is exposed to the solar wind and solar UV radiation, the sail can acquire a differential electrical charge between the front and rear surfaces. This is attributable to the incident proton flux from the solar wind and the photoelectric effect ejecting electrons from the sail. To prevent electrical discharges from the front surface to the rear of the sail, which is a potential source of failure and tearing, both surfaces of the sail must be in electrical contact. This can be achieved through grounding straps passed between neighbouring elements of the sail film and through the use of electrically conducting adhesives, as shown in Fig. 3.3. For sail films coated on both sides, the plastic substrate can be laser drilled before coating to ensure electrical contact. The sail surfaces must also be grounded to the spars, stay lines and any other structural components. A further consequence of electrical charging is that the sail may form a 'bubble' within the solar wind plasma. Within this region, field and particle instruments would not be able to obtain accurate measurements. Clearly, this is a major source of concern for future space science missions and requires further investigation. Finally, although sail film materials have been characterised in detail and samples fabricated, the ultimate performance, degradation and lifetime of sail films will only be assessed through on-orbit testing.

3.3.6   Sail bonding, folding and packing

Given that any sail film will only be produced in relatively narrow strips during

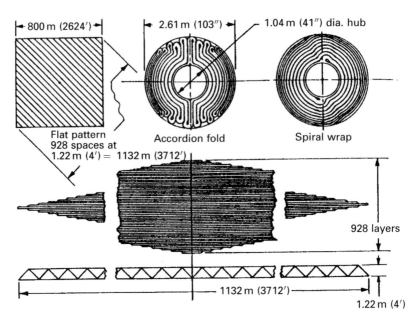

Fig. 3.6. Sail film packing for the JPL comet Halley square sail using spiral and accordion fold (NASA/JPL).

manufacture, individual elements of the sail must be bonded to produce large continuous sheets. The exception to this requirement is the heliogyro where the blades may be equal in width to the sail film produced by the manufacturing process. For large continuous sail sheets the bonded sail film must be folded and packed into a small volume for launch. Since it would be impractical to manufacture the entire sail and then fold it, fabrication processes must be devised to allow sail folding to proceed in parallel with the bonding process.

The bonding of individual elements of the sail film can be achieved through a variety of techniques. The simplest technique requires adjacent elements of the sail to be brought into contact in a butt joint. This joint is then overlaid with a tape of sail film coated with a suitable adhesive. The bonding is completed by a heat-sealing technique whereby the sealer runs along the join with the sail element held stationary on a long table. To limit the size of the manufacturing rig the sail is wound along the table between adjacent rollers after bonding. As the sail element is wound onto the roller it is folded over its neighbour and the process repeated with a new sail element. The use of rollers allows the entire manufacturing rig to be quite compact. For a square solar sail the bonding of individual elements forms seams running diagonally across the sail. This allows the sail to be folded along a diagonal into a long ribbon during manufacture. This ribbon of folded sail can then be spiral wound onto a drum or accordion packed, as shown in Fig. 3.6. While the accordion pack is the simplest to deploy, the spiral warp avoids double folds in the sail and so minimises wrinkling after deployment. For a heliogyro sail folding and packing is considerably easier with each blade of the heliogyro simply wound onto a roller for packing and unwound from the roller during deployment. As discussed in section 3.3.5, to ensure that tears due to failure of the sail film do not propagate, the sail must be provided with rip stops at regular intervals. These can be formed from Kapton tapes joined to the rear surface of the sail during the manufacturing process.

3.4 SOLAR SAIL STRUCTURES

For any solar sail design concept the sail film and structure must be stowed in a small volume consistent with the launch vehicle payload faring or faring volume allocation. Once in orbit the structure must then be able to deploy reliably from the stowed state and rigidise a large area of reflective film. The structure must also be designed so that any redundant mass may be jettisoned after deployment. For example, deployment drive motors, spar canisters and sail packing covers are no longer required after deployment and only degrade the performance of the solar sail. Since the sail film has no resistance to bending or compression, the structure must apply tension to the sail film in order to maintain an approximately flat form. For a square solar sail, spars cantilevered from a central hub are used to keep the sail film in tension. For large square solar sails additional stays are required to prevent buckling of the spars under load. In typical configurations the structural elements are also required to transmit mechanical loads from the sail film to the payload and bus, which represent an inertial load at the centre of the solar sail structure. In contrast, the heliogyro configuration dispenses with the need for cantilevered spars to keep the sail film in tension. The heliogyro achieves a flat sail film by spinning the spacecraft so that the centripetal force acting on the blades provides radial tension. However, the heliogyro has its own structural design considerations, as will be discussed later. Such spin-induced tension methods can also be utilised for disc solar sails using radial compression members either alone or in combination with an outer hoop.

There is extensive on-orbit experience of deployable structures for applications such as solar arrays, antennae and experiment booms. However, as yet none of these applications appears to have been on a scale suitable for solar sails, although there is evidence of classified military satellites using very large deployable antennae. A popular deployable structure with potential application to solar sailing is the storable tubular expandable member (STEM). The STEM structure is a metallic or carbon fibre tube which can be pressed flat and rolled onto a spool for packing. The structure is deployed using the potential energy stored in the prestressed flattened tube, or using a small drive motor for a more controlled deployment. A section of a carbon fibre STEM boom is shown in Fig. 3.7. By using carbon fibre with layers built-up in alternate directions, deployable booms can be manufactured with essentially zero coefficient of thermal expansion. Once deployed the drive motor and associated hardware can be jettisoned in order to reduce to the total mass of the solar sail. Another popular structure is the continuous longeron coilable boom (CLCB). The CLCB is a linear truss structure with a triangular cross-section. Triangular elements are joined to longerons to form the truss, as shown in Fig.

Fig. 3.7. STEM collapsible tubular boom (DLR).

Fig. 3.8. CLCB collapsible boom (AEC – Able Engineering Company, Inc.)

3.8. Pretensioned diagonal elements store enough potential energy to allow self-deployment, although a lanyard cable attached to the end of the truss can be used to control the deployment rate. The CLCB typically has a stowed length of the order of 2% of its final deployed length.

In addition to deployable mechanical structures, inflatables provide an attractive means of reliable deployment. Inflatable structures have long been considered for solar concentrators, antenna reflectors and truss structures. The main benefit of inflatables is the ease and reliability of deployment with few failure modes. The structure consists of a thin membrane which is deployed solely by internal gas pressure, typically using stored nitrogen or warm gas generators (such as sodium azide, used in automobile air bags). Once the structure begins to deploy, the internal gas pressure ensures full deployment and rigidisation. Additional gas can be provided to account for leakage through seams or micrometeorite penetration, although the requirement for such make-up gas varies as the square of the mission duration. As a lower mass alternative, rigidisation may be provided by a space-curing resin enclosed between two films in the inflatable membrane. Following deployment the resin hardens once heated by solar radiation. Other rigidisation methods rely on fabric membranes impregnated with gelatine. On exposure to hard vacuum a water-based solvent is released thus rigidising the membrane. The timescales for deployment and rigidisation must be carefully selected in order to avoid rigidisation prior to full deployment of the structure.

Of great interest for solar sail applications was the inflatable antenna experiment (IAE) flown on the Space Shuttle mission STS-77 in May 1996. As discussed in section 1.2.5, the IAE was primarily designed to test the deployment, rigidisation and surface accuracy of a large inflatable antenna. However, the experiment provided valuable insight into the possibilities of using inflatable structures for solar sails. The 14 m diameter reflector was manufactured from 6.4 µm aluminised Mylar. The reflector was supported by three struts, each 28 m long, manufactured from neoprene-coated Kevlar. The mass of the entire assembly was some 60 kg and demonstrated packing and deployment of an inflatable structure with a dimensional expansion ratio of over 100. Due to unexpected venting of air in the folded structure just after release, the full deployment did not follow the expected sequence, as shown in Fig. 3.9. However, the ultimately successful deployment clearly demonstrated the reliability of inflatable structures since internal gas pressure forced the process to completion.

3.5   SOLAR SAIL CONFIGURATIONS

The choice of solar sail configuration is dependent on mission requirements such as the required solar sail characteristic acceleration and maximum sail turning rate, along with programmatic considerations such as cost and risk. In general, the main configurations are the square solar sail, disc solar sail and heliogyro, as discussed in section 3.1. During the JPL comet Halley studies, the square solar sail and heliogyro configurations were investigated in some detail. Although the heliogyro appears to be the more efficient configuration in terms of structural mass, studies showed that the square solar sail and heliogyro provided remarkably similar performances. This was in part due to the need for load-carrying members in each of the heliogyro blades which offset some of the potential benefits of a spinning

Fig. 3.9. Inflatable antenna experiment deployment (NASA/JPL).

structure. For long-term development, however, the square solar sail was ultimately seen as the better choice. More recent studies, detailed in section 3.6, have considered all three configurations, although the square solar sail is still seen as the optimum choice. This is in part due to its simplicity of operation and its ability to provide the rapid turning rates required for planetary escape and capture spirals. Such spiral trajectories are extremely demanding and are complicated by local gravity gradient torques which scale with the moments of inertia of the solar sail configuration. For heliocentric cruise, where rapid turning rates are not required, the disc solar sail has the potential for significantly higher performance than the square solar sail.

3.5.1   Optimum solar sail configurations

In order to illustrate the potential trade-off between solar sail configurations, a hypothetical spar-supported square solar sail and hoop-supported disc solar sail will be evaluated. The payload of the disc solar sail is assumed to be located at the centre of the disc and connected to the rim using thin stay lines, while the payload of the square solar sail is located at the junction of the spars. These two schematic configurations are only used to provide a simple geometrical evaluation of two competing design concepts. The trade-off between real solar sail configurations will, of course, require a more thorough analysis than the cursory evaluation provided

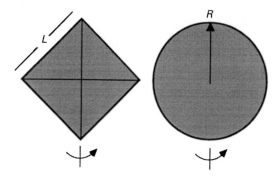

Fig. 3.10. Square solar sail of side L and a disc solar sail of radius R.

here. For example, while minimisation of structural mass is a key consideration, it is constrained by the need for a reasonable load safety factor and reliable deployment.

Firstly, a comparison of the estimated structural mass of the square solar sail and disc solar sail is made based on a simple analysis of the solar sail geometry, as shown in Fig. 3.10. For a square solar sail of side L and area A the total length of the four diagonal spars D_S is $2\sqrt{2}\,L$. Therefore, the total spar length for the square solar sail can be written as a function of the sail area as

$$D_S = 2\sqrt{2A} \qquad (3.7)$$

The disc solar sail is assumed to have an outer hoop structure of radius R so that the total length of structure is just the circumference of the disc $2\pi R$. Therefore, the hoop length of the disc solar sail may be written as a function of the sail area as

$$D_C = 2\sqrt{\pi A} \qquad (3.8)$$

It can be seen that the length of supporting structure varies as $\sqrt{A}$ for both configurations. Therefore, the ratio of the length of the supporting structure of the square solar sail and disc solar sail is independent of the sail area. This ratio is then an approximate measure of the relative efficiency of these two configurations in terms of structural design. From Eqs (3.7) and (3.8) it can be seen that

$$\frac{D_S}{D_C} = \sqrt{\frac{2}{\pi}} \sim 0.8 \qquad (3.9)$$

so that for a given sail area the square solar sail requires a shorter length of supporting structure than the disc solar sail. Although this analysis indicates that the square solar sail configuration is more efficient, it is of course only an approximate comparison as the mass of the structural elements will differ due to the different loads experienced by the two solar sail configurations. The square solar sail will experience loads at the spar tips, inducing buckling, whereas the disc solar sail will experience uniform tension around its circumference. By spinning the disc solar sail a reduction in edge tension is possible, as will be seen in section 3.5.4.

A similar comparison can also be made to evaluate the moments of inertia of the square solar sail and disc solar sail. For a fixed sail turning rate the required control torque is directly proportional to the solar sail moment of inertia. Again a square solar sail of side L will be considered with a spar mass per unit length λ. In this approximate analysis the moment of inertia of the sail film will be ignored relative to that of the solar sail structure. Considering a rotation of the square solar sail along a diagonal, as shown in Fig 3.10, the moment of inertia I_S is found to be

$$I_S = \frac{\lambda}{3\sqrt{2}} A^{3/2} \tag{3.10}$$

It can be seen that as the area of the solar sail increases there is a greater than linear growth in the moment of inertia. Large solar sails will therefore require significant control torques and actuator mass. Conversely, small solar sails offer the possibility of low mass actuation methods due to this scaling law. A similar analysis can now be performed for the disc solar sail. For a solar sail with the same structural mass per unit length as the square solar sail, the moment of inertia about an axis in the plane of the disc is found to be

$$I_C = \frac{\lambda}{\sqrt{\pi}} A^{3/2} \tag{3.11}$$

so that the moment of inertia again scales as $A^{3/2}$. Therefore, the ratio of the moment of inertia of the square solar sail and the disc solar sail is independent of sail area. This ratio then provides an approximate measure of the relative magnitude of the control torques required for attitude control. From Eqs (3.10) and (3.11) it can be seen that

$$\frac{I_S}{I_C} = \frac{1}{3}\sqrt{\frac{\pi}{2}} \sim 0.4 \tag{3.12}$$

so that for a given sail area the square solar sail has a significantly smaller moment of inertia. Again, this approximate analysis does not consider the differing loads experience by the two solar sail configurations, which will lead to a somewhat different structural mass per unit length for each. In addition, if the disc solar sail is spinning then attitude control will be achieved through precession of the spin axis. This precession rate is also a function of the polar moment of inertia about the axis of symmetry of the disc.

As a further illustration of the issues of configuration selection, the optimality of the square solar sail will now be investigated in more detail. The square solar sail has a number of related configurations using either a greater or smaller number of spars. For example, a configuration using three spars will have a triangular form while configurations with more than four spars will have a polygonal form. A measure of the efficiency of each configuration is given by the ratio of sail area to number of spars required to support the sail film. In order to evaluate this measure an arbitrary polygonal solar sail will be considered with N spars each of length R, as shown in Fig. 3.11. From the geometry of the polygon it can then be shown that the area of a

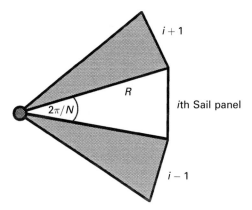

Fig. 3.11. Polygonal solar sail section.

single sail panel is $R^2 \sin(\pi/N) \cos(\pi/N)$. The total area of a polygonal solar sail with N panels is then given by

$$A = NR^2 \sin\left(\frac{\pi}{N}\right) \cos\left(\frac{\pi}{N}\right) \qquad (3.13)$$

As the degree of the polygon increases, $A \to \pi R^2$ as $N \to \infty$ so that a disc solar sail is obtained in this limit. The total length D of the N spars for the polygonal solar sail is simply

$$D = NR \qquad (3.14)$$

Now, using these results, the ratio of solar sail area to the total spar length may be obtained. This ratio is then a measure of the structural efficiency of the solar sail design. Using Eqs (3.13) and (3.14) it can be seen that

$$\frac{A}{D} = \frac{R}{2} \sin\left(\frac{2\pi}{N}\right) \qquad (3.15)$$

It is clear from Eq. (3.15) that the optimum number of spars is $N = 4$, corresponding to a square solar sail. This choice will then maximise the sail area while minimising the number of spars required to support the sail film.

It appears that, from simple geometrical considerations, the square solar sail is the optimum choice of configuration. However, there are many other considerations which must be addressed to synthesise a complete solar sail design which meets all of the required mission objectives and engineering constraints. Even when a configuration has been selected, the solar sail sized and the structure designed, the final concept must be validated through detailed engineering analysis. The sail shape and structural loads must be evaluated in all possible sail orientations over the range of heliocentric distances expected during the mission. As with all structural design problems, an adequate safety factor must also be incorporated to avoid unexpected failure by, for example, spar buckling. The sail film temperature must then be determined, again as a function of both the solar sail orientation and heliocentric distance. Regions of the sail film which may attain elevated temperatures through multiple reflections must be identified and possibly eliminated. Lastly, the interaction of the payload with the solar sail must be determined to assess the impact of the design on camera and antenna viewing angles, disturbances to the local plasma environment and thermal control.

3.5.2   Three-axis stabilised square sail

3.5.2.1   Design

The square solar sail configuration uses single or multiple sheets of sail film which are kept in tension using diagonal spars cantilevered from a central hub. For large solar sails the bending loads on the spars may become excessive, and the spars must then be supported by stays connected to a central boom attached normal to the hub. The stays reduce the structural loads on the solar sail and hence prevent the spars from buckling. Since the spars are sized for bending loads they can contribute significantly to the total mass of the solar sail. The autonomous deployment of such a system of spars and stays represents a large number of potential serial failures and is perhaps the greatest disadvantage of the square solar sail. For the 800 × 800 m JPL comet Halley square solar sail, discussed in section 1.2.3, the spars were to be CLCB lattice structures, constructed of titanium to prevent undue thermal expansion. Owing to the lattice nature of the structure the spars could be simply coiled for storage, as discussed in section 3.4. The stays were designed as flat tapes to prevent snarling during deployment. For automated in-orbit deployment the central boom and diagonal spars were to be simultaneously extended, thus pulling the stays along the spars from storage reels. Once the entire structure was erected the sail itself would be deployed. The sail is firstly pulled into a ribbon along a diagonal spar and then pulled flat along the perpendicular diagonal, as shown in Fig. 3.12. If the sail were constructed from four individual triangular sheets, the same two-stage procedure would take place for each element. Although this deployment scheme has been shown to be feasible, the large number of serial operations leads to many potential failure modes. Since the Jet Propulsion Laboratory comet Halley scheme could not even be partially tested prior to implementation, the deployment had a high element of risk. This element of risk was the main consideration when the heliogyro was selected over the square sail for the comet Halley mission.

3.5.2.2   Control

Three-axis attitude control of a square solar sail may be accomplished by several means. One method is to displace the centre-of-mass of the solar sail relative to the centre-of-pressure. This can be achieved by mounting the payload on an articulated boom and utilising boom rotations to control the centre-of-mass location. Alternatively, the entire sail may be translated across the solar sail structure to displace the centre-of-pressure. The sail may be displaced by reeling in outboard support lines while reeling out the opposite set of lines. While these displacement methods can provide full three-axis attitude control, control authority is weak when the solar sail

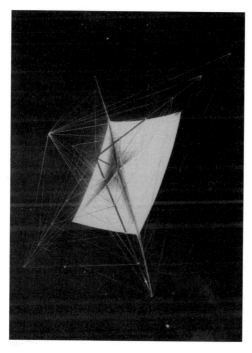

Fig. 3.12. Square solar sail deployment (NASA/JPL).

is almost edgewise to the Sun. Better control authority can be achieved through the use of small reflective vanes at the spar tips. Using combinations of vane rotations, arbitrary pitch, roll and yaw torques may be generated, even when the main sail is in a null attitude edgewise to the Sun-line. It can be shown that suitable control can in fact be achieved with vanes which only rotate about axes along the spars to which they are attached. In practice, combinations of more than one method may provide the optimal solution for good three-axis attitude control. The required control authority is, however, a function of the mission phase. Planet-centred escape or capture spiral trajectories place greater demands on the solar sail attitude control actuators than Sun-centred cruise trajectories.

Firstly, centre-of-mass offset control will be investigated. An ideal, flat solar sail will be considered with body axes (x, y, z), as shown in Fig. 3.13. Pitch rotations will be defined about the $+y$ axis, yaw rotations about the $+x$ axis and roll rotations about the $+z$ axis. It will be assumed that the payload of mass m_P is located at the end of a boom of fixed length l so that the position of the payload is defined in body axes as

$$\mathbf{r}_P = l\cos\theta\cos\phi\,\mathbf{e}_1 + l\cos\theta\sin\phi\,\mathbf{e}_2 + l\sin\theta\,\mathbf{e}_3 \qquad (3.16)$$

where θ is the boom elevation and ϕ is the boom azimuth relative to the solar sail body axes. These two boom angles then provide the controls for modulating the

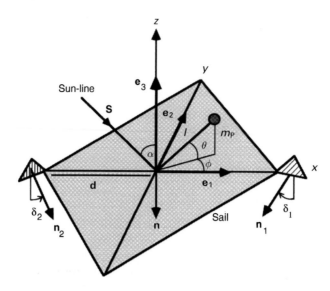

Fig. 3.13. Square solar sail attitude control actuators.

torque exerted on the solar sail. If the mass of the solar sail excluding the payload is m_S, the position of the centre-of-mass $\mathbf{r}_C$ is defined in body axes as

$$\mathbf{r}_C = \frac{m_P}{m_P + m_S}\mathbf{r}_P \tag{3.17}$$

It can be seen that if the boom length is fixed, the centre-of-mass location can only be displaced over a hemispherical surface of radius $m_P/(m_P + m_S)l$ on the sunward face of the solar sail. Since the solar sail is assumed to be ideal, the solar radiation pressure force acts in direction $\mathbf{n}$ normal to the sail surface. Then, the force f exerted on the solar sail is given by

$$f = \tilde{f}_S(\mathbf{S} \cdot \mathbf{n})^2 \mathbf{n} \tag{3.18}$$

where $\mathbf{S}$ is the Sun-line vector and $\tilde{f}_S$ is the force exerted on the solar sail when facing the Sun. For ease of illustration it will be assumed that the Sun-line is located in the x–z plane at an angle α relative to the sail normal. Since the force vector does not act through the centre-of-mass of the solar sail, the torque $\mathbf{M}$ generated by the centre-of-mass offset is now defined to be

$$\mathbf{M} = -\mathbf{r}_C \times f \tag{3.19}$$

where $-\mathbf{r}_C$ is the point of action of the solar radiation pressure force vector relative to the solar sail centre-of-mass. Writing Eq. (3.19) in component form, the torque generated about each body axis of the solar sail is found to be

$$M_x = \tilde{f}_S \frac{m_P}{m_P + m_S} l \cos^2 \alpha \cos \theta \sin \phi \qquad (3.20a)$$

$$M_y = -\tilde{f}_S \frac{m_P}{m_P + m_S} l \cos^2 \alpha \cos \theta \cos \phi \qquad (3.20b)$$

$$M_z = 0 \qquad (3.20c)$$

It can be seen from Eqs (3.20) that body pitch and yaw torques may be generated by a suitable choice of boom orientation, but that roll torques cannot be generated. However, it can be shown that, for a non-ideal solar sail, small roll torques are in fact generated since the solar radiation pressure force vector is no longer directed normal to the sail surface. In addition, it can also be seen that control torques cannot be generated when the solar sail is oriented edgewise to the Sun.

Normalised pitch and yaw torques are shown in Fig. 3.14 for an ideal solar sail in a Sun-facing attitude. It can be seen that the magnitude of the torque may be controlled using the boom elevation while the sign of the torque may be controlled using the boom azimuth. For a required set of control torques, Eqs (3.20) may then be used to obtain the necessary boom azimuth and elevation angles. In addition, Eqs (3.20) may also be used to size the boom length to generate the necessary maximum torques expected during the solar sail mission. It is clear then that the centre-of-mass offset can provide useful control torques for attitude control. However, an unintentional offset of the centre-of-mass and centre-of-pressure can easily generate bias torques which can dominate the nominal torques required for attitude manoeuvres. Such an unintentional offset can occur due to thermal bending of the spars, asymmetries in the mass and reflectivity properties of the sail film, and incomplete deployment of the solar sail structure. The control boom length must therefore be sized to compensate for any expected disturbance torques along with the manoeuvring torques required for attitude control.

As an alternative means of three-axis attitude control, the use of spar tip-mounted vanes will now be considered. The tip vanes are small reflective panels attached to the end of the spars through drive motors. Although the force generated by a tip vane is small, the large moment arm relative to the solar sail centre-of-mass can provide suitable control torques. For ease of illustration the solar sail will be assumed to have only two articulated vanes located at the tips of the x-axis spar structure, as shown in Fig. 3.13. While two vanes will only provide control in pitch and roll, it will be seen that yaw control may in fact be achieved through an appropriate combination of pitch and roll rotations of the entire solar sail. It will now be assumed that the vanes may only be rotated about the spar axis through a clockwise rotation δ. Then, the unit normal vector to each vane $\mathbf{n}$ may be written in the solar sail body axes as

$$\mathbf{n}_1 = \sin \delta_1 \, \mathbf{e}_2 - \cos \delta_1 \, \mathbf{e}_3 \qquad (3.21a)$$

$$\mathbf{n}_2 = -\sin \delta_2 \, \mathbf{e}_2 - \cos \delta \, \mathbf{e}_3 \qquad (3.21b)$$

Each tip vane will generate a torque which can be superimposed to provide a useful

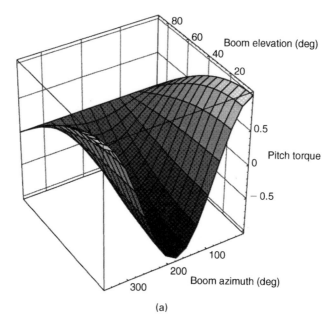

(a)

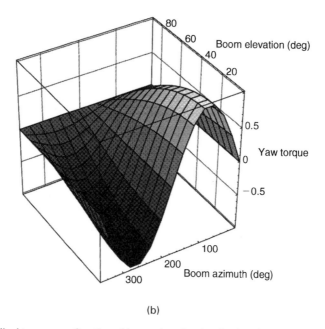

(b)

Fig. 3.14. Normalised torque as a function of boom elevation θ and azimuth ϕ: (a) pitch torque; (b) yaw torque.

total control torque. Then, the control torque generated by the ith vane is obtained from

$$\mathbf{M}_i = \mathbf{d}_i \times \tilde{f}_V (\mathbf{S} \cdot \mathbf{n}_i)^2 \mathbf{n}_i \tag{3.22}$$

where $\mathbf{d}_i$ is the location of the ith vane relative to the solar sail centre-of-mass and $\tilde{f}_V$ is the tip vane force obtained when facing the Sun. For spars of length d the control torque generated about each body axis of the solar sail may then be written in component form as

$$M_x = 0 \tag{3.23a}$$

$$M_y = \tilde{f}_V \, d \cos^2 \alpha (\cos^3 \delta_1 - \cos^3 \delta_2) \tag{3.23b}$$

$$M_z = \tilde{f}_V \, d \cos^2 \alpha (\cos^2 \delta_1 \sin \delta_1 + \cos^2 \delta_2 \sin \delta_2) \tag{3.23c}$$

Therefore, through appropriate combinations of vane rotations, arbitrary pitch and roll torques may be generated with the roll torques generated through a collective rotation of both vanes. In addition to determining the required vanes angles, Eqs (3.23) may also be used to size the area of the tip vanes to generate the necessary maximum torques expected during the solar sail mission.

Normalised pitch and roll torques generated for an ideal solar sail facing the Sun are shown in Fig. 3.15. It can be seen that full control authority can now be achieved in both pitch and roll. If yaw control is only required infrequently, it can be demonstrated that full three-axis control can in fact be obtained with only two vanes located at the tips of the x-axis spar. The sail is firstly rolled by 90° using collective rotations of the two vanes. Then, the sail is pitched to the desired attitude and finally rolled back through 90°. This full sequence of rotations then corresponds to a change in the sail yaw attitude.

Tip vanes and centre-of-mass displacement each have their own particular advantages for solar sail attitude control. While tip vanes provide excellent control authority, actuators are required at the end of each spar. This may present additional demands for the sail structural design and pose problems for packing and deployment. In addition, the vane orientations must be accurately controlled to avoid the generation of unintentional disturbance torques, particularly in roll. While an articulated central boom does not provide a control authority as efficient as tip vanes, only a single gimbal actuator is required near the centre-of-mass of the solar sail. Boom rotations may not be problematic for passive experiments, but continual rotation of the payload may pose additional pointing difficulties for optical payloads and communication antennae.

3.5.3   Spin-stabilised heliogyro

3.5.3.1   Design

Whereas the square solar sail relies on a rigid structure to provide tension at the edges of the sail film, the heliogyro has several long blades of film, rotating to provide tension and spin stabilisation. Invented in the late 1960s by Richard

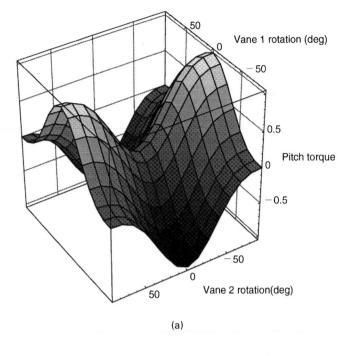

(a)

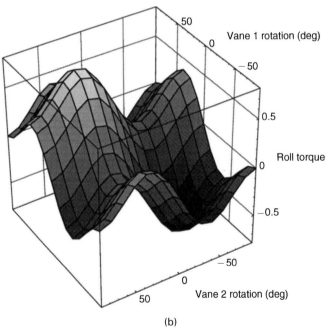

(b)

Fig. 3.15. Normalised torque as a function of vane 1 rotation δ_1 and vane 2 rotation δ_2: (a) pitch torque; (b) roll torque.

McNeal, the heliogyro offers the possibility of deleting structural mass and using spin-induced tension to maintain a large flat reflecting surface. Although this appears an attractive option, it has been found that for a large heliogyro the additional mass required to provide blade torsional stiffness may in fact reduce some of the apparent advantages of the concept. Smaller heliogyro designs do, however, appear to provide performance advantages. The Jet Propulsion Laboratory comet Halley heliogyro, discussed in section 1.2.3, consisted of twelve blades, each 8 m wide and 7.5 km in length. The blades were to be joined to a central hub containing the payload and rotated in pitch by drive motors to provide attitude control. Centripetal loads can be carried in tension members at the edges of the blades, fabricated from tapes of high-tension graphite polyimide fibres. The tapes require periodic bonded crossovers to provide multiple load paths, and this gives each tension member the ability to withstand several failures. If a blade were to become detached, there is a high probability of impact with the remaining blades, leading to catastrophic failure. The blades must also have transverse battens in compression, spaced along the blade length to provide torsional stiffness. Without these battens the blades have essentially no torsional stiffness and so could not be controlled in pitch.

The deployment sequence for the heliogyro is a much simpler and lower risk procedure than for a square solar sail. Firstly, the central hub of the solar sail is spun using small thrusters; then, a set of rollers allow the heliogyro blades to unwind, with the central hub directed along the Sun-line and the blades held at the same collective pitch angle. The radiation pressure torque generated by the blades then adds angular momentum to the heliogyro allowing the blades to fully extend, as shown in Fig. 3.16. The deployment must be carefully controlled in order to avoid undesired interactions between the various modes of the heliogyro blade dynamics. Such interactions and dynamic instability, either during deployment or during operation, represent perhaps the most serious problem for the heliogyro concept. Various instabilities due to thermoelastic and purely mechanical effects can occur, although blade flutter appears potentially the most serious. This is due to a coupling between the blade pitch and blade bending. Flutter may, however, be avoided by separating the frequencies of the first few torsional and bending modes of the blades. Although the heliogyro configuration offers many advantages, the uncertain dynamics of long, high aspect ratio blades may still limit its utility in practice.

3.5.3.2   *Blade shape*

In this section the approximate shape of a single rectangular heliogyro blade will be determined. The blade will be considered to be a tensioned film so that thin membrane theory can be applied. It will be assumed that the blade may be perturbed from a uniformly flat state by vertical flapwise motion, in-plane chordwise motion, and blade twist. A full investigation of the blade shape requires detailed analysis of the coupling between these modes. However, for simplicity it will be assumed that the modes are uncoupled.

A blade of length R, chord C and thickness h will now be considered rotating about one end with angular velocity Ω, as shown in Fig. 3.17. An element of the

Fig. 3.16. Heliogyro deployment (NASA/JPL).

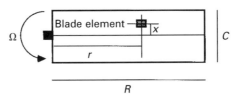

Fig. 3.17. Heliogyro blade element.

blade at radial distance r and distance x from the mid-chord will then be considered. The centripetal force density (centripetal force experienced by a unit volume of the blade) experienced by this blade element in the radial and chordwise directions is given by

$$f_r = \rho \Omega^2 r \tag{3.24a}$$

$$f_x = \rho \Omega^2 x \tag{3.24b}$$

where ρ is the density of the sail film. From this definition the tensile stress in the radial direction σ_r and chordwise direction σ_x experienced by the sail film are obtained by integration along the blade, viz.

$$\sigma_r = \int_r^R \rho\Omega^2 r \, dr \tag{3.25a}$$

$$\sigma_x = \int_x^{C/2} \rho\Omega^2 x \, dx \tag{3.25b}$$

Therefore, integrating Eqs (3.25) provides the radial and chordwise stresses experienced by the blade element as

$$\sigma_r(r) = \tfrac{1}{2}\rho\Omega^2(R^2 - r^2) \tag{3.26a}$$

$$\sigma_x(x) = \tfrac{1}{2}\rho\Omega^2\left[\left(\frac{C}{2}\right)^2 - x^2\right] \tag{3.26b}$$

As expected, the stresses are always positive and are greatest at the blade root and at the mid-chord. Since the blade is always in tension, vertical flapwise displacements can now be considered. For small displacements, the vertical deflection w of the sail film normal to its surface is obtained from the membrane equation

$$\frac{\partial}{\partial r}\left(\sigma_r \frac{\partial w}{\partial r}\right) + \frac{\partial}{\partial x}\left(\sigma_x \frac{\partial w}{\partial x}\right) + \frac{P_n}{h} = 0 \tag{3.27}$$

where P_n is the pressure normal to the blade surface. It should be noted that since the sail is accelerating under the action of solar radiation pressure, P_n is obtained by subtracting the inertial reaction of the sail film from the direct solar radiation pressure. If it is now assumed that the vertical deflection w is uncoupled from any blade twist so that $dw/dx = 0$, Eq. (3.27) may be written as

$$\frac{d}{dr}\left(\sigma_r \frac{dw}{dr}\right) + \frac{P_n}{h} = 0 \tag{3.28}$$

where the vertical flapwise deflection is only a function of the radial distance along the blade. The so-called coning angle of the blade ϑ is now defined to be dw/dr. The coning angle then describes the blade curvature as a function of the distance from the root of the blade. Integrating Eq. (3.28) gives the local coning angle as

$$\vartheta = \frac{1}{\sigma_r}\int_r^R \frac{P_n}{h} \, dr \tag{3.29}$$

In general the coning angle is non-uniform, although a uniform coning angle can be achieved by tapering the blade. For uniform solar radiation pressure and a uniform blade thickness, Eq. (3.29) can be integrated and, using Eq. (3.26a), yields the coning angle as

$$\vartheta(r) = \frac{2P_n}{\rho h \Omega^2(R + r)} \tag{3.30}$$

As expected, the coning angle decreases radially along the blade as centripetal tension flattens the sail film. The blade deflection is also reduced as the heliogyro spin

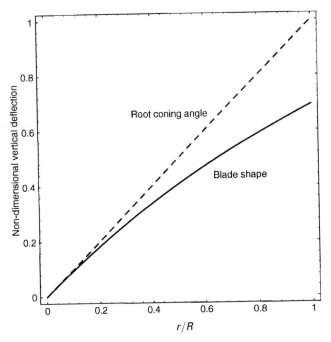

Fig. 3.18. Normalised heliogyro blade shape $w(r)/\vartheta(0)R$.

rate increases. It can be seen from Eq. (3.30) that the root coning angle is twice as large as the tip coning angle since

$$\vartheta(R) = \tfrac{1}{2}\vartheta(0) \tag{3.31}$$

which is a general property for uniform heliogyro blades. Even for long blades the tip deflection is, however, small relative to the chord. By a further integration of Eq. (3.30) the true shape of the heliogyro blade can now be obtained as

$$w(r) = \vartheta(0)R\ln\left(1 + \frac{r}{R}\right) \tag{3.32}$$

A characteristic blade shape is shown in Fig. 3.18 in dimensionless units. In practice, large coning angles are to be avoided owing to the reduction in solar radiation pressure force with increasing Sun aspect angle. Also, if the heliogyro spin axis is not parallel with the Sun-line, a constant precessional torque is induced.

The blade twist θ may now be obtained from Eq. (3.27) by defining $w = \theta x$, assuming that the blade is flat along the chord. Then, integrating Eq. (3.27) over the entire blade chord, it can be shown that

$$\frac{\partial}{\partial r}\left(I\sigma_r\frac{\partial\theta}{\partial r}\right) - \rho\Omega^2 I\theta + t_\theta = 0 \tag{3.33}$$

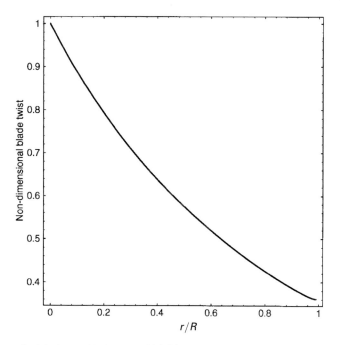

Fig. 3.19. Normalised heliogyro blade twist $\theta(r)/\theta(0)$.

where I is the area moment of inertia of the blade in cross-section and t_θ is the twisting moment applied by the solar radiation pressure, viz.

$$I = \tfrac{1}{12} C^3 h \tag{3.34a}$$

$$t_\theta = \int_{-C/2}^{+C/2} P_n x \, dx \tag{3.34b}$$

It can be seen from Eq. (3.33) that there is a restoring moment proportional to the blade twist which is due to the chordwise centripetal tension. For control authority over the heliogyro it is clear that the entire blade must be able to twist by applying a torque at the root of the blade. If $t_\theta = 0$ it is found that the blade twist equation may now be written as

$$\tfrac{1}{2}(R^2 - r^2)\frac{d^2\theta}{dr^2} - r\frac{d\theta}{dr} - \theta = 0 \tag{3.35}$$

Interestingly, in this case the blade twist is independent of the mechanical properties of the blade and the solution to Eq. (3.35) defines a universal curve describing the blade twist. The form of this solution is shown in Fig. 3.19. It is found that the twist at the tip of the blade is 36.4% of the twist at the root. Therefore, by applying a torque to the root of the blade the entire blade can be twisted, as required for

Table 3.4. Required root torque for a 7.6 μm thick 1×300 m heliogyro blade.

Blade root pitch angle (deg)	Root torque: $\Omega = 0.2$ rpm (N m)	Root torque: $\Omega = 0.3$ rpm (N m)	Root torque: $\Omega = 0.5$ rpm (N m)
10°	1.25×10^{-5}	2.81×10^{-5}	7.80×10^{-5}
20°	2.50×10^{-5}	5.61×10^{-5}	1.56×10^{-4}
30°	3.74×10^{-5}	8.42×10^{-5}	2.34×10^{-4}

control. The torque M_0 required to twist the blade at its root is defined to be

$$M_0 = I\sigma_0 \frac{\partial \theta}{\partial r}\bigg|_{r=0} \tag{3.36}$$

where σ_0 is the radial stress at the blade root, obtained from Eq. (3.26a). Then, from Eq. (3.36) it is found that this torque may be written as

$$M_0 = \frac{1.208\theta_0}{R} I\sigma_0 \tag{3.37}$$

Even for large blades the torque required at the blade root is extremely small due to the low moment of inertia of slender blades. For illustration, the torque required to generate a given blade pitch is listed in Table 3.4 for a Kapton blade of 300 m span, 1 m chord and 7.6 μm thickness. It can be seen that the required torque is extremely low, although there is some increase with blade spin rate. For large blades the required torques are easily generated using electric drive motors. However, for smaller blades innovative designs using piezoelectric actuators appear possible. Such novel actuation methods will be discussed later.

3.5.3.3   *Control*

The heliogyro may be controlled by twisting the blades in pitch in a collective and cyclic manner. The collective pitch is a constant twist applied to each blade whereas cyclic pitch is time varying and is modulated every rotation period. While collective pitch may be used to change the spin rate of the heliogyro, cyclic pitch will create torques across the blade disc causing the heliogyro spin axis to precess. A general pitch control law may be written as

$$\theta(t) = \theta_0 + A\sin(\Omega t - \psi_0) + B\cos(\Omega t - \psi_0) \tag{3.38}$$

where θ_0 is the collective pitch, A and B are the components of the cyclic pitch and ψ_0 is a phase angle. Assuming that the heliogyro spin axis is directed along the Sun-line, pure collective pitch with $A = 0$ and $B = 0$ will generate a torque about the polar axis of the heliogyro therefore increasing or decreasing its spin rate. Similarly, pure cyclic pitch can be generated by a control law of the form

$$\theta(t) = A\sin(\Omega t - \psi_0) \tag{3.39}$$

which will induce a lateral component of force in the plane of the blades, as shown in

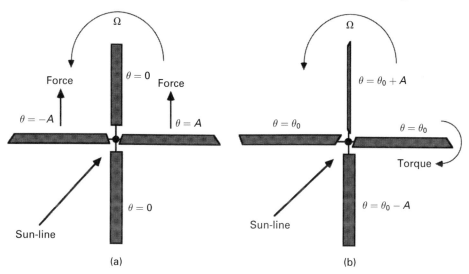

Fig. 3.20. Heliogyro control laws: (a) lateral force control; (b) torque control.

Fig. 3.20(a). Such lateral forces may be useful for planetary escape or capture spirals, particularly in polar orbit where the heliogyro spin axis is directed along the Sun-line. When the collective pitch is combined with a cyclic pitch of the form

$$\theta(t) = \theta_0 + A\sin(\Omega t - \psi_0) \qquad (3.40)$$

a torque which will precess the heliogyro spin axis is generated, as shown in Fig. 3.20(b). Although precessional torques are induced, a torque about the spin axis is also induced which will alter the spin rate of the heliogyro. More complex schemes can be devised to form other standing patterns in the blade pitch. By periodically altering the sign of the collective and cyclic pitch angles the standing patterns can generate torques across the blade disc while leaving the spin rate unchanged. Therefore, by appropriate combinations of collective and cyclic pitch the heliogyro attitude can be fully controlled. It should be noted, however, that the blade pitch must be modulated on the timescale of the heliogyro rotation period, requiring significant actuator activity compared with a square solar sail. Cruise operations for the square solar sail are therefore somewhat simpler than for the heliogyro.

3.5.4 Spin-stabilised disc sail

3.5.4.1 Design

An intermediate concept between the three-axis stabilised square solar sail and the spin-stabilised heliogyro is the disc solar sail. This concept allows deletion of structural mass by using spin-induced tension to maintain a flat sail film, but without the need for long, high aspect ratio blades. The attitude of the rotating disc solar sail is controlled through torques induced by an offset of the centre-of-mass and centre-of-pressure. These induced torques then lead to precession of the spin axis of the solar sail. In order to transmit loads during precession, radial spars are required and perhaps a short boom normal to the hub of the disc to attach stay lines to an outer hoop structure. As will be discussed in section 3.5.6, the spinning disc solar sail appears to be an attractive option for the manufacture of high-performance solar sails.

3.5.4.2   Sail shape

As discussed earlier, the disc solar sail uses spin-induced tension to maintain a flat sail film. In this section the shape of a spinning sail will be obtained and the effect of spin rate and edge tension investigated. As with the heliogyro, it will be assumed that the sail is a tensioned film to enable thin membrane theory to be applied. The solar sail will now be defined as a uniform disc of mass per unit area σ with radius R and rotational angular velocity Ω. The disc of sail film is supported by a hoop structure which applies a radial tension T_0 to the edge of the disc, as shown in Fig. 3.21. The payload is assumed to be located at the centre of the disc and attached to the hoop via radial stays lines which are not considered here. Then, if the sail film experiences a load due to solar radiation pressure P, the vertical displacement of the sail film w is obtained from

$$T\frac{d^2 w}{dr^2} + \frac{T}{r}\frac{dw}{dr} + \frac{dT}{dr}\frac{dw}{dr} = P \tag{3.41}$$

In order to integrate Eq. (3.41) it is necessary to obtain the radial tension T experienced by the sail film (since the disc solar sail is a circular domain the tension is strictly a surface tension). Considering the equilibrium of forces in the radial direction it can be shown that

$$T(r) = T_0\left(\frac{R}{r}\right) + \frac{\sigma\Omega^2 R^3}{3r}\left[1 - \left(\frac{r}{R}\right)^3\right] \tag{3.42}$$

Then, substituting for the radial tension in Eq. (3.41), and integrating, yields the

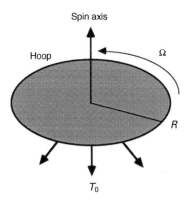

Fig. 3.21. Spinning disc solar sail with hoop structure.

vertical displacement of the sail film as

$$w(r) = \frac{P}{2\sigma\Omega^2} \ln\left\{1 + \frac{1}{3}\frac{\sigma\Omega^2 R^2}{T_0}\left[1 - \left(\frac{r}{R}\right)^3\right]\right\}$$
(3.43)

The profile of a 100 m radius disc solar sail with a mass per unit area of $15\,\mathrm{g\,m^{-2}}$ is shown in Fig. 3.22 for a range of spin rates. Firstly, when it is not rotating it can be seen that the disc sail film will billow under load. In this case the only support is due to the tension provided at the edge of the disc by the hoop structure. If the sail billowing is severe, the solar sail will suffer a loss of efficiency compared to a flat sail due to the oblique incidence of photons at the edge of the disc. When the sail is rotating, the billowing is reduced, leading to a flatter film profile. Increasing the tension at the edge of the disc also leads to a reduction in billowing. However, this increases the compression load on the hoop and so requires an increase in structural mass. It can be seen then that a spinning solar sail can in principle lead to a reduction in edge tension over a non-spinning sail by using spin-induced tension in the sail film. This reduction in edge tension then leads to a reduction in the solar sail structural mass and, hence, an increase in the solar sail performance.

3.5.4.3   Control

As discussed earlier, the attitude of the disc solar sail may be controlled through precession of the spin axis. Precession may be induced through the action of torques arising from a centre-of-mass offset. Free precession of the solar sail may also be used for orbit-raising purposes. For example, if the precession rate is equal to the orbit rate of the solar sail, a resonance condition occurs leading to an increase in semi-major axis. A more sophisticated concept is the dual spin solar sail which utilises a counter-rotating disc at the sail hub to nullify the solar sail angular momentum. In this case spin is only used to provide tension in the sail film. Although control is somewhat easier, there is a substantial mass and power penalty due to the need for a heavy rotor disc.

3.5.5   Solar photon thruster

The conventional flat solar sail suffers from a loss of efficiency due to the cosine squared reduction in solar radiation pressure force as the Sun aspect angle increases. In addition, since it is the transverse component of force tangent to the solar sail orbit which does useful work, only 38% of the available solar radiation pressure force is of use when the sail orientation is optimised. However, by separating the functions of collecting and directing the solar radiation, significant performance improvements over conventional flat solar sails are possible. The concept of separating these two functions was originally proposed in the Soviet literature in the early 1970s and was subsequently re-invented by physicist Robert Forward who coined the term 'solar photon thruster'. Although the concept appears attractive in principle, as yet no detailed engineering design studies have been undertaken for this type of solar sail.

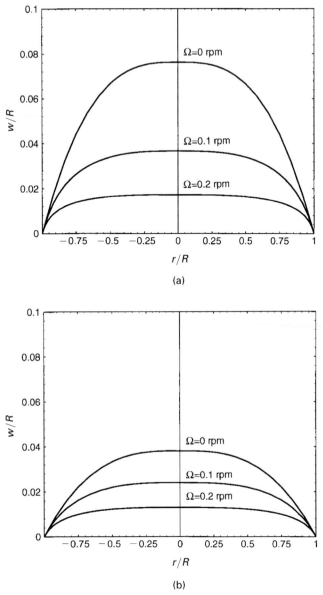

Fig. 3.22. Normalised disc solar sail profile w/R: (a) $T_0 = 2 \times 10^{-3}\,\mathrm{N\,m^{-1}}$; (b) $T_0 = 4 \times 10^{-3}\,\mathrm{N\,m^{-1}}$.

The solar photon thruster is constructed from a large Sun-pointing reflector which directs incoming solar radiation onto a small collimating mirror. This collimating mirror then directs a uniform beam of radiation onto an articulated directing mirror, as shown in Fig. 3.23. This articulated mirror is used to direct the collimated solar radiation and so control the orientation of the force vector acting on the solar sail.

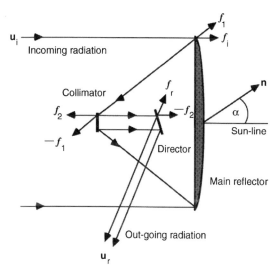

Fig. 3.23. Solar photon thruster optical path.

Since the main collector always faces the Sun, the solar photon thruster is not subjected to the reduction in projected area that a flat solar sail suffers when directing the solar radiation pressure force away from the Sun-line. In addition, since the force vector is directed by articulating a small mirror, the solar photon thruster can perhaps be more easily controlled than a flat solar sail where the entire structure must be manoeuvred. In order to minimise undesired torques, the sail design can be configured such that the total force acting on the solar sail is directed through its centre-of-mass. By translating the relative position of the mirrors and payload, and so displacing the centre-of-mass, control torques can be generated as required.

The force exerted on a solar photon thruster will now be determined by ray tracing a single photon through the entire optical system. It will be assumed that all the optical surfaces are perfectly reflecting and that there are no other losses in the system. The photon will firstly be reflected from the collector generating an incident force f_i and a reaction force f_1, as shown in Fig. 3.23. After being reflected from the collector, the photon is incident on the collimating mirror generating a force $-f_1$. Since it is assumed that the optical surfaces are perfectly reflective, these internal forces at the collector and collimating mirror will cancel. This cancellation of internal forces appears in even more complex optical paths. On reflection from the collimating mirror the photon exerts a reaction force f_2. The photon is then incident of the directing mirror and exerts a force $-f_2$, again assuming no losses. On reflection from the directing mirror the photon finally exerts a reaction force f_r on the directing mirror and leaves the optical system. Therefore, owing to internal cancellation within the optical path, the total force exerted on the solar sail is given simply by the sum of f_i and f_r. For a collector of area A the total force exerted due to incident solar

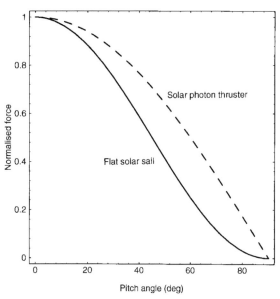

Fig. 3.24. Comparison of the force exerted on a solar photon thruster ($\cos \alpha$) and flat solar sail ($\cos^2 \alpha$) as a function of pitch angle α.

radiation is given by

$$f_i = PA\mathbf{u}_i \tag{3.44}$$

where P is the solar radiation pressure and the unit vector $\mathbf{u}_i$ is directed along the Sun-line. Similarly, assuming perfect reflectivity, the radiation reflected from the directing mirror exerts a force given by

$$f_r = -|f_i|\mathbf{u}_r \tag{3.45}$$

where the unit vector $\mathbf{u}_r$ is directed along the path of photons leaving the optical system. The total force acting on the solar sail is now defined to act in direction $\mathbf{n}$ and is obtained by the vector summation of Eqs (3.44) and (3.45). Using the identity $\mathbf{u}_i - \mathbf{u}_r = 2(\mathbf{u}_i \cdot \mathbf{n})\mathbf{n}$, the total force f exerted on the solar sail is given by

$$f = 2PA(\mathbf{u}_i \cdot \mathbf{n})\mathbf{n} \tag{3.46}$$

If the resultant force vector is defined to be directed at an angle α relative to the Sun-line, then

$$f = 2PA \cos \alpha \, \mathbf{n} \tag{3.47}$$

It can be seen that since the force magnitude only varies as the cosine of the angle between the Sun-line and force vector, the solar photon thruster has a significant advantage over a conventional flat solar sail, as shown in Fig 3.24. This advantage is particularly pronounced when the force vector is directed at large angles from the Sun-line.

In practice the collector of the system would be parabolic in order to focus the incident radiation onto the collimating mirror. This geometric shape needs to be maintained accurately to avoid spillover of reflected radiation at the collimator. Alternatively, the collimator would need to be large enough to capture all the radiation directed from an imperfect collector. In addition, for a small articulated directing mirror the concentration ratio provided by the collector would be large, perhaps of order 100 or greater. Therefore, the incident energy flux on these mirrors would be increased by the same ratio. For realistic, non-perfect reflectors elevated temperatures would be expected. Although such difficulties are not insurmountable, the solar photon thruster concept has yet to be subjected to the detailed engineering scrutiny given to other design concepts. If it is subjected to such scrutiny and appears feasible, then the advantages provided over a conventional flat solar sail are significant. Not only is the concept more efficient at generating transverse forces for orbit transfer, but control is greatly simplified since the main reflecting surface is fixed in a Sun-facing attitude.

3.5.6   High-performance solar sails

For many future mission applications high-performance solar sails are required with a lightness number of order unity. This corresponds to a solar sail loading of $1.53 \, \mathrm{g \, m^{-2}}$ and a characteristic acceleration of $5.9 \, \mathrm{mm \, s^{-2}}$. As discussed earlier in this chapter, square solar sail designs use a spar structure to provide tension to hold the sail film flat after deployment. Such structures are always subject to bending loads and hence contribute significantly to the total solar sail mass. The most efficient means of fabricating a high-performance solar sail is to use spin-induced tension to delete a significant fraction of the solar sail structural mass. Then, the structural elements of the solar sail will be in tension and so can be much lighter than would be in compression. In addition to deleting structural mass, a high-performance solar sail must also have a greatly reduced sail film mass. Again, for a conventional solar sail design, a plastic substrate is required to allow handling, packing and deployment of the sail film. Once deployed, however, the substrate is no longer required for a well-designed structure and only lowers the solar sail performance. One means of increasing the solar sail performance, while allowing safe packing and deployment, is to use a substrate which vaporises or detaches in vacuum or under the action of solar UV radiation, as discussed in section 3.3.4. Kapton strips may be left to provide rip stops and secondary load paths in the sail film. Small-scale production of $0.05 \, \mu\mathrm{m}$ all-metal films has already been demonstrated, as discussed in section 3.3.4.

A surface of characteristic acceleration is shown in Fig. 3.25 for a 100 m radius disc solar sail with an efficiency of 0.85. As expected, only a small payload mass is possible, even with a low sail assembly loading. An approximate mass breakdown for such a high-performance sail is listed in Table 3.5 for a payload mass of 25 kg. The solar sail would have a total mass of 41 kg and a lightness number of order unity. If a $2 \, \mu\mathrm{m}$ UV-sensitive plastic substrate is included, this adds 90 kg of mass (assuming similar mass properties to Kapton) giving a total launch mass of 131 kg. This is well within the Earth escape capacity of many small, low-cost launch vehicles. Small displacements of the centre-of-mass of the solar sail relative to the centre-of-pressure may be used to generate torques to precess the sail spin axis for attitude

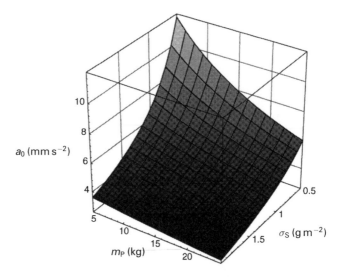

Fig. 3.25. Surface of characteristic acceleration for a 100 m radius disc solar sail with an efficiency of 0.85.

Table 3.5. High-performance solar sail (HPSS) and micro-solar sail (μ-SS) mass properties and performance estimates.

	HPSS	μ-SS
Total sail area (m^2)	31,400	16
Total solar sail mass (kg)	41	0.24
Sail film mass (kg)	7	0.024
Structural mass (kg)	9	0.016
Payload mass (kg)	25	0.2
Assembly loading (g m^{-2})	0.51	2.5
Total loading (g m^{-2})	1.31	15.0
Characteristic acceleration (mm s^{-2})	5.9	0.52
Λ_1	− 0.39	− 0.17
Λ_2	− 0.61	− 0.83
Λ_3	0.39	0.17

control. Although such precession is slow for a large disc solar sail, rapid turning rates are not required for many of the future applications of high-performance solar sails. A schematic layout of the solar sail is shown in Fig. 3.26. It is assumed that Kevlar tapes are used to provide load-bearing paths. Since these tapes are in tension, significant loads can be accommodated without unduly stressing the sail film.

As discussed in section 3.3.4, concepts exist to increase solar sail performance yet further by perforating the sail film with holes smaller than the mean wavelength of visible sunlight. Such perforation is possible using techniques well established in the semiconductor industry for micro-fabrication. These perforations can reduce the sail

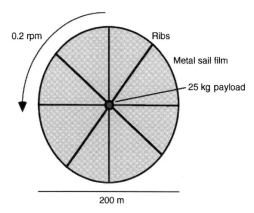

Fig. 3.26. Schematic high-performance solar sail.

mass by an order of magnitude while reducing the sail reflectivity only marginally. Using such techniques sail lightness numbers of order 10 appear possible. However, to benefit from perforated sails films the payload mass fraction of the solar sail must be kept low. It is clear that for a large payload, where the payload mass dominates the total mass of the solar sail, the characteristic acceleration of the solar sail becomes insensitive to reductions in the sail film mass. Therefore, to use a perforated sail film effectively the payload mass must be minimised.

3.5.7   Micro-solar sails

As discussed in section 1.2.6, one of the primary motivations for the renewed interest in solar sailing has been the significant reduction in the mass of useful payloads obtained in recent years. Projecting such mass reductions further leads to the possibility of extremely small solar sails with a payload mass of order 1 kg or less. Some of these possibilities will now be investigated and the downward scaling of conventional solar sail concepts explored. Firstly, the effect of greatly reducing sail area will be considered. A surface of characteristic acceleration for a 4×4 m solar sail is shown in Fig. 3.27 for an efficiency of 0.85. It can be seen that a characteristic acceleration of order 1 mm s^{-2} can be achieved with a low sail assembly loading if the payload mass is reduced below 0.25 kg. An approximate mass breakdown for a micro-solar sail is listed in Table 3.5 with a payload mass of 0.2 kg. A payload of this size is characteristic of recent developments in micro-spacecraft technologies using microelectromechanical systems (MEMS). A schematic layout of the solar sail is shown in Fig. 3.28. Owing to the small sail area the design assumes a 1 μm aluminised substrate tensioned by 1 mm diameter collapsible carbon fibre rods. The sail orientation is controlled by inducing a centre-of-pressure offset using thermoelastic wires which distort the sail film profile.

A benefit of reducing the scale of the solar sail is that the principal moments of inertia of the sail are significantly reduced, leading to small control torque

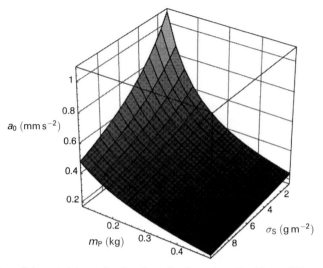

Fig. 3.27. Surface of characterisic acceleration for a $4 \times 4\,\mathrm{m}$ solar sail with an efficiency of 0.85.

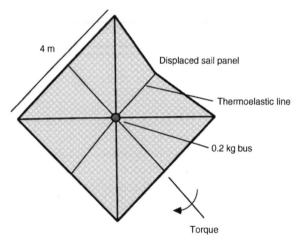

Fig. 3.28. Schematic micro-solar sail.

requirements, as seen in section 3.5.1. For such small control torques, novel low-mass methods of actuation are possible. For example, piezoelectric films may be considered for rotating small control vanes or altering the sail geometry. Other possibilities include windows of nickel oxide which can be switched from transparent to opaque by a small electrical current and thermoelastic wires which can be used to shift the centre-of-pressure of the sail relative to the centre-of-mass, as described above. All of these forms of actuation require very low levels of electrical power and are highly reliable due to the absence of mechanical moving parts.

Other concepts for micro-solar sails utilise recent developments in nanorover technology developed for planetary surface exploration. Nanorovers with a mass of under 100 grams appear possible using MEMS techniques. These rovers are in fact self-contained spacecraft with a suite of miniaturised sensors, actuators and telemetry systems. Recent concepts at NASA/JPL have considered using a micro-solar sail with conventional sail films to provide a characteristic acceleration of order $0.5 \, \mathrm{mm \, s^{-2}}$ for delivery of a self-contained micro-rover to a near Earth asteroid. The rover wheel drive motors would reel shroud lines attached to the corners of the sail to alter its shape for attitude control while attitude-sensing functions would be provided using the rover camera to detect pairs of bright stars. On arrival at the asteroid the wheel drive motors would unwind the shroud lines, allowing the rover to descend in free-fall to the asteroid surface.

3.6   RECENT CASE STUDIES IN SOLAR SAIL DESIGN

3.6.1   World Space Foundation (WSF)

The World Space Foundation has been a long-standing proponent of solar sailing through design and hardware-testing activities. Initial work from 1979 centred on a $28 \times 28 \, \mathrm{m}$ engineering development mission (EDM), designed to test solar sail deployment in low Earth orbit. Subsequently in 1989 a $55 \times 55 \, \mathrm{m}$ solar sail race vehicle (SSRV) was designed for the ill-fated 1992 Mars race, as discussed in section 1.2.4. As shown in Fig. 3.29, the SSRV is again a square solar sail with a number of enhancements over the EDM required for interplanetary operations. With a cruise mass of 139 kg the SSRV has a total loading of $46.3 \, \mathrm{g \, m^{-2}}$, providing an estimated characteristic acceleration of $0.17 \, \mathrm{mm \, s^{-2}}$, as listed in Table 3.6.

The SSRV comprises six main functional elements which combine to form the solar sail spacecraft. Firstly, the propulsion module provides mechanical interfaces to the launch vehicle and contains a solid kick motor to raise the perigee of the initial orbit (assumed to be geostationary transfer orbit (GTO)) to avoid excessive air drag. Cold gas thrusters are also provided for attitude control during the apogee burn and prior to the sail deployment. The module supports the spar deployment mechanisms, storage for the spar tip attitude control vanes and the primary batteries required to power the deployment sequence. Once the sail is deployed this redundant mass is jettisoned.

The sail film is manufactured from $2.5 \, \mu\mathrm{m}$ Kapton aluminised on both sides. The dual reflective coating avoids differential electrical charging of the sail prior to deployment and during operation. Individual sail elements are formed from 38 cm strips of film which are bonded in seams with aluminised polyester tape. The strips run along the z axis of the sail for ease of packing. Rip stops of aluminised tape are bonded to the sail at 1.5 m intervals perpendicular to the seams. In order to prevent damage to the packaged sail by out-gassing of trapped air during ascent, the sail film has perforations to provide pathways for venting. The sail is packed by folding along the z axis of the sail film with each fold equal in width to the individual sail strips. The partially folded sail is further accordion folded separately along the $+y$ and $-y$

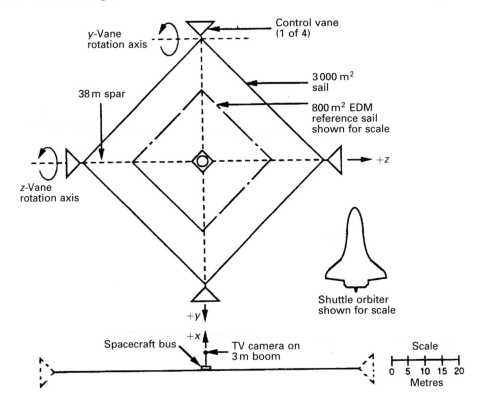

Fig. 3.29. World Space Foundation square solar sail (Hoppy Price/WSF).

Table 3.6. Solar sail design mass properties and performance estimates.

	WSF	U3P	JHU	MIT	CCL	DLR/JPL
Total sail area (m^2)	3000	4000	22,700	1000	60,000	1600
Total solar sail mass (kg)	139	227	180	15	300	77
Sail film mass (kg)	10	55	23	11	120	25
Structural mass (kg)	49	91	77	1	120	16
Payload mass (kg)	80	81	80	3	60	36
Assembly loading (g m^{-2})	19.7	36.5	4.4	12.0	4.0	25.6
Total loading (g m^{-2})	46.3	56.8	7.9	15.0	5.0	48.1
Characteristic acceleration (mm s^{-2})	0.17	0.14	0.98	0.52	1.55	0.16
Λ_1	-0.42	-0.64	-0.55	-0.80	-0.80	-0.53
Λ_2	-0.58	-0.36	-0.44	-0.20	-0.20	-0.47
Λ_3	0.42	0.64	0.55	0.80	0.80	0.53

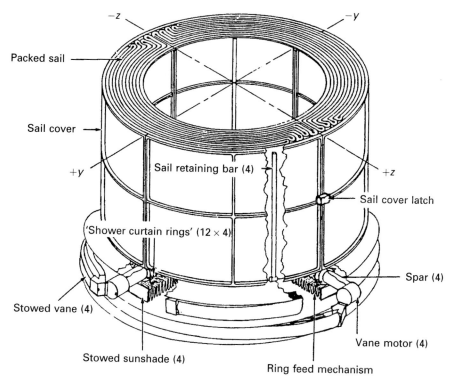

Fig. 3.30. Stowed solar sail (Hoppy Price/WSF).

axes. The folded sail is then restrained by four retaining bars and covered by a protective aluminium shell, as shown in Fig. 3.30. After deployment the sail film is held in tension by four diagonal spars, cantilevered from the central hub. The spars, which provide tension at the corners of the sail, are also subject to bending loads once deployed. Owing to the relatively small area of the solar sail, stay lines are not required to support the spars. The preferred option for the spars is commercially available prestressed STEM tubular spars which can deploy from a spool. The spars pack flat onto a spool, but attain a tubular cross-section on deployment. The spool deployment mechanism is attached to the propulsion module which is jettisoned after deployment. The sail film is attached to the spars using rings at 3 m intervals which help maintain a flat sail shape even if the sail is illuminated along the $-x$ axis. A 10 cm wide sunshade strip is also attached to the rings to prevent thermal distortion of the spars. In addition, the sail tips are attached to the spars with constant force springs to limit the maximum load on the spars to 0.4 N, providing a structural load safety factor of 3.

The attitude of the SSRV is controlled using $15\,\mathrm{m}^2$ vanes mounted at the tip of each of the four spars. Each vane is actuated by a drive motor which connects the vane and spar. Pitch control of the spacecraft is achieved by rotations of the two vanes on the $+y$ and $-y$ spars about the z axis, as shown in Fig. 3.29. Roll and yaw

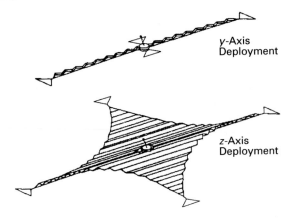

Fig. 3.31. Solar sail deployment (Jerome Wright/WSF).

control is achieved using the remaining two vanes on the $+z$ and $-z$ spars which rotate about the spar axis. The triangular vanes are supported by collapsible beams which wrap around the main hub prior to deployment, as shown in Fig. 3.30. Deployment of the vanes is achieved through the release of restraining latches which allow the strain energy stored in the collapsed beams to drive the deployment. A 3 m deployable boom, which extends from the $+x$ face of the solar sail, supports a TV camera to monitor the sail deployment. In addition, by rotating the boom about the y and z axes, the centre-of-mass of the solar sail can be displaced to provide additional pitch and yaw torques.

Prior to and during the apogee burn the spacecraft stack is spin-stabilised. After the apogee burn, the $+x$ axis of the spacecraft is rotated to face the Sun and allow battery charging through the solar arrays. The stack is then de-spun and held in a three-axis stabilised mode for deployment of the sail. Firstly, the four vanes are deployed as described above. Following the vane deployment the protective shell is released by unlatching a restraining pin. Then, the main sail is deployed by extending the diagonal spars in pairs. Initially the sail is deployed along the y axis pulling the sail from its accordion fold. Then, the z axis spars are deployed fully unfolding the sail, as shown in Fig. 3.31. The spars are deployed at a rate of $0.04\,\mathrm{m\,s}^{-1}$, providing full deployment in approximately 15 minutes for each of the y and z axes. After a successful deployment the propulsion module remains attached to the solar sail to provide additional attitude control during initial on-orbit testing. The propulsion module is then finally jettisoned by releasing a marmon clamp and applying a small separation impulse with the propulsion module cold gas thrusters.

3.6.2   Union pour la Promotion de la Propulsion Photonique (U3P)

The U3P solar sail was initially designed in the 1980s for the proposed solar sail Moon race, again discussed in section 1.2.4. The design has been refined since, consisting of a $4000\,\mathrm{m}^2$ square sail film with four diagonal spars each of 45 m length.

The spars are based on flight-proven coilable technologies using carbon epoxy composites. The sail film uses a 7.6 μm Kapton substrate of mass 55 kg, aluminised on both sides to reduce differential charging and to minimise sail turning rates during orbit raising from GTO. An 81 kg bus provides support functions for the solar sail. The total solar sail mass is 227 kg and provides a loading of 56.8 g m^{-2}, with an estimated characteristic acceleration of order 0.14 mm s^{-2}, as listed in Table 3.6. The solar sail attitude is controlled using eight triangular flaps, two attached to each edge of the sail. The flaps have an area of 240 m^2, some 6% of the sail area, and provide a full 180° rotation of the solar sail in 200 minutes. This turning rate, combined with a dual-sided aluminised sail film, is sufficient for orbit raising from GTO to a lunar fly-past.

3.6.3   Johns Hopkins University (JHU)

The Johns Hopkins University solar sail was designed in the late 1980s, again for the proposed 1992 solar sail race to Mars. The baseline solar sail is a 170 m diameter disc with a total mass of 180 kg. This total mass comprises an 80 kg bus with science payload, a 77 kg sail structure and a high-performance aluminised Kapton sail film of mass 23 kg. With an assumed sail efficiency of 0.85 a characteristic acceleration of order 1 mm s^{-2} is estimated, as listed in Table 3.6. The high-performance sail film is manufactured by etching 85% of a standard 7.6 μm Kapton substrate to an areal density of 1 g m^{-2} using atomic oxygen in vacuum. This manufacturing process has been demonstrated in small-scale laboratory experiments. Patterns of unetched Kapton at the edges of the sail elements would carry almost all of the required loads.

The solar sail design concept is innovative in that it seeks to incorporate the best of both the heliogyro and standard uniform, flat sail designs. The sail is a flat disc composed of large numbers of triangular 'petals' which are manufactured and packaged individually. The petal concept has the simplicity of the heliogyro manufacture and packing processes, while the fully deployed flat disc has the smaller dimensions and simpler structural dynamics of a conventional flat solar sail. In addition, the use of petals avoids the bonding of long seams, as required for a continuous flat sail. The radius of the sail determines the length of each petal while the number of petals is determined by the width of available film. For the 170 m diameter solar sail design, some 480 petals would be required.

Each petal is rolled and packaged in a small, thin tube for launch. These tubes are attached end-to-end in a circular chain and packed around the central spacecraft bus. The tubes are linked by springs forming hinge joints between each member of the chain. During deployment, the restoring force in the springs pushes the connected tubes outwards in an expanding hoop. As the hoop expands the petals of sail film are drawn out of their packing tubes forming the flat disc sail, as shown in Fig. 3.32. When fully deployed the tubes also form an outer hoop structure with which to support the sail film. Some of the tubes carry a mechanism to twist their petals, assisting in attitude control of the solar sail, principally in roll. The spacecraft bus is mounted at the end of a lightweight 50 m deployable boom attached to the rim of the hoop via three or more stay lines. By using these lines to displace the boom tip,

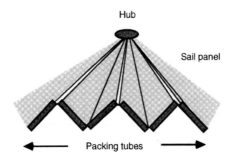

Fig. 3.32. Sail film petal deployment.

pitch and yaw control is provided through displacement of the solar sail centre-of-mass. The long boom also provides an excellent vantage point from which to image the deployment of the hoop and sail.

3.6.4   Massachusetts Institute of Technology (MIT)

Again designed for the 1992 Mars race, the Massachusetts Institute of Technology solar sail is one of the few low-mass designs yet proposed. With a total mass of only 15 kg and no moving mechanical parts, the design appears attractive in terms of both cost and reliability. The penalty for the low launch mass, however, is that the bus can only support a low bandwidth downlink and can only deliver a small payload. However, due to advances in micro-spacecraft technologies since the original design, its key features still prove attractive.

The design concept is essentially a heliogyro to simplify manufacture and to provide reliable deployment on-orbit. In order to increase confidence in the design, high aspect ratio blades are avoided to minimise uncertainties in the heliogyro dynamics and control. A novel feature of the design, apart from its low mass, is the use of solid-state piezoelectric actuators and distributed control electronics on each blade. These actuators eliminate the need for bearings and drive motors for each blade and significantly increase the reliability of the design. The heliogyro utilises eight blades, each of which is 83 m in length and 1.5 m in width with a total mass of 11.5 kg. Using Kapton film a characteristic acceleration of order $0.5\,\mathrm{mm\,s}^{-2}$ is possible, as listed in Table 3.6, although Lexan film was also considered for higher performance.

The actuators are located at the root of each blade and connect the blades to the central hub, as shown in Fig. 3.33. The actuators are only required to generate torques of order $10^{-5}\,\mathrm{N\,m}$ and can rapidly induce pitch angles of up to 45°. These low-mass actuators utilise the piezoelectric effect of polyvinyldifluoride (PVDF) film in a manner similar to a bimetallic strip. An added advantage is that the actuator film can be rolled up along with the blades for ease of packing. Two solid-state Sun sensors on each blade provide blade pitch information for a fast feedback loop to

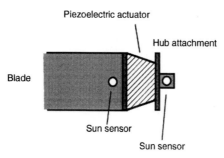

Fig. 3.33. Solid-state heliogyro blade actuator.

maintain cyclic control. A slower control loop maintains the direction of the heliogyro spin axis.

Owing to its low mass, the spacecraft can be launched on a range of small satellite launchers. The launcher upper stage would be required to spin the packaged solar sail to 30 rpm to provide the angular momentum to passively deploy the heliogyro blades. After pyrotechnic release from the upper stage, the individual blades would be released from their stowed configuration by a series of hot wires. The blades are then freely deployed in only 40 seconds, slowing the spin of the spacecraft to the operational rate of 0.1 rpm. During the initial stages of deployment the hub is expected to de-spin rapidly, thus inducing wrinkle in shear at the root of each blade. Although deployment of the blades cannot be tested pre-launch the actuators can be tested, by suspending a small test article in a vacuum chamber. Such pre-launch testing is possible due to the narrow 1.5 m blades and the extremely low control torques required at the blade root.

3.6.5   Cambridge Consultants Ltd (CCL)

The Cambridge Consultants Ltd solar sail was designed principally for the 1992 Mars race, although wider application was ultimately envisaged. In common with many of the designs proposed for this event, the concept displays a high degree of engineering innovation. The solar sail is a 276 m diameter disc of aluminised 2 μm film, supported on a structure of thirty-six radial carbon fibre-reinforced plastic (CFRP) profile spars. Each of the spars is individually cantilevered from a central load-bearing hub, also manufactured from CFRP. A unique spiral fold allows the sail and spars to be wrapped around the hub and packaged into a compact form for launch. This warp-rib concept is both reliable and well understood through prior experience with large satellite communication antenna deployment mechanisms, albeit on a smaller scale. The CFRP profile spars fold flat for packing, but take on an elliptical cross-section once deployed to provide bending stiffness for the sail structure. The 35 mm diameter CFRP profile spars require a mass of 120 kg with a further 120 kg for the sail film. A payload and bus mass of 60 kg provides a total mass of some 300 kg. Using this baseline design, accelerations of order 1.6 mm s^{-2}

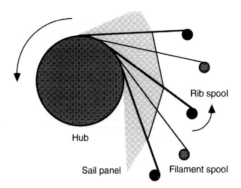

Fig. 3.34. Wrap-rib solar sail manufacture.

may be achieved, as listed in Table 3.6. However, since the solar sail structural design leads to low stresses on the sail film, thinner materials than the 2 μm film proposed may be considered. By plasma etching the non-aluminised rear surface of the sail film large improvements in sail performance and characteristic acceleration appear possible.

During manufacture the spars would be wound onto the hub from 100 mm diameter rollers mounted in a 15 m diameter circle around the hub, as shown in Fig. 3.34. The hub, manufactured using conventional pre-preg lay up techniques, would be mounted on a rotating jig. In addition to the thirty-six spars, two filaments of polyaramid fibres are wound between the spars to provide additional support for the sail film. As the spars are wound from rollers onto the hub, panels of sail film are attached by adhesive bonding allowing the manufacture process to be both relatively simple and compact. However, approximately 5000 individual, and unique, panels would be required, necessitating ink jet printing of alignment marks and computer integrated manufacture. The panels would be formed by automated laser cutting. When one tour of the circumference of the disc is complete, the hub is rolled to pack that section of the sail. This process is repeated until the spars have been fully wound from the rollers onto the hub. Each rib is 138 m long and wraps around the hub 11 times during manufacture. An added feature of the process is that deployment may be continually tested during manufacture by allowing the spars to partially unwind from the hub, drawing out the sail film. After manufacture the completed assembly is a compact cylinder 4 m in diameter and 4 m in height.

Deployment of the sail is achieved by the pyrotechnic release of a clamping ring, thus allowing the thirty-six spars to unwind elastically from the central hub, progressively deploying the sail film, as shown in Fig. 3.35. Therefore, the deployment is fully passive with the elastic energy added to the spars during manufacture providing the tension to deploy the sail film. Since significant potential energy is stored in the spars, the rate of deployment is controlled by the viscous forces of initially unstretched polymer threads linking the spars at regular intervals. These threads stretch viscoelastically as the sail deploys, limiting the deployment rate of the spars and also the angular velocity of the hub. This is particularly important

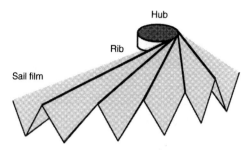

Fig. 3.35. Wrap-rib solar sail deployment.

otherwise the hub would gain significant rotational energy which would create a mechanical shock as it dissipated through the sail once deployment was complete.

The sail attitude may be controlled by inducing small distortions in the entire sail shape through actuation of the spars. The spar actuation is achieved by applying bending moments at the sail hub using thermoelastic bracing wires connected to the spars and small pillars on the hub. By applying differential electrical heating to these wires, the sail geometry can be altered from a flat disc to a cone for passive stability, or a saddle for active attitude manoeuvring. The solar sail spin rate may also be controlled by bending alternate spars to form a turbine configuration. With spar bending of less than 1° only, the solar sail can be rotated by 180° in approximately 3 hours. In addition to controlling the overall geometry of the solar sail, the bracing wires may be used to damp vibration modes in the thirty-six CFRP spars. Such active structural control would require optical sensors or strain gauges to determine accurately the sail shape in real-time.

3.6.6   ODISSEE mission (DLR/JPL)

One of the most recent solar sail design studies has been performed as a joint undertaking between the German Aerospace Research Establishment (DLR) and NASA/JPL. The ODISSEE (orbital demonstration of an innovative solar sail driven expandable structure experiment) study has defined a practical low-cost solar sail demonstration mission. The goals of the mission are to demonstrate fabrication, packing and deployment of a solar sail along with attitude control through centre-of-mass displacement. Central to the mission concept is the use of the Ariane 5 ASAP (Ariane structure for auxiliary payload) ring to provide a low-cost piggy-back launch to GTO. The ASAP ring can accommodate up to eight spacecraft with a strict mass limit of 100 kg and a volume limit of $0.6 \times 0.6 \times 0.8$ m. It has been found that the main design driver for a solar sail ASAP is in fact the allowable volume for sail packing, rather than the total mass limit.

The concept selected is a 40×40 m square solar sail with a separate sail module and bus, as shown in Fig. 3.36. The sail is manufactured from commercially available 7.6 μm Kapton coated with an 0.1 μm film of aluminium on the front surface and an 0.015 μm film of chromium on the rear surface for thermal control. Individual sheets

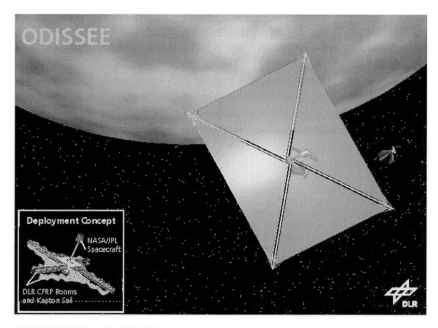

Fig. 3.36. ODISSEE solar sail (DLR).

of film are bonded with Kapton tapes to form quarter panels which are then individually folded and stored in the sail module. The sail film is held in tension using STEM-type diagonal spars manufactured from two laminated sheets of CFRP, bonded at the edges to give the spars a tubular form which can be pressed flat and rolled for packing. The spar design results in a low-mass structure of less than $0.1 \, \text{kg m}^{-1}$ with essentially zero coefficient of thermal expansion.

The sail module is connected to the bus through a commercially available 10 m CLCB coilable boom. The boom is articulated with a two-degrees-of-freedom gimbal to provide an offset of the centre-of-mass and centre-of-pressure for attitude control. The spacecraft bus provides power and telemetry along with additional attitude control using cold gas thrusters. The thrusters are mounted in four blocks of three thrusters with the nitrogen propellant tanks sized for six months of manoeuvring. The mass of the bus is 36 kg (including a 1 kg ASAP adapter) giving a total launch mass of 77 kg. This is well within the ASAP ring mass limit of 100 kg and provides a comfortable mass margin. This total mass yields a loading of $48.1 \, \text{g m}^{-2}$ and an estimated characteristic acceleration of $0.16 \, \text{mm s}^{-2}$, as listed in Table 3.6.

The baseline mission plan requires a standard Ariane 5 midnight launch to GTO. The midnight launch mode results in the apogee of the initial solar sail orbit being directed sunward. The sail is deployed firstly by ejection of the entire spacecraft from the ASAP ring. The coilable boom is then deployed and latched. The sail booms are deployed by rotating a central hub within the sail module which also unfolds the packaged sail quarter panels. The entire deployment sequence is imaged by wide and narrow field cameras on the spacecraft bus. There is also an opportunity for an additional 1 kg free-flying camera to observe the deployment sequence. After deployment a careful orbit control strategy is required to raise the perigee of the initial orbit to avoid excessive air drag. The perigee is raised to a safe limit of 1400 km during the first 110 days of the mission. The control strategy then requires the rate of change of the orbit semi-major axis to be maximised with a fly-past of the lunar pole 550 days later and a final escape from Earth orbit after approximately 630 days.

3.7 SUMMARY

It is clear from this chapter that effective solar sail design is a synthesis of a number of independent engineering design problems. For example, the selection and coating of a low-mass sail film which can also withstand the rigours of packing and deployment is just one of the many challenges encountered. Such competing requirements are, of course, common in conventional spacecraft design, but are perhaps more acute with solar sailing. With conventional spacecraft design, mass growth can lead the spacecraft mass budget towards the upper limit of the launch vehicle capacity. However, for solar sails any mass growth leads directly to a reduction in characteristic acceleration and a potentially degraded mission.

While solar sailing presents many engineering challenges, none of these challenges are insurmountable. What is lacking is practical on-orbit experience, and it is this experience which will allow solar sail design concepts to be refined and sail performance to be enhanced. Inevitable failures will also provide valuable experience in the unseen difficulties involved with solar sail deployment and control. While conventional design concepts can enhance many current solar sail mission applications, future developments in high-performance sail films and micro-solar sails offer exciting possibilities which can truly enable new and exciting mission opportunities. As will be discussed later in this book, high-performance solar sails are capable of exotic new orbits that are impossible for conventional spacecraft, while micro-solar sails may enable autonomous, open-ended exploration of the solar system.

3.8 FURTHER READING

Non-spinning solar sails

Wiley, C. (pseudonym: Sanders, R.), 'Clipper Ships of Space', *Astounding Science Fiction*, 136–143, May 1951.

Williams, T. & Collins, P., 'Design Considerations for an Amateur Solar Sail Spacecraft', IAF-83-395, 34th *International Astronautical Congress*, Budapest, October 1983.

Forward, R.L., 'Solar Photon Thrustor', *Journal of Spacecraft and Rockets*, **27**, 4, 411–416, 1990.

Jack, C. & Welch, C.S., 'Solar Kites: Small Solar Sails with no Moving Parts', IAF-96-S.4.03, *47th International Astronautical Congress*, Beijing, October 1996.

Leipold, M., 'ODISSEE – A Proposal for Demonstration of a Solar Sail in Earth Orbit', IAA-L98-1005, *3rd International Academy of Astronautics Conference on Low Cost Planetary Missions*, Pasadena, April 1998.

Spinning solar sails

MacNeal, R.H., 'The heliogyro, an Interplanetary Flying Machine', *NASA CR-84460*, Astro Research Corporation, June 1967.

MacNeal, R.H., 'Structural Dynamics of the Heliogyro', *NASA CR-1745*, MacNeal-Schwendler Corporation, May 1971.

Friedman, L.D. *et al.*, 'Solar Sailing – The Concept Made Realistic', AIAA-78-82, *16th AIAA Aerospace Sciences Meeting*, Huntsville, January 1978.

Svitek, T. *et al.*, 'Solar Sail Concept Study', IAF-ST-82-12, *33rd International Astronautical Congress*, Paris, October 1982.

Mitsugi, J., Natori, M. & Miura, K., 'Preliminary Evaluation of the Spinning Planar Solar Sail', AIAA-87-0742, *28th AIAA/ASME/ASCE/AHS Structural Dynamics Conference*, Monterey, April 1987.

High-performance solar sails

Drexler, K.E., 'High Performance Solar Sails and Related Reflecting Devices', AIAA-79-1418, *4th Princeton/AIAA Conference on Space Manufacturing Facilities*, Princeton, May 1979.

Uphoff, C., 'Very Fast Solar Sails', *International Conference Space Missions and Astrodynamics III*, Turin, June 1994.

Genta, G. & Brusa, E., 'The Aurora Project: A New Sail Layout', *2nd IAA Symposium on Realistic Near-Term Advanced Scientific Space Missions*, Aosta, Italy, June 1998.

Solar sail technologies

Lippman, M.E., 'In-Space Fabrication of Thin-Film Structures', *NASA CR-1969*, February 1972.

Weis, R., 'Preliminary Design Fabrication Assessment for Two Solar Sail Candidates', *NASA CR-155617*, ILC Dover Industries, August 1977.

Rowe, W.M., Luedke, E.F. & Edwards, D.K., 'Thermal Radiative Properties of Solar Sail Film Materials', AIAA-78-852, *2nd AIAA/ASME Thermophysics and Heat Transfer Conference*, Palo Alto, May 1978.

Hill, J.R. & Whipple, E.C., 'Charging of Large Structures in Space with Application to the Solar Sail Spacecraft', *Journal of Spacecraft and Rockets*, **22**, 3, 245–253, 1985.

Bernasconi, M.C. & Reibaldi, G.C., 'Inflatable, Space-Rigidised Structures: Overview of Applications and their Technology Impact', *Acta Astronautica*, **14**, 455–465, October, 1986.

Scaglione, S. & Vulpetti, G., 'The Aurora Project: Removal of Plastic Substrate to Obtain an All-Metal Solar Sail', *2nd IAA Symposium on Realistic Near-Term Advanced Scientific Space Missions*, Aosta, Italy, June 1998.

Attitude control

Angrilli, F. & Bortolami, S., 'Attitude and Orbital Modelling of Solar Sail Spacecraft', *European Space Agency Journal*, **14**, 4, 431–446, 1990.

Williams, T., 'Attitude Control Requirements for Various Solar Sail Missions', N91-22150, *NASA Lewis Research Centre Vision 21: Space Travel for the Next Millennium*, Cleveland, June 1991.

4

Solar sail orbital dynamics

4.1 INTRODUCTION

The orbital dynamics of solar sail spacecraft are similar in many respects to the orbital dynamics of other spacecraft utilising low thrust propulsion. That is, a small continuous thrust is used to modify the spacecraft orbit over an extended period of time. However, a solar-electric propulsion system may orient its thrust vector in any direction, whereas solar sails are constrained to thrust vector orientations within 90° of the Sun-line. For some mission applications this constraint leads to significant differences between the spacecraft trajectories. For example, to transfer from a prograde to a retrograde orbit, a solar-electric propulsion system may direct its thrust vector perpendicular to the Sun-line to lose prograde angular momentum and then gain retrograde angular momentum. However, for solar sails the transfer is made by increasing the spacecraft ecliptic inclination to greater than 90° by alternately orienting the solar radiation pressure force vector above and below the ecliptic plane. The analysis of such cranking orbit manoeuvres is of importance for some mission applications for both initial mission design and sail sizing. It is this type of preliminary orbit analysis that will be addressed in this chapter. In particular, closed form analytical solutions will be derived wherever possible to provide physical insight. Such analytical solutions also provide a simple and effective means of generating trajectory data for preliminary mission design. More specialised optimal trajectories which require a numerical solution will also be discussed later in this chapter.

The equations of motion for a solar sail in a Sun-centred orbit will firstly be derived in vector form. This vector equation of motion may then be resolved into any suitable co-ordinate system, such as spherical polar co-ordinates. As an alternative means of representing the dynamics of solar sails, the Lagrange variational equations will also be presented. These provide the equations of motion in terms of the osculating orbital elements of the solar sail. In general, the orbital elements are slowly varying so that this representation has significant benefits for some applications.

Following the derivation of the solar sail equations of motion, solutions are obtained for Sun-centred conic section orbits. These orbits are classified and simple elliptical transfer orbits are considered. Logarithmic spiral trajectories are then investigated as a potential form of interplanetary trajectory. Although it will be found that these trajectories are not particularly practical, they do provide valuable insight into solar sail dynamics. The Lagrange variational equations are then used to explore locally optimal solar sail trajectories where the sail attitude is chosen to maximise the rate of change of some particular orbital element or combination of elements. Although these trajectories are not globally optimal, they provide simple, suboptimal (although in some cases near optimal) orbit manoeuvring schemes. The calculation of true minimum time trajectories is also outlined using optimal control theory.

Following the discussion of Sun-centred orbits, planet-centred orbits are then considered. Various orbit-raising schemes are investigated using simple sail-steering laws. In particular, a locally optimal steering law is derived which maximises the instantaneous rate of increase of orbit energy. These laws are then compared and their relative merits discussed. It is also shown that for an inertially fixed solar sail attitude, the equations of motion have a closed form solution obtained using the Hamilton–Jacobi method. Lastly, minimum time planet-centred escape trajectories are discussed, again using optimal control theory. However, it will be found that for planetary escape, locally optimal steering laws provide excellent performance without the complexity of numerically generated optimal laws.

4.2   EQUATIONS OF MOTION

4.2.1   Vector equation of motion

The equations of motion for a solar sail in a Sun-centred orbit may be obtained by considering the Sun (mass M) and solar sail (mass m) in an inertial frame of reference I, as shown in Fig. 4.1. In this frame of reference the centre-of-mass of the Sun–sail system C is located at position $\mathbf{R}$, defined by

$$\mathbf{R}\frac{M\mathbf{r}_1 + m\mathbf{r}_2}{M + m} \tag{4.1}$$

Using Eq. (2.24) from section 2.3.3, the equation of motion of the Sun and a perfectly reflecting solar sail in the inertial frame I are given by

$$M\frac{d^2\mathbf{r}_1}{dt^2} = \frac{GMm}{r^2}\hat{\mathbf{r}} \tag{4.2a}$$

$$m\frac{d^2\mathbf{r}_2}{dt^2} = -\frac{GMm}{r^2}\hat{\mathbf{r}} + \beta\frac{GMm}{r^2}(\hat{\mathbf{r}}\cdot\mathbf{n})^2\mathbf{n} \tag{4.2b}$$

where $\mathbf{r} = \mathbf{r}_2 - \mathbf{r}_1$ and $\hat{\mathbf{r}}$ is the unit vector directed along the Sun–sail line. Subtracting Eq. (4.2b) and Eq. (4.2a) and using the definition of centre-of-mass from Eq. (4.1),

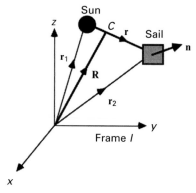

Fig. 4.1. Inertial frame of reference I with centre-of-mass C.

the acceleration of the centre-of-mass of the system C is obtained as

$$\frac{d^2\mathbf{R}}{dt^2} = \beta \frac{GMm}{M+m} \frac{(\hat{\mathbf{r}} \cdot \mathbf{n})^2}{r^2} \mathbf{n} \tag{4.3}$$

Since $(d^2\mathbf{R}/dt^2) \neq 0$ it is clear that the centre-of-mass of the Sun–sail system accelerates relative to the inertial frame of reference. This is due to the gravitational acceleration of the Sun towards the solar sail as the solar sail is accelerated by solar radiation pressure. However, since $m \ll M$ the acceleration of the centre-of-mass is, of course, negligible. This is quite unlike the conventional two-body problem where the centre-of-mass forms the origin of a truly inertial frame of reference.

Assuming that the centre-of-mass does not accelerate, a linear transformation is now used to transform to a new inertial frame I', with an origin at the centre-of-mass C, so that $\mathbf{R} = 0$, viz.

$$M\mathbf{r}_1 + m\mathbf{r}_2 = 0 \tag{4.4}$$

Then, using the definition $\mathbf{r} = \mathbf{r}_2 - \mathbf{r}_1$, the relative acceleration of the Sun and the solar sail is given by

$$\frac{d^2\mathbf{r}}{dt^2} = \left(1 + \frac{m}{M}\right) \frac{d^2\mathbf{r}_2}{dt^2} \tag{4.5}$$

Substituting from Eq. (4.2b) the equation of motion for a perfectly reflecting solar sail in a Sun-centred orbit is then obtained as

$$\frac{d^2\mathbf{r}}{dt^2} + \frac{\mu}{r^2}\hat{\mathbf{r}} = \beta \frac{\mu}{r^2} (\hat{\mathbf{r}} \cdot \mathbf{n})^2 \mathbf{n} \tag{4.6}$$

where $\mu = G(M + m)$, with the approximation that $\mu \approx GM$ since $m \ll M$. This vector equation of motion may now be transformed into scalar components in any convenient frame of reference.

4.2.2 Sail force vector

Now that the vector equation of motion has been derived, the orientation of the solar sail force vector will be defined. The magnitude of the force vector has already been investigated in Chapter 2 for two non-perfect solar sail models. In this section the orientation of the force vector will be defined through the sail clock angle and cone angle. Usually, these angles refer explicitly to the orientation of the true solar sail force vector rather than the orientation of the unit vector normal to the sail surface. Recall from section 2.6 that these two vectors are not coincident for a solar sail with non-perfect reflectivity. However, for a perfect solar sail these two vectors are coincident. For such a perfect solar sail, the sail normal vector **n** will therefore be defined in terms of a cone angle α and clock angle δ, as shown in Fig. 4.2. The sail cone angle is defined to be the angle between the sail normal and the Sun-line, while the sail clock angle is defined to be the angle between the projection of the sail normal and some reference direction onto a plane normal to the Sun-line. In the planar case the cone angle is equivalent to the sail pitch angle used earlier. Using these definitions and resolving along the radial, orbit normal and transverse directions it is found that

$$\mathbf{n} = \cos\alpha\,\hat{\mathbf{r}} + \sin\alpha\cos\delta\,\hat{\mathbf{p}} + \sin\alpha\sin\delta\,\hat{\mathbf{p}}\times\hat{\mathbf{r}} \qquad (4.7)$$

where in this instance $\hat{\mathbf{p}}$ is the unit vector normal to the orbit plane.

As will be seen, an important consideration in many orbit transfer manoeuvres is the optimisation of the sail attitude to maximise the component of solar radiation pressure force in a given direction. For example, for orbit-raising purposes an efficient strategy is to maximise the component of the solar radiation pressure force directed along the solar sail velocity vector. This optimisation process will now be considered. An arbitrary unit vector **q** will be defined to represent some required direction along which the solar radiation pressure force is to be maximised, viz.

$$\mathbf{q} = \cos\tilde{\alpha}\,\hat{\mathbf{r}} + \sin\tilde{\alpha}\cos\tilde{\delta}\,\hat{\mathbf{p}} + \sin\tilde{\alpha}\sin\tilde{\delta}\,\hat{\mathbf{p}}\times\tilde{\mathbf{r}} \qquad (4.8)$$

where $\tilde{\alpha}$ and $\tilde{\delta}$ are the cone and clock angles of the required force direction. The force

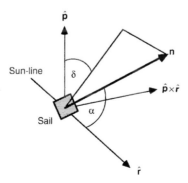

Fig. 4.2. Solar sail of cone and clock angles.

magnitude f_q in this direction is therefore given by

$$f_q = 2PA(\mathbf{n}\cdot\hat{\mathbf{r}})^2(\mathbf{n}\cdot\mathbf{q}) \tag{4.9}$$

where P is again the local solar radiation pressure and A is the sail area. Using Eqs (4.7) and (4.8) the force component may be written in terms of the cone and clock angles as

$$f_q = 2PA\cos^2\alpha[\cos\alpha\cos\tilde{\alpha} + \sin\alpha\sin\tilde{\alpha}\cos(\delta - \tilde{\delta})] \tag{4.10}$$

It is clear that the force component in direction $\mathbf{q}$ is maximised if $\delta = \tilde{\delta}$, so that the sail clock angle is aligned with that of the vector $\mathbf{q}$. In order to maximise f_q it is then required that $\partial f_q/\partial\alpha = 0$. Therefore, setting the derivative of the force component with respect to the sail cone angle to zero yields

$$2\sin\alpha\cos(\alpha - \tilde{\alpha}) + \cos\alpha\sin(\alpha - \tilde{\alpha}) = 0 \tag{4.11}$$

After some reduction, the optimal sail cone angle α^* which maximises the component of solar radiation pressure force in direction $\mathbf{q}$ is found to be

$$\tan\alpha^* = \frac{-3 + \sqrt{9 + 8\tan^2\tilde{\alpha}}}{4\tan\tilde{\alpha}} \tag{4.12}$$

The variation of the optimal sail cone angle α^* with the required cone angle $\tilde{\alpha}$ is shown in Fig. 4.3. In general, the optimal sail cone angle lags behind the required cone angle due to the reduction in total force magnitude with increasing cone angle.

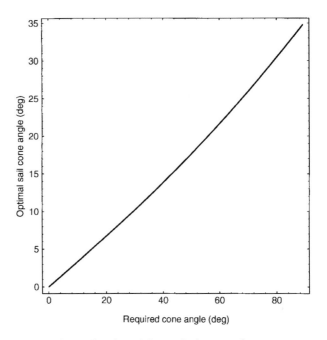

Fig. 4.3. Optimal sail cone angle as a function of the required cone angle.

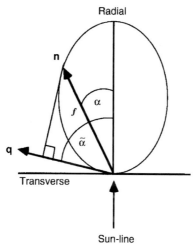

Fig. 4.4. Optimisation of the sail cone angle.

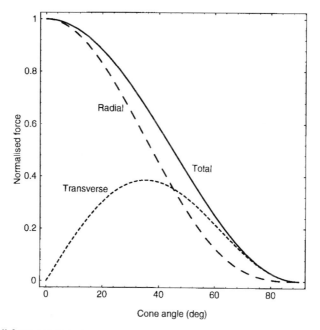

Fig. 4.5. Solar sail force components.

It can also be seen that the sail cone angle is limited to approximately $35°$ as the required cone angle reaches $90°$. This is again due to the reduction in total force magnitude as the sail cone angle increases. The effect is illustrated in Fig. 4.4 which shows a schematic polar plot of the solar radiation pressure force magnitude as a function of the sail cone angle, and the projection of the solar radiation pressure

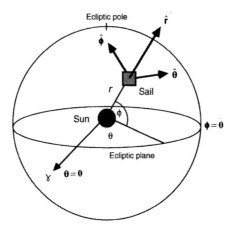

Fig. 4.6. Definition of spherical polar co-ordinates.

force vector onto the required vector **q**. As the required cone angle increases there is a sail cone angle which maximises the product of the total force magnitude and its projection in the direction **q**. From Eq. (4.12) it is found that with $\tilde{\alpha} = 90°$ the sail cone angle which maximises the force in the transverse direction is $\tan \alpha^* = 1/\sqrt{2}$, so that $\alpha^* = 35.26°$. This can clearly be seen by considering the radial and transverse components of the solar radiation pressure force, as shown in Fig. 4.5. Whereas the radial force component monotonically decreases with increasing cone angle, the transverse component has a turning point with a single maximum. This optimum cone angle, which maximises the transverse force, will arise many times in this chapter.

4.2.3   Polar equations of motion

The vector equation of motion may now be resolved into scalar components in any suitable co-ordinate system, such as spherical polar co-ordinates shown in Fig. 4.6. The plane $\phi = 0$ will be defined as the ecliptic with $\phi = \pi/2$ directed to the north ecliptic pole. The direction $\theta = 0$ will be defined as the first point of Aries ♈, an astronomical reference direction in the ecliptic plane. Resolving Eq. (4.6) into radial $\hat{\mathbf{r}}$, transverse $\hat{\boldsymbol{\theta}}$ and normal $\hat{\boldsymbol{\phi}}$ components and using Eq. (4.7) it is found that

$$\frac{d^2 r}{dt^2} - r\left(\frac{d\phi}{dt}\right)^2 - r\left(\frac{d\theta}{dt}\right)^2 \cos^2 \phi = -\frac{\mu}{r^2} + \beta\frac{\mu}{r^2}\cos\alpha^3 \tag{4.13a}$$

$$\frac{1}{r}\cos\phi\frac{d}{dt}\left(r^2\frac{d\theta}{dt}\right) - 2r\left(\frac{d\theta}{dt}\right)\left(\frac{d\phi}{dt}\right)\sin\phi = \beta\frac{\mu}{r^2}\cos\alpha^2\sin\alpha\sin\delta \tag{4.13b}$$

$$\frac{1}{r}\frac{d}{dt}\left(r^2\frac{d\phi}{dt}\right) + r\left(\frac{d\theta}{dt}\right)^2\sin\phi\cos\phi = \beta\frac{\mu}{r^2}\cos\alpha^2\sin\alpha\cos\delta \tag{4.13c}$$

These equations of motion may be used to simulate three-dimensional solar sail

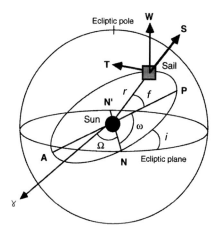

Fig. 4.7. Definition of orbital elements.

trajectories with a fixed sail attitude or with a predefined sail attitude time history. As will be seen, the planar problem can be solved in closed form in a number of particular cases.

4.2.4   Lagrange variational equations

An alternative means of representing the solar sail dynamics is through the Lagrange variational equations. These equations represent the rate of change of the osculating solar sail orbital elements. These are the instantaneous elliptical orbital elements which would be obtained if the solar sail were to be jettisoned. Since the orbital elements are in general slowly changing, the variational equations may provide a more appropriate means of representing solar sail orbital dynamics in certain cases. The instantaneous osculating orbital elements of the solar sail are defined in Fig. 4.7. The orbit may be defined with respect to a celestial sphere centred on the Sun, the ecliptic plane and the first point of Aries ♈. The orbit intersects the ecliptic plane along the line of nodes defining the ascending node N and descending node N'. Then, the longitude of the ascending node Ω is defined as the angle from ♈ to the ascending node N measured along the ecliptic plane. Using the inclination i between the orbit plane and the ecliptic plane completes the definition of the orbit plane orientation. The solar sail orbit itself is oriented within the orbit plane using the argument of perihelion ω, which defines the angle between N and P. The perihelion P and aphelion A are the closest and furthest points on the orbit from the Sun respectively. The line from P to A is the line of apses. Finally, the solar sail is located along the orbit using the orbit radius r and true anomaly f. The true anomaly is the polar angle of the solar sail in the orbit plane measured from P. The size of the orbit is parameterised by the semi-major axis a, defined as one half of the distance from P to A, while the shape of the orbit is parameterised by the eccentricity e. Using these definitions it can be shown that the variational equations may be written as

$$\frac{da}{df} = \frac{2pr^2}{\mu(1-e^2)^2}\left[Se\sin f + T\frac{p}{r}\right] \tag{4.14a}$$

$$\frac{de}{df} = \frac{r^2}{\mu}\left[S\sin f + T\left(1+\frac{r}{p}\right)\cos f + T\frac{r}{p}e\right] \tag{4.14b}$$

$$\frac{di}{df} = \frac{r^3}{\mu p}\cos(f+\omega)W \tag{4.14c}$$

$$\frac{d\Omega}{df} = \frac{r^3}{\mu p \sin i}\sin(f+\omega)W \tag{4.14d}$$

$$\frac{d\omega}{df} = -\frac{d\Omega}{df}\cos i + \frac{r^2}{\mu e}\left[-S\cos f + T\left(1+\frac{r}{p}\right)\sin f\right] \tag{4.14e}$$

$$\frac{dt}{df} = \frac{r^2}{\sqrt{\mu p}}\left\{1 - \frac{r^2}{\mu e}\left[S\cos f - T\left(1+\frac{r}{p}\right)\sin f\right]\right\} \tag{4.14f}$$

where $p = a(1 - e^2)$ is the semi-latus rectum, $n = \sqrt{\mu/a^3}$ is the mean motion and $M = n(t - \tau)$ is the mean anomaly, where τ is the time of perihelion passage. In addition, using Eq. (4.7) the radial, transverse and orbit normal force components S, T and W may be written as

$$S = \beta\frac{\mu}{r^2}\cos\alpha^3 \tag{4.15a}$$

$$T = \beta\frac{\mu}{r^2}\cos\alpha^2\sin\alpha\sin\delta \tag{4.15b}$$

$$W = \beta\frac{\mu}{r^2}\cos\alpha^2\sin\alpha\cos\delta \tag{4.15c}$$

where S is directed along the Sun-line, T is normal to S and in the orbit plane, while W is directed normal to the orbit plane and completes the triad. The variational equations may again be used for simulation purposes or, as will be seen, to determine the sail cone and clock angles required for locally optimal solar sail trajectories. Lastly, it can be seen that this set of variational equations contain singularities for circular orbits and orbits within the ecliptic plane. These singularities may, however, be removed by using sets of equinoctial orbital elements which provide a uniformly valid representation of the solar sail orbit.

4.3 SUN-CENTRED ORBITS

4.3.1 Introduction

Now that the solar sail equations of motion have been presented, families of Sun-centred orbits will be investigated. Firstly, it will be shown that a radially oriented sail can be used to reduce the effective solar gravity experienced by the solar sail, thereby generating conic section orbits with interesting new properties. Then, logarithmic spiral trajectories will be considered as a simple means of interplanetary transfer. While these trajectories are appealing owing to their simplicity, they do not satisfy the arbitrary boundary conditions required for general interplanetary transfer. Locally optimal sail-steering laws are then derived which maximise the instantaneous rate of change of selected solar sail orbital elements. Used in combination, these locally optimal methods provide an effective means of generating preliminary solar sail trajectories. Lastly, in order to provide true minimum time interplanetary trajectories which satisfy arbitrary boundary conditions, optimal control methods are outlined.

4.3.2   Conic section orbits

4.3.2.1   Orbit classification

In this section the classical two-body problem will be considered with the addition of the solar radiation pressure force from a radially oriented solar sail. With the sail normal directed along the Sun-line, the effect of solar radiation pressure is to modify the effective solar gravity experienced by the solar sail. The nett effective force may be reduced from its usual value of μ/r^2 to zero by increasing the sail lightness number from zero to unity. In addition, for high-performance solar sails with a sail lightness number greater than unity, the effective force switches sign. In this case the solar sail orbit is hyperbolic, but the Sun is no longer at the prime focus of the orbit, as will be seen later. Therefore, when classifying solar sail conic section orbits the sail lightness number provides an additional free parameter to consider.

If the solar sail attitude is now chosen such that the sail normal is directed along the Sun-line, the vector equation of motion Eq. (4.6) may be written as

$$\frac{d^2\mathbf{r}}{dt^2} + (1 - \beta)\frac{\mu}{r^2}\hat{\mathbf{r}} = 0 \tag{4.16}$$

Again, this is a statement of the classical two-body problem with the effective solar gravity determined by the solar sail lightness number. An effective gravitational parameter may now be defined as

$$\tilde{\mu} = \mu(1 - \beta) \tag{4.17}$$

With this effective gravitational parameter the usual relations for the classical two-body problem apply. Therefore, the orbit equation for the solar sail is given by

$$r(f) = \frac{h^2/\tilde{\mu}}{1 + e\cos f} \tag{4.18}$$

where h is the orbit angular momentum per unit mass. Kepler's third law can also be stated for the solar sail conic section orbit as

$$\left(\frac{2\pi}{T}\right)^2 a^3 = \tilde{\mu} \tag{4.19}$$

so that the orbit period T is a function of both semi-major axis and the solar sail lightness number. In addition, the solar sail orbit speed and the total orbit energy may be written as

$$v^2 = \tilde{\mu}\left(\frac{2}{r} - \frac{1}{a}\right) \tag{4.20a}$$

$$E = \tfrac{1}{2}v^2 - \frac{\tilde{\mu}}{r} \tag{4.20b}$$

Using Eq. (4.20a) it can be seen that the total orbit energy may also be written as

$$E = -\frac{\tilde{\mu}}{2a} \tag{4.21}$$

The conic section orbits will now be classified in the usual manner using total orbit energy:

(i) *Elliptical orbit*

$$E < 0 \quad \Rightarrow \quad v < \sqrt{\frac{2\tilde{\mu}}{r}} \quad \Rightarrow \quad a > 0 \tag{4.22a}$$

(ii) *Parabolic orbit*

$$E = 0 \quad \Rightarrow \quad v = \sqrt{\frac{2\tilde{\mu}}{r}} \quad \Rightarrow \quad a \to \infty \tag{4.22b}$$

(iii) *Hyperbolic orbit*

$$E > 0 \quad \Rightarrow \quad v > \sqrt{\frac{2\tilde{\mu}}{r}} \quad \Rightarrow \quad a < 0 \tag{4.22c}$$

Of particular interest is the condition for a parabolic orbit which defines the local escape speed, separating closed elliptical orbits from open hyperbolic orbits. The escape speed is, however, a function of the solar sail lightness number, which leads to a more complex classification scheme. In particular, Eq. (4.22b) shows that for a solar sail lightness number of $\frac{1}{2}$ the local escape speed becomes the local Keplerian circular orbit speed $\sqrt{\mu/r}$.

Consider now a solar sail deployed on a circular heliocentric orbit of radius r with circular orbit speed $\sqrt{\mu/r}$ and the sail normal directed along the Sun-line. For a lightness number $\beta = 0$ the orbit is clearly a circular Keplerian orbit of radius r, as shown in Fig. 4.8. For a lightness number in the range $0 < \beta < \frac{1}{2}$, the solar sail will be moving faster than the local circular orbit speed $\sqrt{\tilde{\mu}/r}$ so that the orbit is then elliptical. For a lightness number $\beta = \frac{1}{2}$ the solar sail is moving at escape speed and so will move along a parabolic orbit. Similarly, for a lightness number in the range $\frac{1}{2} < \beta < 1$ the solar sail will be moving along a hyperbolic orbit. For a lightness number of unity, the solar radiation pressure force exactly balances solar gravity so that the solar sail moves along a rectilinear trajectory with speed $\sqrt{\mu/r}$. Lastly, for a lightness number greater than unity, the effective force exerted on the solar sail

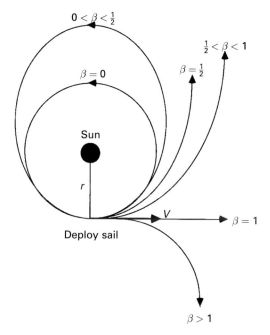

Fig. 4.8. Orbit type as a function of solar sail lightness number.

switches sign and the solar sail again moves along a hyperbolic orbit, but with positive effective solar gravity.

4.3.2.2   Circular orbits (e = 0)

Firstly, a simple circular orbit will be considered with zero eccentricity. For a circular orbit, the solar sail orbit period is a function of both the orbit radius and the solar sail lightness number. Therefore, using Eq. (4.19) it is found that

$$T = \frac{2\pi}{\sqrt{\mu}}(1 - \beta)^{-1/2}r^{3/2} \tag{4.23}$$

so that the solar sail orbit period is decoupled from its orbit radius. By an appropriate choice of solar sail lightness number the orbit period can then be chosen arbitrarily with the constraint that

$$T > \frac{2\pi}{\sqrt{\mu}}r^{3/2} \tag{4.24}$$

since $0 < \beta < 1$. Therefore, the solar sail orbit period is always longer than the usual Keplerian orbit period at that orbit radius. This decoupling of orbit period from orbit radius leads to a number of interesting applications which will be discussed in Chapter 6.

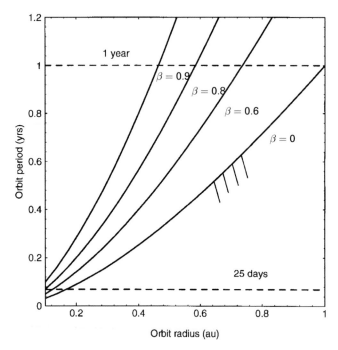

Fig. 4.9. Orbital period variation with solar sail lightness number.

In order to synchronise the solar sail orbit period T at some required value, the required solar sail lightness number at radius r must be chosen as

$$\beta = 1 - \frac{4\pi^2}{\mu} \frac{r^3}{T^2} \qquad (4.25)$$

Two examples of particular interest are 1-year Earth synchronous orbits and orbits with a period in the order of 25 days, which are synchronous with the solar rotation. For circular Keplerian orbits the required orbit radii are 1 and 0.167 au (~ 36 solar radii) respectively. The variation of the solar sail orbit period with orbit radius is shown in Fig. 4.9 with these two orbit periods indicated. Again, the solar sail orbit period is always greater than the Keplerian orbit period at the same orbit radius since the effect of solar radiation pressure is to reduce the effective solar gravity experienced by the solar sail. It can also be seen that for orbits close to the Sun, the required solar sail lightness number quickly approaches unity for both the 1-year and 25-day orbits.

4.3.2.3   *Elliptical orbits* $(0 < e < 1)$

For elliptical orbits, the relationships for Keplerian orbits will again apply but with a modified gravitational parameter. Of particular interest is the case of a solar sail deployed on an Earth escape trajectory with an orbit radius R of 1 au. After

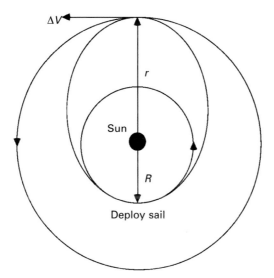

Fig. 4.10. Solar sail single impulse transfer.

deployment, the solar sail will be moving in a Sun-centred orbit with the local circular orbit speed $\sqrt{\mu/R}$. However, owing to the reduced gravitational parameter this speed is greater than that required for a circular orbit. Therefore, after deployment the solar sail will move along an elliptical orbit, as shown in Fig. 4.10. Using the relationship for the speed on an elliptical orbit given by Eq. (4.20a), the following identity is seen to hold, viz.

$$\frac{\mu}{R} = \tilde{\mu}\left(\frac{2}{R} - \frac{1}{a}\right)$$

(4.26)

Therefore, the semi-major axis of the Sun-centred orbit obtained after deployment of the solar sail is given by

$$a = R\left(\frac{1-\beta}{1-2\beta}\right)$$

(4.27)

For bound orbits it can be seen that the solar sail lightness number is constrained such that $0 < \beta < \frac{1}{2}$, as discussed in section 4.3.2.1. In addition, the perihelion distance R and aphelion distance r of the elliptical orbit are given by

$$R = a(1 - e)$$

(4.28a)

$$r = a(1 + e)$$

(4.28b)

Therefore, using Eq. (4.27) to eliminate the orbit semi-major axis, the orbit eccentricity is found to be

$$e = 1 - \left(\frac{1-2\beta}{1-\beta}\right)$$

(4.29)

Interestingly, it can be seen that the eccentricity of the orbit is independent of the initial circular orbit radius. Then, using Eq. (4.28b), the aphelion distance r may be written as

$$r = \frac{R}{1 - 2\beta} \qquad (4.30)$$

Again, it can be seen that the sail lightness number is constrained to $0 < \beta < \frac{1}{2}$ for bound orbits.

The use of elliptical orbits also provides the opportunity for Hohmann-type transfers, but using only a single impulse. Such a manoeuvre is shown in Fig. 4.10 for a transfer between circular co-planar orbits of radius R and r. The solar sail is again deployed on the initial circular orbit of radius R and, with the correct choice of lightness number, is transferred to an elliptical orbit with the aphelion located at the final circular orbit radius r. At the aphelion of the transfer ellipse the sail is then jettisoned and a single impulse used to circularise the orbit. From the orbit geometry the required semi-major axis of the transfer ellipse is given by

$$a = \tfrac{1}{2}(R + r) \qquad (4.31)$$

Then, using Eq. (4.27) the solar sail lightness number required to attain an aphelion distance r may be written as

$$\beta = \tfrac{1}{2}\left(1 - \frac{R}{r}\right) \qquad (4.32)$$

In addition, the duration of the transfer may also be calculated from Eq. (4.19) as one half of the period of the elliptical transfer orbit, viz.

$$T_S = \pi\left[\frac{(R + r)^3}{8\mu(1 - \beta)}\right]^{1/2} \qquad (4.33)$$

Finally, the Δv required to circularise at the final orbit may be obtained from the difference between the solar sail speed and the circular orbit speed at the aphelion distance r as

$$\Delta v_S = \sqrt{\frac{\mu}{r}}\left(1 - \sqrt{\frac{R}{r}}\right) \qquad (4.34)$$

The performance of a single impulse transfer using a solar sail will now be compared with a standard Hohmann transfer. For the Hohmann transfer two impulses are required. The first impulse injects the spacecraft onto the transfer ellipse while the second impulse circularises the spacecraft at the final orbit. The solar sail transfer mode therefore eliminates the need for an initial impulse. The required Δv, transfer time and solar sail lightness number are shown in Fig. 4.11 for a range of final orbits up to 20 au. It can be seen that the cost for the solar sail transfer Δv_S has a maximum at 4 au and then tends asymptotically to zero. This limit corresponds to an escape orbit using a solar sail with a lightness number of 0.5. However, the Hohmann cost Δv_H has a maximum at 15.58 au and tends asymptotically to 12.34 km s^{-1}, which is

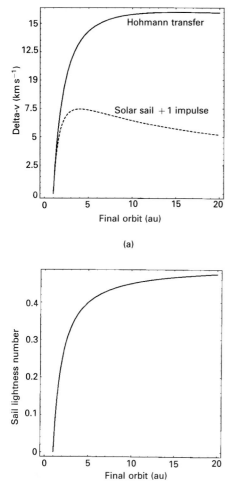

(a)

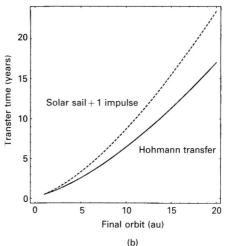

(b)

Fig. 4.11. Solar sail elliptical orbit transfer: (a) comparison of the required Δv for Hohmann and solar sail transfer; (b) comparison of the transfer time for Hohmann and solar sail transfer; (c) required solar sail lightness number.

(c)

the solar system escape Δv from 1 au. Using the above analysis the requirements for particular transfers can be found, as listed in Table 4.1. It can be seen that the solar sail transfer can provide a significant reduction in Δv, but at the expense of an increased transfer time, with the solar sail transfer time T_S always greater than that for the Hohmann transfer time T_H. This reduction in Δv must, of course, be offset against the required solar sail mass.

4.3.2.4 Rectilinear orbits ($e \rightarrow \infty$)

If the solar sail lightness number is now chosen as $\beta = 1$, no nett force will be exerted on the solar sail. The solar sail will completely remove the effect of solar gravity. Since solar radiation pressure and solar gravity both scale as inverse square laws

Table 4.1. Single impulse solar sail transfer

	Mars	Jupiter	Saturn	Uranus	Escape
a (au)	1.52	5.20	9.55	19.13	∞
Δv_{H} (km s^{-1})	5.60	14.44	15.74	15.95	12.34
T_{H} (years)	0.71	2.73	6.06	15.97	∞
Δv_{S} (km s^{-1})	4.58	7.34	6.52	5.26	0
T_{S} (years)	0.78	3.54	8.15	22.01	∞
β	0.17	0.40	0.45	0.47	0.5

(apart from close to the Sun, as discussed in section 2.5) the solar sail can then remain at rest in any location relative to the Sun. In addition, if the solar sail has some initial velocity relative to the Sun, it will move along a rectilinear trajectory with constant speed. For example, if a solar sail of unit lightness number is injected into an Earth escape trajectory and then deployed in a circular orbit at 1 au, it will move along a rectilinear trajectory away from the Sun at 29.78 km s^{-1}, the circular orbit speed of the Earth. If the solar sail is deployed on a circular Sun-centred orbit at some radius R, the speed and linear distance traversed are given by

$$v = \sqrt{\frac{\mu}{R}} \tag{4.35a}$$

and

$$s = \sqrt{\frac{\mu}{R}}(t - t_0) \tag{4.35b}$$

so that the radial distance of the solar sail from the Sun is given by

$$r(t) = R\left[1 + \frac{\mu}{R^3}(t - t_0)^2\right]^{1/2} \tag{4.36}$$

Since the efficiency of such a trajectory is a function of the initial solar sail speed, there is a strong motivation to manoeuvre the solar sail close to the Sun before rotating the sail to face the Sun and so begin the rectilinear trajectory.

As the solar sail moves along the rectilinear trajectory its osculating semi-major axis and eccentricity will continually increase. These are the instantaneous orbital elements which would be obtained if the sail were to be jettisoned. In particular, the solar sail will quickly reach a distance from the Sun where its speed v is equal to the local escape speed $\sqrt{2\mu/r}$. This transition to escape occurs at a heliocentric distance of $2R$. For a rectilinear trajectory starting from 1 au it is found from Eq. (4.36) that the transition occurs after only 100.86 days. After this time, even of the sail is jettisoned or fails, the payload will be moving on a solar system escape trajectory.

4.3.2.5   Escape orbits (e ≥ 1)

As discussed in section 4.3.2.1, escape orbits are possible directly through the choice of solar sail lightness number, with $\beta = \frac{1}{2}$ providing parabolic escape. Of particular

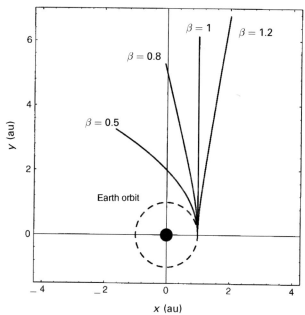

Fig. 4.12. Families of escape orbits.

interest are the family of hyperbolic orbits obtained with a sail lightness number greater than unity. In this case the effective force exerted on the solar sail follows an inverse square law, but the force is directed radially outwards, away from the Sun. Since the effective force is inverse square, the orbit is still a conic section.

A selection of escape trajectories are shown in Fig. 4.12 for a duration of one year. Although the fabrication of a solar sail with a high lightness number is a significant challenge, the performance leverage apparent from Fig. 4.12 is quite remarkable. It can be seen that for a solar sail lightness number of greater than unity the hyperbolic orbit inverts so that the Sun occupies the opposite focus of the conic section. The mission applications of such high-performance solar sails will be considered in Chapter 6.

4.3.3 Logarithmic spiral trajectories

Logarithmic spiral trajectories may in principle be generated for any type of spacecraft using continuous low-thrust propulsion. However, it is found that this type of trajectory requires the spacecraft thrust-induced acceleration to have an inverse square variation with orbit radius. Since this is clearly the case for solar sails, logarithmic spiral trajectories appear to be an attractive option for interplanetary transfer. In addition, since the sail attitude must be fixed relative to the Sun-line, only a simple steering law is required. A principal characteristic of the logarithmic spiral trajectory is that the solar sail velocity vector maintains a fixed angle with respect to the instantaneous radius vector throughout the transfer. As will be seen, this characteristic of the trajectory leads to difficulties in the practical application of logarithmic spirals for interplanetary transfer. For transfers between circular orbits, the initial and final solar sail velocity vector will not match that of the initial and final orbits. Therefore, a significant hyperbolic excess is required at launch to inject the solar sail onto the logarithmic spiral trajectory. Similarly, on arrival the solar sail again has a significant hyperbolic excess relative to the final orbit. The hyperbolic excess is a measure of the energy required from a kick-stage to inject the solar sail onto the logarithmic spiral, above that required for an Earth escape trajectory. The arrival hyperbolic excess can be removed again by using a kick-stage or, if appropriate, jettisoning the solar sail and aerobraking the payload.

In order to investigate the dynamics of logarithmic spiral trajectories the solar sail equations of motion will now be considered in the ecliptic plane with a clock angle δ of $90°$. Therefore, the sail attitude is defined solely by the cone angle α. In this planar case the sail cone angle will also be referred to as the sail pitch angle, as discussed earlier. From Eqs (4.13), the Sun-centred equations of motion in plane polar co-ordinates may be written as

$$\frac{\mathrm{d}^2 r}{\mathrm{d}r^2} - r\left(\frac{\mathrm{d}\theta}{\mathrm{d}t}\right)^2 = -\frac{\mu}{r^2} + \beta\frac{\mu}{r^2}\cos^3\alpha \qquad (4.37\mathrm{a})$$

$$r\frac{\mathrm{d}^2\theta}{\mathrm{d}t^2} + 2\left(\frac{\mathrm{d}r}{\mathrm{d}t}\right)\left(\frac{\mathrm{d}\theta}{\mathrm{d}t}\right) = \beta\frac{\mu}{r^2}\cos\alpha^2\sin\alpha \qquad (4.37\mathrm{b})$$

It can be seen that the effect of the radial component of the solar radiation pressure force is to reduce the effective gravitational force exerted on the solar sail while the transverse component changes the angular momentum of the orbit. For a fixed sail pitch angle α particular solution to these equations of motion is found to be

$$r(\theta) = r_0 \exp(\theta \tan\gamma) \qquad (4.38)$$

where the spiral angle γ is the angle between the solar sail velocity vector and the transverse direction, as shown in Fig. 4.13. In order to investigate the consequences of this logarithmic spiral solution, Eq. (4.38) is firstly differentiated to obtain the radial velocity and acceleration as

$$\frac{\mathrm{d}r}{\mathrm{d}t} = r\tan\gamma\frac{\mathrm{d}\theta}{\mathrm{d}t} \qquad (4.39\mathrm{a})$$

$$\frac{\mathrm{d}^2 r}{\mathrm{d}t^2} = \left[\tan\gamma\frac{\mathrm{d}^2\theta}{\mathrm{d}t^2} + \tan^2\gamma\left(\frac{\mathrm{d}\theta}{\mathrm{d}t}\right)^2\right]r \qquad (4.39\mathrm{b})$$

These relations can now be used to eliminate the radial velocity and acceleration terms in Eqs (4.37) to obtain

$$r\left[\frac{\mathrm{d}^2\theta}{\mathrm{d}t^2}\tan\gamma + \left(\frac{\mathrm{d}\theta}{\mathrm{d}t}\right)^2\tan^2\gamma\right] - r\left(\frac{\mathrm{d}\theta}{\mathrm{d}t}\right)^2 = -\frac{\mu}{r^2} + \beta\frac{\mu}{r^2}\cos^2\alpha \qquad (4.40\mathrm{a})$$

$$r\left[\frac{\mathrm{d}^2\theta}{\mathrm{d}t^2} + 2\left(\frac{\mathrm{d}\theta}{\mathrm{d}t}\right)^2\tan\gamma\right] = \beta\frac{\mu}{r^2}\cos\alpha^2\sin\alpha \qquad (4.40\mathrm{b})$$

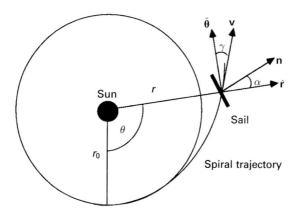

Fig. 4.13. Logarithmic spiral trajectory.

Therefore, combining Eqs (4.40) to eliminate the angular acceleration term it is found that

$$r^3 \left(\frac{d\theta}{dt} \right)^2 = \mu [1 - \beta \cos^2 \alpha (\cos \alpha - \tan \gamma \sin \alpha)] \cos^2 \gamma \qquad (4.41)$$

This equation represents a modified form of Kepler's third law, which is recovered when $\beta = 0$. For $\beta \neq 0$ it is clear that the logarithmic spiral is non-Keplerian, with a reduced effective gravitational parameter.

Using Eq. (4.41) the transverse component of the solar sail velocity vector $v_\theta = r\dot\theta$ may now be obtained as

$$v_\theta = \sqrt{\frac{\mu}{r}} [1 - \beta \cos^2 \alpha (\cos \alpha - \tan \gamma \sin \alpha)]^{1/2} \cos \gamma \qquad (4.42)$$

However, $\dot\theta$ may also be related to $\dot r$ using Eq. (4.39a) so that the radial component of the solar sail velocity vector $v_r = \dot r$ is obtained as

$$v_r = \sqrt{\frac{\mu}{r}} [1 - \beta \cos^2 \alpha (\cos \alpha - \tan \gamma \sin \alpha)]^{1/2} \sin \gamma \qquad (4.43)$$

Therefore, the magnitude of the solar sail velocity is then easily obtained from Eqs (4.42) and (4.43) as

$$v(r) = \sqrt{\frac{\mu}{r}} [1 - \beta \cos^2 \alpha (\cos \alpha - \sin \alpha \tan \gamma)]^{1/2} \qquad (4.44)$$

It can be seen that the solar sail speed is always less than the local circular orbit speed so that there will indeed be a discontinuity at the boundary conditions for transfers between circular orbits, as discussed earlier.

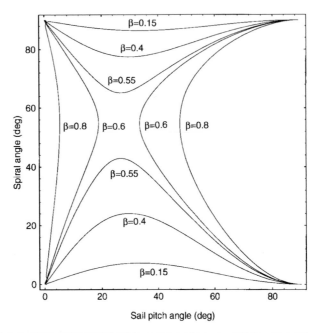

Fig. 4.14. Sail pitch angle and spiral angle with contours of equal sail lightness number.

By combining the solar sail velocity components with the equations of motion it may be shown that there is an implicit relationship between the sail pitch angle α and the spiral angle γ, viz.

$$\frac{\sin \gamma \cos \gamma}{2 - \sin^2 \gamma} = \frac{\beta \cos^2 \alpha \sin \alpha}{1 - \beta \cos^3 \alpha} \tag{4.45}$$

Therefore, for a given solar sail lightness number Eq. (4.45) may be solved for the required sail pitch angle and spiral angle, as shown in Fig. 4.14. For small values of lightness number there are two classes of admissible solution corresponding to spiral angles close to $0°$ or $90°$. Owing to launch energy requirements, only the lower branch of solutions are of interest. Similarly, for large values of lightness number there are again two classes of solution corresponding to sail pitch angles close to $0°$ or $90°$. Again, only solutions with small spiral angles will be of interest.

Using Eq. (4.45) to simplify Eq. (4.43), the transfer time between the initial orbit radius r_0 and some radial distance r may be obtained by integration of the radial component of the solar sail velocity as

$$\int_{r_0}^{r} \sqrt{r}\, dr = \int_{t_0}^{t} \left(2\beta\mu \sin \alpha \cos^2 \alpha \tan \gamma\right)^{1/2} dt \tag{4.46}$$

Therefore, integrating Eq. (4.46) the transfer time between any two orbits is obtained as

$$t - t_0 = \tfrac{1}{3}(r^{3/2} - r_0^{3/2})\left(\frac{2\cot\gamma}{\beta\mu\sin\alpha\cos^2\alpha}\right)^{1/2} \tag{4.47}$$

For a given solar sail lightness number, optimal logarithmic spiral trajectories may be obtained by calculating the value of the sail pitch angle which minimises the transfer time between orbits. However, this optimisation must also use Eq. (4.45) to obtain the correct spiral angle for a given solar sail lightness number and pitch angle. Since Eq. (4.45) is implicit, the optimisation requires numerical methods.

However, since only small spiral angles are of interest due to launch energy requirements, it is found that Eq. (4.45) may be approximated to obtain

$$\tan\gamma = \frac{2\beta\cos^2\alpha\sin\alpha}{1 - \beta\cos^3\alpha} \tag{4.48}$$

so that the spiral angle may be obtained as an explicit function of the solar sail lightness number and the sail pitch angle. In addition, using this approximation the transfer time is then given by

$$t - t_0 = \tfrac{1}{3}(r^{3/2} - r_0^{3/2})\left(\frac{1 - \beta\cos^3\alpha}{\beta^2\,\mu\cos^4\alpha\sin^2\alpha}\right)^{1/2} \tag{4.49}$$

so that the calculation of minimum time transfers now corresponds to minimising the function

$$f(\alpha) = \frac{(1 - \beta\cos^3\alpha)^{1/2}}{\beta\cos^2\alpha\sin\alpha} \tag{4.50}$$

with respect to the sail pitch angle. Setting $f'(\alpha) = 0$ it can be shown that there is an implicit relationship between the solar sail lightness number and the optimal sail pitch angle α^* which minimises the transfer time, viz.

$$\beta - \frac{2}{\cos\alpha^*}\frac{1 - 2\tan^2\alpha^*}{2 - \tan^2\alpha^*} = 0 \tag{4.51}$$

Therefore, for a given solar sail lightness number, Eq. (4.51) may be solved to obtain the required sail pitch angle, with the required spiral angle provided by Eq. (4.48). The variation of the optimal sail pitch angle and spiral angle is shown in Fig. 4.15 for small values of solar sail lightness number. It can be seen that the optimal sail pitch angle is always close to 35.26°, the sail pitch angle which maximises the transverse component of the solar radiation pressure force, as discussed in section 4.2.2. In addition, the required spiral angle slowly increases with solar sail lightness number. However, this increase in spiral angle corresponds to a significant increase in launch energy, as will be seen below.

In order to evaluate the utility of the logarithmic spiral trajectory an Earth–Mars transfer will be investigated. It will be assumed that the transfer is between circular, co-planar orbits from the Earth at 1 au to Mars at 1.524 au. Solar sail lightness numbers of 0.05, 0.1, 0.125 and 0.15 will be considered to illustrate the trajectories

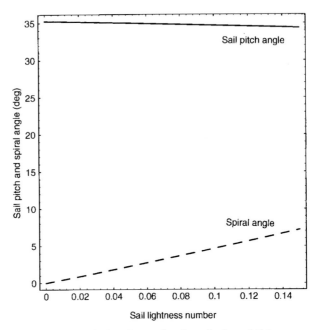

Fig. 4.15. Sail pitch and trajectory spiral angle as a function of solar sail lightness number.

Table 4.2. Earth–Mars logarithmic spiral trajectories

β	Hohmann	0.05	0.1	0.125	0.15
a_0 (mm s^{-2})	–	0.30	0.60	0.75	0.89
α (deg)	–	34.98	34.68	34.53	34.37
γ (deg)	–	2.66	4.66	5.90	7.18
T (days)	259	875	431	342	283
Δv_1 (km s^{-1})	2.95	1.54	2.52	3.17	3.84
Δv_2 (km s^{-1})	2.65	1.25	2.10	2.57	3.11
Δv_T (km s^{-1})	5.60	2.79	4.62	5.74	6.95

generated for a range of sail performances. Firstly, the optimal sail pitch angle and spiral angle may be calculated from Eqs (4.51) and (4.48). The resulting parameters are listed in Table. 4.2. Again, it can be seen that the optimal sail pitch angle is always close to $35.26°$ with little variation as the solar sail lightness number increases. The transfer time T may then be calculated from Eq. (4.47) with the resulting times again listed in Table 4.2. It can be seen that for a lightness number of 0.15 the spiral trajectory still has a longer duration than a conventional two-impulse Hohmann transfer. In addition, the variation of transfer time with sail pitch angle is shown in Fig. 4.16, where it can be seen that the transfer time is relatively insensitive to small variations in the sail pitch angle. A typical logarithmic spiral trajectory is shown in Fig. 4.17 for a lightness number of 0.05.

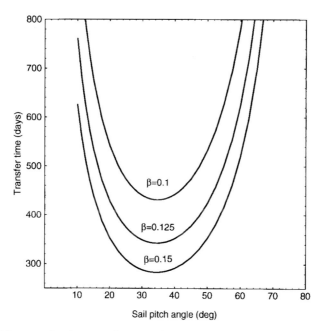

Fig. 4.16. Earth–Mars transfer time as a function of solar sail lightness number and sail pitch angle.

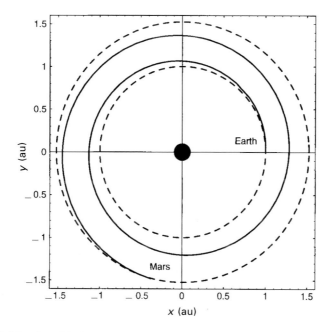

Fig. 4.17. Earth–Mars logarithmic spiral trajectory.

In order to evaluate fully the logarithmic spiral trajectory for interplanetary transfer, the hyperbolic excess required to inject the solar sail onto a logarithmic spiral trajectory is required. Since the spiral angle is constant the initial hyperbolic excess may be calculated using

$$\Delta v_1^2 = v_E^2 + v_0^2 - 2v_E v_0 \cos \gamma \qquad (4.52a)$$

where v_E is the circular orbit speed of the Earth and v_0 is the speed of the solar sail at the start of the spiral, obtained from Eq. (4.44). Similarly, the hyperbolic excess on arrival at Mars is given by

$$\Delta v_2^2 = v_M^2 + v_f^2 - 2v_M v_f \cos \gamma \qquad (4.52b)$$

where v_M is the circular orbit speed of Mars and v_f is the speed of the solar sail at the end of the spiral. The resulting initial and final hyperbolic excess and the total hyperbolic excess Δv_T for the complete transfer are listed in Table. 4.2. It can be seen that the logarithmic spiral trajectory does not appear attractive in comparison to the conventional Hohmann transfer owing to the large hyperbolic excess required at the boundaries of the transfer. Only trajectories with low solar sail lightness numbers and long transfer times show any benefit. As will be seen later in this chapter, the initial and final hyperbolic excess can be removed completely using optimal control theory to generate minimum time trajectories which accurately satisfy the boundary conditions of the transfer. Such optimal trajectories do, however, incur significantly more complex steering laws than the simple logarithmic spiral.

Although the logarithmic spiral does not appear attractive for interplanetary transfer with boundary conditions to enforce, it does provide a simple trajectory with a monotonically increasing or decreasing orbit radius. This property, along with the simple sail steering law, may be attractive for some mission applications to map, for example, spatial variations in the properties of the solar wind. Such mission applications will be considered in Chapter 6.

4.3.4 Locally optimal trajectories

Although optimal control theory can generate true minimum time trajectories, there are many instances when only simple manoeuvre strategies are required. For example, transferring a solar sail to a close heliocentric orbit, or increasing its inclination using a cranking orbit. Such manoeuvres can be efficiently generated using simple sail-steering laws. To demonstrate this it will now be shown that a large class of locally optimal trajectories can be generated through the use of such simple steering laws. These control laws maximise the instantaneous rate of change of a particular orbital element, or combination of elements, and provide the required sail orientation in closed, analytic form. It should be noted that locally optimal control laws do not guarantee global optimality. Global optimality over an entire trajectory requires numerical methods, which will be discussed in section 4.3.5. However, in some instances these locally optimal laws can generate near optimal results when used in appropriate combination.

Firstly, the variational equations defined in section 4.2.4 will be written in the following form for some arbitrary orbital element Z, viz.

$$\frac{dZ}{df} = \lambda(Z) \cdot f \qquad (4.53)$$

where $\lambda = (\lambda_1, \lambda_2, \lambda_3)$ is a vector of functions of the solar sail orbital elements and $f = (S, T, W)$ is the solar radiation pressure force exerted on the solar sail. In order to maximise the rate of change of some particular orbital element it is clear that the projection of the solar radiation pressure force vector along the vector λ must be maximised. The component of solar radiation pressure force in this desired direction is given by

$$f_\lambda = 2PA(\mathbf{n} \cdot \hat{\mathbf{r}})^2 \, \mathbf{n} \cdot \lambda \qquad (4.54)$$

The optimal sail cone and clock angles are then obtained using the same procedure as discussed in section 4.2.2 – that is, the optimal solar sail attitude angles (α^*, δ^*) are chosen to maximise the force in direction $(\tilde{\alpha}, \tilde{\delta})$, where $\tilde{\alpha}$ and $\tilde{\delta}$ are the cone and clock angles of the vector λ.

For changes to the solar sail semi-major axis and eccentricity, only in-plane forces are required so that $W = 0$. The cone and clock angles of the required force direction are then given by

$$\tan \tilde{\alpha} = \left(\frac{\lambda_2}{\lambda_1} \right) \qquad (4.55a)$$

$$\tilde{\delta} = \frac{\pi}{2} \qquad (4.55b)$$

as shown in Fig. 4.18. The optimal solar sail cone and clock angles to maximise the component of solar radiation pressure force in this direction are then obtained using

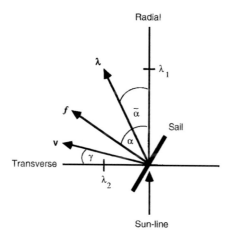

Fig. 4.18. Required force vector for locally optimal trajectories.

Eq. (4.12) as

$$\tan \alpha^* = \frac{-3 + \sqrt{9 + 8 \tan^2 \tilde{\alpha}}}{4 \tan \tilde{\alpha}} \tag{4.56a}$$

$$\delta^* = \frac{\pi}{2} \tag{4.56b}$$

Using Eqs (4.55) and (4.56) a sail steering law is then obtained in closed, analytic form. This steering law provides the required solar sail attitude as a function of the osculating orbital elements of the solar sail and maximises the instantaneous rate of change of any desired orbital element, or indeed combination of orbital elements. While these equations provide a locally optimal control law, operational constraints may limit the range of allowed sail orientations. For example, it may be desirable that the sail normal is constrained to some finite cone about the Sun-line in order to provide a good Sun aspect angle for the payload solar arrays, Sun sensors or other instruments. In order to evaluate the effect of such constraints the sail cone angle will now be chosen such that

$$\alpha = \begin{cases} \alpha^* & \text{if} \quad \alpha^* \le \tan^{-1}(1/\sqrt{2}) \\ \tan^{-1}(1/\sqrt{2}) & \text{if} \quad \alpha^* > \tan^{-1}(1/\sqrt{2}) \end{cases} \tag{4.57}$$

so that the sail normal is constrained to lie within 35° of the Sun-line. As will be seen, this hypothetical constraint does not qualitatively change the locally optimal trajectories obtained, but it does affect the time required to execute certain manoeuvres.

It can be seen from Eq. (4.14) that for changes to the solar sail orbit inclination and ascending node angles only out-of-plane forces are required. Then, the cone and clock angles of the required force direction are given by

$$\tan \tilde{\alpha} = \frac{\pi}{2} \tag{4.58a}$$

$$\tilde{\delta} = \begin{cases} 0 \\ \pi \end{cases} \tag{4.58b}$$

corresponding to a force directed above or below the instantaneous orbit plane. Then, the optimal solar sail cone and clock angles are given by

$$\tan \alpha^* = \frac{1}{\sqrt{2}} \tag{4.59a}$$

$$\delta^* = \begin{cases} 0 \\ \pi \end{cases} \tag{4.59b}$$

so that the cone angle maximises the out-of-plane component of the solar radiation pressure force while the clock angle determines the side of the instantaneous orbit plane along which the force vector is to be directed. As can be seen, these steering laws always satisfy the hypothetical constraint which has been imposed on the solar sail orientation. Using these locally optimal steering laws it will now be shown that changes to individual solar sail orbital elements or combinations of elements may be easily generated.

4.3.4.1 Semi-major axis

In this first manoeuvre a sail steering law will be generated to maximise the rate of change of the solar sail semi-major axis at all points along the trajectory. Using the variational equations in section 4.2.4 and ignoring the common scaling factors it can be seen from Eq. (4.14a) that

$$\lambda_1 = e \sin f \tag{4.60a}$$

$$\lambda_2 = \frac{p}{r} \tag{4.60b}$$

Therefore, the cone angle of the required force direction is given by

$$\tan \tilde{\alpha} = \frac{p}{re \sin f} \tag{4.61}$$

However, using the orbit equation from the two-body problem, viz.

$$r = \frac{p}{1 + e \cos f} \tag{4.62}$$

it can be seen that

$$\tan \tilde{\alpha} = \frac{1 + e \cos f}{e \sin f} \tag{4.63}$$

The optimal sail cone angle is then obtained from Eq. (4.56a) and constraints on the sail pointing enforced through Eq. (4.57). In addition, from the two-body problem the angle γ between the solar sail velocity vector and the transverse direction is given by

$$\tan \gamma = \frac{e \sin f}{1 + e \cos f} \tag{4.64}$$

as defined in Fig. 4.18. Therefore, it can be seen from Eqs (4.63) and (4.64) that the constraint $\tilde{\alpha} + \gamma = \pi/2$ must always hold. The locally optimal control law for changes to the solar sail semi-major axis therefore corresponds to maximising the component of the solar radiation pressure force directed along the solar sail velocity vector. It can be shown that this strategy also corresponds to maximising the instantaneous rate of change of the total orbit energy.

In the current and following sections the locally optimal steering laws will be illustrated using a solar sail with a lightness number of 0.05 (corresponding to a characteristic acceleration of $0.3 \, \mathrm{mm \, s^{-2}}$). A low-performance solar sail is used to avoid rapid changes in the orbital elements which would mask the underlying dynamics of the steering laws. The orbit manoeuvre will begin from a planar Sun-centred orbit with a perihelion of 1 au and aphelion of 1.5 au. This corresponds to an initial orbit with a semi-major axis of 1.25 au and an eccentricity of 0.2. An example trajectory is shown in Fig. 4.19 where the instantaneous rate of change of the solar sail semi-major axis has been maximised. Here the orbit number, defined as $2\pi/f$, is used as the independent variable to highlight the periodic nature of the steering laws.

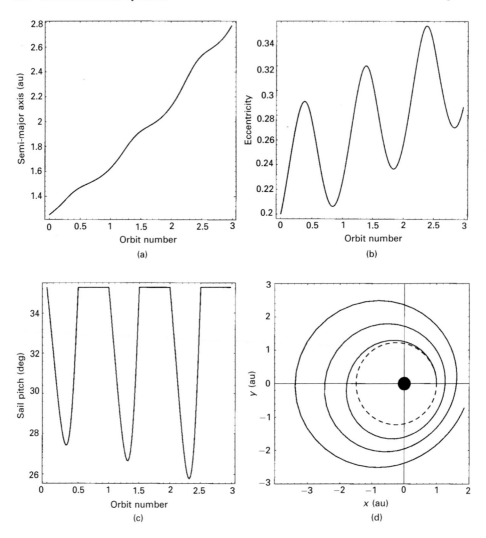

Fig. 4.19. Optimal semi-major axis increase: (a) semi-major axis; (b) eccentricity; (c) sail pitch angle; (d) solar sail orbit.

It can be seen that there is a rapid monotonic increase in semi-major axis along with a slow increase and periodic variation in eccentricity. Initially the solar sail velocity vector is normal to the Sun-line so that the sail is pitched at $35°$ when the manoeuvre begins at perihelion. This maximises the transverse component of the solar radiation pressure force since the required force direction has a cone angle of $90°$ at this point. Then, as the solar sail moves out along the elliptical orbit its velocity vector is no longer normal to the Sun-line. Therefore, the sail pitch angle decreases in order to continually maximise the force component along the solar sail velocity vector.

After aphelion, the velocity vector makes an angle of greater than 90° with the Sun-line so that the sail pitch angle required by the control law is greater than 35°. Since a constraint has been imposed whereby the sail pitch angle is limited, the pitch angle is fixed at 35°, maximising the transverse component of solar radiation pressure force. The sail then remains in this fixed pitch attitude until the next perihelion passage. It can also be seen that as the eccentricity of the orbit slowly increases, the deviation of the sail pitch angle from 35° increases in order to track the solar sail velocity vector. In addition, due to an accumulated change in the argument of perihelion, it can be seen that the solar sail does not return to perihelion after the total change in true anomaly of 6π radians. Lastly, the hypothetical sail-pointing constraint which has been enforced does not qualitatively affect the locally optimal solar sail trajectory. The constraint only reduces the semi-major axis gain somewhat as the solar sail moves sunward from aphelion to perihelion.

4.3.4.2 Eccentricity

A second locally optimal steering law will now be generated to maximise the instantaneous rate of change of orbit eccentricity. Again, using the variational equations in section 4.2.4 it can be seen from Eq. (4.14b) that

$$\lambda_1 = \sin f \qquad (4.65a)$$

$$\lambda_2 = \left(1 + \frac{r}{p}\right)\cos f + e\frac{r}{p} \qquad (4.65b)$$

Therefore, the cone angle of the required force direction is given by

$$\tan\tilde{\alpha} = \frac{2 + e\cos f}{1 + e\cos f}\cot f + \frac{e\operatorname{cosec} f}{1 + e\cos f} \qquad (4.66)$$

where the optimal sail cone angle is obtained as before. An example manoeuvre is shown in Fig. 4.20 where it can be seen that the steering law results in a rapid increase in eccentricity with an asymmetric pitch angle profile. This rapid increase in eccentricity then leads to a reduction in the perihelion radius and an increase in the aphelion radius, as expected. In this case the sail-pointing constraint has little influence on the solar sail trajectory.

4.3.4.3 Aphelion radius

It will now be shown that combinations of orbital elements may be varied using locally optimal control laws. In principle any function of the solar sail orbital elements may be optimised. For example, the aphelion radius r_A is defined as

$$r_A = a(1 + e) \qquad (4.67)$$

so that

$$\frac{dr_A}{df} = \frac{da}{df}(1 + e) + a\frac{de}{df} \qquad (4.68)$$

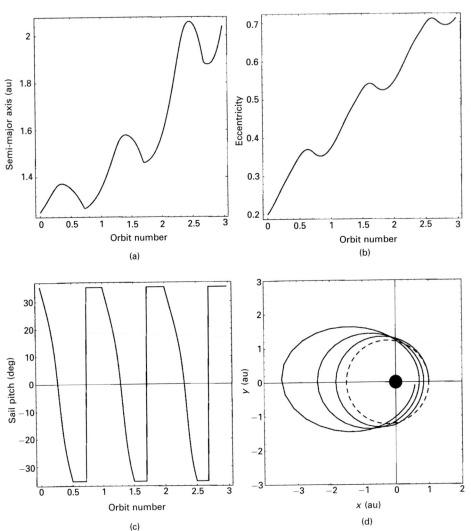

Fig. 4.20. Optimal eccentricity increase: (a) semi-major axis; (b) eccentricity; (c) sail pitch angle; (d) solar sail orbit.

Then, substituting for the appropriate variational equations, the functions λ_1 and λ_2 may be identified and a control law formed as above. The resulting trajectory is shown in Fig. 4.21. In order to maximise the rate of change of aphelion radius both the semi-major axis and eccentricity rapidly increase, as expected from Eq. (4.68). Again, an asymmetric pitch angle profile is generated which leads to an increase in the perihelion radius along with the aphelion radius. Although this steering law is locally optimal for aphelion changes it is found that the eccentricity steering law also generates near optimal changes in aphelion radius.

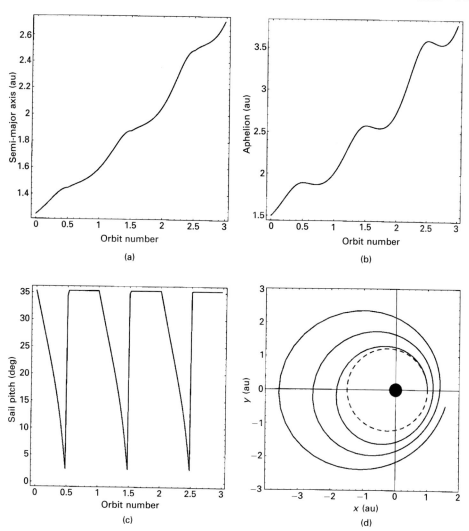

Fig. 4.21. Optimal aphelion radius increase: (a) semi-major axis; (b) aphelion radius; (c) sail pitch angle; (d) solar sail orbit.

4.3.4.4 *Orbit inclination*

Along with in-plane steering laws, locally optimal laws which change the orientation of the orbit plane may also be generated. These steering laws only require control over the out-of-plane component of the solar radiation pressure force using a simple switching function. For example, it can be seen from Eq. (4.14c) that in order to maximise the rate of change of inclination, a simple switching control law of the following form is required, viz.

$$\text{sign } W = \text{sign}[\cos(f + \omega)] \tag{4.69}$$

This steering law then ensures that the rate of change of orbit inclination is maximised at all points along the trajectory. Then, in order to maximise the out-of-plane solar radiation pressure force, the sail cone and clock angles are given by

$$\alpha^* = \tan^{-1}\left(\frac{1}{\sqrt{2}}\right) \tag{4.70a}$$

$$\delta^* = \frac{\pi}{2}\{1 - \text{sign}[\cos(f + \omega)]\} \tag{4.70b}$$

so that there is a 180° sail rotation every half orbit when the out-of-plane force is alternately directed above and below the orbit plane. This steering law then leads to a monotonic increase in orbit inclination with the maximum rate of change obtained close to the orbit nodes when $\cos(f + \omega) = 1$. Combined steering laws which alter both the orbit plane orientation and the orbit shape are also possible. Out-of-plane steering may be used close to the line of nodes and other in-plane steering laws close to the apse line.

An example trajectory is shown in Fig. 4.22 for a circular orbit initially in the ecliptic plane at 0.25 au. A solar sail with a lightness number of 0.05 will be considered. It can be seen that the effect of the control law is to maintain the initial circular orbit size and shape while efficiently rotating the orbit plane about the line of nodes. As will be seen, this cranking manoeuvre is more efficient closer to the Sun. However, thermal limitations on the sail film will limit the minimum cranking orbit radius in practice.

The change in inclination per orbit which can be achieved with this control law will now be obtained in closed form for a circular cranking orbit. Firstly, substituting for W in the variational equations it is found that

$$\frac{di}{df} = \beta \cos^2 \alpha^* \sin \alpha^* \cos \delta^*(f) \cos f \tag{4.71}$$

Then, the change in inclination per orbit Δi is obtained from

$$\Delta i = \beta \cos^2 \alpha^* \sin \alpha^* \int_0^{2\pi} \cos \delta^*(f) \cos f \, df \tag{4.72}$$

Evaluating this integral with the control law for the sail clock angle gives

$$\Delta i = 4\beta \cos^2 \alpha^* \sin \alpha^* \tag{4.73}$$

so that the change in inclination per orbit is independent of the orbit radius. Then, substituting for the optimal sail cone angle gives

$$\Delta i = 88.2\beta \text{ [degrees per orbit]} \tag{4.74}$$

Although the inclination change per orbit is independent of the solar sail orbit radius, the total time required to achieve a given change inclination is not. Since the solar sail orbit period diminishes with orbit radius, the quickest inclination changes

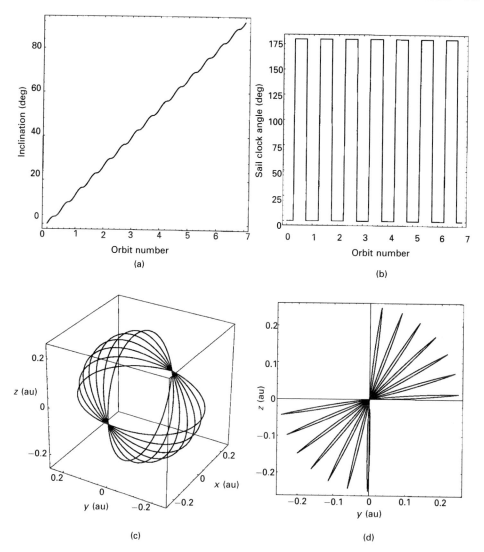

Fig. 4.22. Optimal inclination increase: (a) orbit inclination; (b) sail clock angle; (c) inclination cranking orbit (x–y–z); (d) inclination cranking orbit (y–z).

are clearly achieved using close heliocentric orbits, as shown in Fig. 4.23. It can be seen that, owing to the long orbit period at large heliocentric distances, variations in solar sail performance have little consequence far from the Sun. However, for close cranking orbits there is a significant gain to be achieved with increased lightness numbers. It should also be noted that the solar sail orbit period will be increased somewhat due to the constant radial force which arises from this steering law. As the out-of-plane force is switched above and below the orbit plane, the radial force

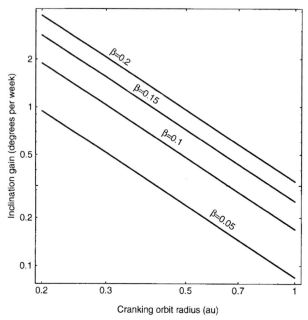

Fig. 4.23. Inclination gain per week as a function of cranking orbit radius and solar sail lightness number.

component remains constant. Therefore, the effective gravitational parameter becomes $\mu(1 - \beta \cos^3 \alpha^*)$ leading to slightly longer orbit periods.

4.3.4.5   Ascending node

Manoeuvres to change the solar sail ascending node angle are similar to those to change the orbit inclination. It can be seen from Eq. (4.14d) that in order to maximise the rate of change of the ascending node angle another switching law of the following form is required, viz.

$$\text{sign}\, W = \text{sign}[\sin(f + \omega)] \tag{4.75}$$

Then, the sail clock angle switches according to

$$\alpha^* = \tan^{-1}\left(\frac{1}{\sqrt{2}}\right) \tag{4.76a}$$

$$\delta^* = \frac{\pi}{2}\{1 - \text{sign}[\sin(f + \omega)]\} \tag{4.76b}$$

with the sail cone angle again chosen to maximise the out-of-plane component of the solar radiation pressure force. Using a similar analysis to section 4.3.4.4 this switching control law for the sail clock angle gives the change in ascending node angle per orbit as

$$\Delta\Omega = \frac{4\beta}{\sin i} \cos^2 \alpha^* \sin \alpha^* \tag{4.77}$$

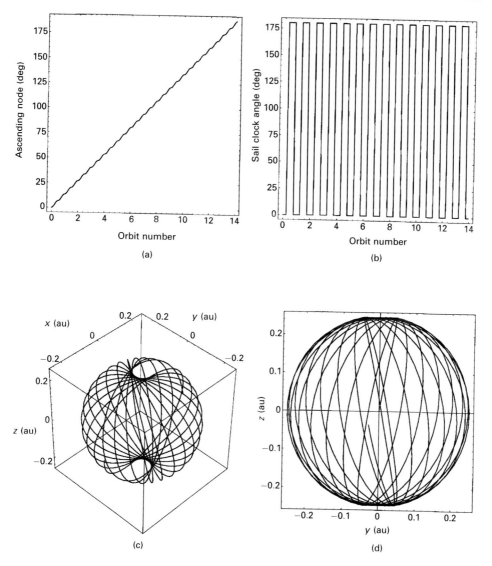

Fig. 4.24. Optimal ascending node increase: (a) ascending node; (b) sail clock angle; (c) node cranking orbit $(x–y–z)$; (d) node cranking orbit $(y–z)$.

Again, substituting for the optimal sail cone angle gives

$$\Delta\Omega = \frac{88.2\beta}{\sin i} \text{ [degrees per orbit]} \tag{4.78}$$

An example trajectory is shown in Fig. 4.24 for an orbit radius of 0.25 au and an orbit inclination of 80°. In this example the steering law is used to rotate the near polar orbit by 180°. Again, such manoeuvres are carried out more rapidly using close solar orbits with a short orbit period. The change in ascending node angle per orbit greatly increases for small orbit inclinations owing to the geometric definition of this particular orbit element.

4.3.5   Globally optimal trajectories

For practical mission analysis purposes locally optimal trajectories can be used only as a guide to finding true globally optimal trajectories. Such trajectories require the use of optimal control techniques to obtain the required solar sail clock and cone angles as a function of time. Since the sail attitude is time varying, the boundary conditions required for interplanetary transfer may be met without the use of initial and final impulses, as required for the logarithmic spiral trajectory. While the use of optimal control methods will produce true minimum time trajectories, operational and engineering constraints are likely to limit how closely they can be tracked in practice. For example, the optimum sail attitude may be forbidden during certain mission phases owing to antenna visibility constraints, constraints on the payload solar array Sun aspect angle or thermal limitations on the sail film. Optimum trajectories do, however, provide a benchmark for mission design and for the evaluation of less complex, suboptimal control schemes which incorporate inevitable operational constraints.

The application of optimal control theory is a complex field in its own right; therefore, the application of these methods to the solar sail orbit transfer problem will only be outlined. Firstly, the vector equation of motion derived in section 4.2.1 will be recast as two first-order equations, viz.

$$\dot{\mathbf{r}} = \mathbf{v} \tag{4.79a}$$

$$\dot{\mathbf{v}} = -\frac{\mu}{r^2}\hat{\mathbf{r}} + \beta\frac{\mu}{r^2}(\hat{\mathbf{r}}\cdot\mathbf{n})^2\mathbf{n} \tag{4.79b}$$

with boundary conditions imposed on the solar sail trajectory such that the solar sail state is defined by $(\mathbf{r}_0, \mathbf{v}_0)$ at initial time t_0 and $(\mathbf{r}_f, \mathbf{v}_f)$ at final time t_f, where $t_f = t_0 + T$. The goal is now to minimise the transfer time T subject to the constraints imposed by the boundary conditions and the equations of motion. To proceed with the optimisation process the Hamiltonian function for the problem is formed. To ensure optimisation, the Hamiltonian must be maximised at all points along the trajectory through an appropriate choice of solar sail cone and clock angles. It should be noted that this is not the Hamiltonian from classical mechanics, but another function of the same name. The Hamiltonian is defined as

$$H = \mathbf{p}_r\cdot\mathbf{v} - \frac{\mu}{r^2}\mathbf{p}_v\cdot\hat{\mathbf{r}} + \beta\frac{\mu}{r^2}(\hat{\mathbf{r}}\cdot\mathbf{n})^2\mathbf{p}_v\cdot\mathbf{n} \tag{4.80}$$

where $\mathbf{p}_r$ and $\mathbf{p}_v$ are the co-states for position and velocity. The velocity co-state is also referred to as the primer vector and defines the optimal direction for the solar radiation pressure force vector. Note that, unlike other low-thrust propulsion systems, there is no co-state for the solar sail mass since it is clearly constant. The rate of change of the co-states is then obtained from the Hamiltonian as

$$\dot{\mathbf{p}}_r = -\frac{\partial H}{\partial \mathbf{r}} \tag{4.81a}$$

$$\dot{\mathbf{p}}_v = -\frac{\partial H}{\partial \mathbf{v}} \tag{4.81b}$$

so that

$$\dot{\mathbf{p}}_r = \frac{\mu}{r^3}\mathbf{p}_v - \frac{3\mu}{r^5}(\mathbf{p}_r\cdot\mathbf{r})\mathbf{r} + 2\beta\frac{\mu}{r^3}(\hat{\mathbf{r}}\cdot\mathbf{n})(\mathbf{p}_v\cdot\mathbf{n})[\mathbf{n} + 2(\hat{\mathbf{r}}\cdot\mathbf{n})\hat{\mathbf{r}}] \tag{4.82a}$$

$$\dot{\mathbf{p}}_v = -\mathbf{p}_r \tag{4.82b}$$

In order to maximise the Hamiltonian it is found that the clock angle of the solar sail must be aligned with the clock angle of the primer vector, as expected from the analysis of section 4.2.2. In addition, it is found that the solar radiation pressure force vector lies in the plane defined by the position vector and primer vector such that

$$\mathbf{n} = \frac{\sin(\alpha - \tilde{\alpha})}{\sin\tilde{\alpha}}\hat{\mathbf{r}} + \frac{\sin\alpha}{\sin\tilde{\alpha}}\hat{\mathbf{p}}_v \tag{4.83}$$

where $\tilde{\alpha}$ is the cone angle of the primer vector. The optimal solar sail cone angle is then found by maximising the component of the solar radiation pressure force vector along the primer vector. Then, using Eq. (4.12) from section 4.2.2, the optimal solar sail cone angle is given by

$$\tan\alpha^* = \frac{-3 + \sqrt{9 + 8\tan^2\tilde{\alpha}}}{4\tan\tilde{\alpha}} \tag{4.84}$$

It is clear that in order to obtain the optimal sail clock and cone angles the co-states are required as a function of time. The co-states can be obtained by integrating Eqs (4.82) along with the solar sail equations of motion. However, unlike the equations of motion, the co-state equation boundary conditions are unknown. In addition, the co-states must also satisfy the so-called transversality condition

$$\left[\beta\frac{\mu}{r^2}(\hat{\mathbf{r}}\cdot\mathbf{n})^2\mathbf{p}_v\cdot\mathbf{n}\right]_{t=t_0} = \left[\beta\frac{\mu}{r^2}(\hat{\mathbf{r}}\cdot\mathbf{n})^2\mathbf{p}_v\cdot\mathbf{n}\right]_{t=t_f} \tag{4.85}$$

which is required to ensure the optimality of the launch date. The co-state boundary conditions must be obtained by iterative methods to generate a trajectory which satisfies the orbit boundary conditions and the transversality conditions. In general the convergence to a true optimal solution is difficult owing to the insensitivity of the transfer time to small variations in the solar sail attitude time history. In addition, the initial guess of the co-states required to begin the iteration process must be close to the true values to ensure convergence. Locally optimal steering laws may be used to provide an initial guess and to start the convergence procedure. While the use of optimal control theory provides extremely accurate trajectories, other methods are

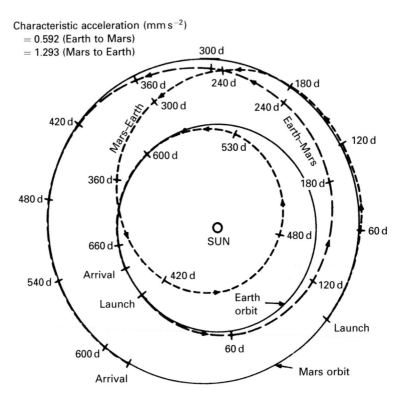

Fig. 4.25. Minimum time Earth–Mars sample return trajectory (NASA/JPL).

somewhat easier to implement. In particular, gradient methods which divide the trajectory into a number of segments may be used. The sail cone and clock angles are then parameterised as, for example, linear functions of time within each trajectory segment. The optimal trajectory is then obtained by choosing these parameters to minimise the transfer time while enforcing the boundary conditions of the problem. This parameter optimisation problem is solved using standard gradient type methods and circumvents the need for repeated iteration of the co-states.

An example time optimal Mars sample return trajectory is shown in Fig. 4.25. This trajectory has been designed for a 1988 launch with an outbound solar sail characteristic acceleration of 0.59 mm s^{-2} and a return characteristic acceleration of 1.29 mm s^{-2}. This particular launch date and trajectory are representative of the regular opportunities for transfer to Mars, with optimum launch dates somewhat prior to opposition. In general it is found that rendezvous is significantly easier prior to the aphelion of the target body rather than after aphelion The outbound and return transfer times are found to be approximately 620 and 680 days respectively, with the longer return trajectory requiring a loop inside the orbit of the Earth. It should be noted that these transfer times assume zero departure and arrival hyperbolic excess so that the total transfer times will increase significantly if Earth escape and Mars capture spirals are included. It can be seen from Fig. 4.25 that the outbound trajectory crosses the orbit of Mars rather quickly. Therefore, if a rendezvous is not required the solar sail can quickly deliver payloads to Mars with some hyperbolic excess velocity. This hyperbolic excess can then be removed, for example by aerobraking. However, for a rendezvous trajectory it can be seen that the final arc of the transfer is outside the orbit of Mars. Therefore, the solar sail approaches Mars moving towards the Sun. This is attributable to the constraint that the solar radiation pressure force vector cannot be directed sunward.

4.4 PLANET-CENTERED ORBITS

4.4.1 Introduction

Now that Sun-centred solar sail trajectories have been discussed, planet-centred trajectories will be considered. In particular Earth escape trajectories will be investigated using a variety of simple sail-steering laws. Since the solar radiation pressure force cannot be directed sunward, solar sail escape trajectories are distinctly different from escape trajectories using other forms of low-thrust propulsion. In particular, almost no energy can be gained while the solar sail is moving sunward or, of course, while the solar sail is in eclipse. These problems can be partially alleviated by using polar escape trajectories which are normal to the Sun-line. Such orbits can have an almost fixed Sun aspect angle and can be chosen to be free of eclipse periods under certain conditions. However, it will be shown that polar escape trajectories are in fact still less efficient than equatorial trajectories. Clearly, the same steering laws which will be derived for escape trajectories can also be used for capture spirals.

The general equation of motion for a perfectly reflecting solar sail in a planet-centred orbit is given by

$$\frac{d^2\mathbf{r}}{dt^2} + \mu\frac{\mathbf{r}}{r^3} = \kappa(\mathbf{l}\cdot\mathbf{n})^2\mathbf{n} \tag{4.86}$$

where the parameter κ is the magnitude of the solar sail acceleration and $\mathbf{l}$ is the unit vector directed along the Sun-line. For Earth-centred trajectories κ is just the solar sail characteristic acceleration. Owing to the scale of the spiral orbit relative to the Earth–Sun distance, the magnitude of the solar sail acceleration can be assumed to be constant. In addition, the Sun-line will have a slow annual rotation of $0.986°$ per day owing to the motion of the Earth about the Sun. Since the Earth's orbit about the Sun is slightly elliptical, this rate will itself vary during the year. For orbits beginning at low altitudes, air drag and orbit precession owing to Earth oblatness are important effects, as are eclipse condition which are a strong function of orbit inclination. In particular, orbit precession has an important bearing on polar escape trajectories and can bring an initially sunlit orbit into eclipse. Air drag is a function of the initial orbit altitude and the solar sail attitude time history. In general, air drag is negligible above approximately 600–900 km, although the effect is strongly modulated by solar activity. In addition, as the solar sail spirals outwards, third-body effects, such as lunar and solar gravity, will become important.

Finally, owing to the high local gravitational acceleration in Earth orbit, escape trajectories are found to require long spiral times. In addition, for effective escape manoeuvres it will be shown that rapid solar sail rotation rates may be required, placing demands on the sail attitude control actuators. For these reasons Earth escape trajectories are in fact best avoided if possible by using direct injection into an interplanetary trajectory using a chemical kick-stage. Clearly such direct escape strategies have cost implications. However, by avoiding long escape spirals and rapid rotations of the sail, solar sail designs can be optimised for the more benign conditions of interplanetary cruise.

4.4.2   Suboptimal trajectories

4.4.2.1   On–off switching

The first method of orbit manoeuvring to be considered is a simple on–off switching of the solar radiation pressure force using the sail pitch attitude. The sail will be oriented to face the Sun for one half of the orbit and will then be oriented edgewise to the Sun for the other half orbit, as shown in Fig. 4.26. Although the switching profile is simple, rapid 90° rotation manoeuvres are required twice per orbit. The duration of these rotations must be short relative to the orbit period to avoid loss of efficiency. Therefore, low-altitude starting orbits are undesirable owing to the long spiral time, short orbit period and the presence of long periods of eclipse. It will now be shown that by using this simple switching law during each orbit, orbit energy is gained and the semi-major axis of the orbit increases. A further consequence of the

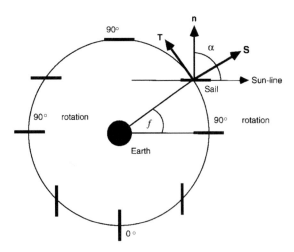

Fig. 4.26. On–off steering law.

asymmetry of the switching law is an increase in orbit eccentricity with the solar sail orbit expanding normal to the Sun-line.

The increase in semi-major axis per orbit can be estimated using the variational equations of section 4.2.4. For a small orbit eccentricity the variational equation for semi-major axis, Eq. (4.14a), becomes

$$\frac{da}{df} = \frac{2a^3}{\mu} T(f) \tag{4.87}$$

Neglecting any periods of eclipse the transverse acceleration experienced by the solar sail is given by

$$T(f) = 0 \quad (0 \le f < \pi) \tag{4.88a}$$

$$T(f) = \kappa \cos\left(f - \frac{3\pi}{2}\right) \quad (\pi \le f < 2\pi) \tag{4.88b}$$

Therefore, the change in semi-major axis over one orbit is obtained by integrating the transverse acceleration using

$$\Delta a = \frac{2a^3}{\mu} \int_0^{2\pi} T(f)\, df \tag{4.89}$$

Then, integrating over the half orbit when the solar sail is facing the Sun it is found that

$$\Delta a = \frac{4\kappa}{\mu} a^3 \tag{4.90}$$

Since the semi-major axis gain is a strong function of the initial semi-major axis, the benefit of a high starting orbit is again seen. In the current and following sections the steering laws will be illustrated using a solar sail with a lightness number of 0.17 (corresponding to a characteristic acceleration of $1\,\mathrm{mm\,s^{-2}}$). The spiral manoeuvre will begin from geostationary orbit (GEO), corresponding to an initial semi-major axis of 42,241 km and an eccentricity of zero. A moderate performance solar sail and a high starting orbit are used to obtain a reasonably quick increase in semi-major axis. Each steering law will be used for a period 72 hours to illustrate the variation of the solar sail orbital elements and the sail pitch angle as a function orbit number. An example orbit-raising manoeuvre is shown in Fig. 4.27 using the simple switching law. It can be seen that rapid sail rotations are required and that energy is gained during only half of each orbit. The method is therefore inefficient in gaining energy, but requires only a simple Sun-facing attitude for half of each orbit.

4.4.2.2 Orbit rate steering

Given the limitations of the switching law described above, a more sophisticated orbit-raising scheme will now be considered. In this scheme the solar sail will be forced to rotate at one half of the orbit rate, allowing some energy to be gained during the half orbit when the solar sail is moving sunward. Since the orbit rate will

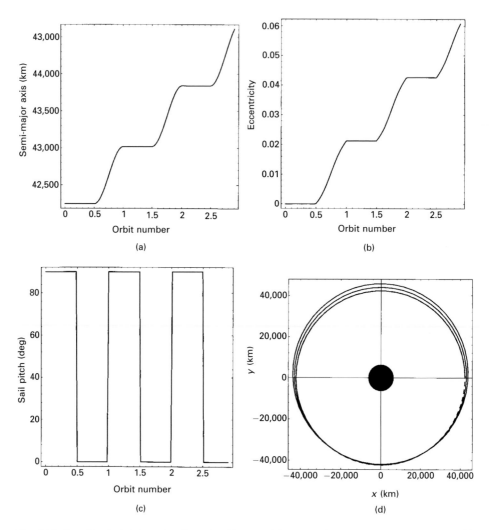

Fig. 4.27. On–off spiral: (a) semi-major axis; (b) eccentricity; (c) sail pitch angle; (d) solar sail orbit.

vary as the semi-major axis and eccentricity of the orbit increases, the sail pitch rate will not in general be constant. The sail pitch rate is then related to the orbit rate by

$$\frac{d\alpha}{dt} = \frac{1}{2}\frac{df}{dt} \tag{4.91}$$

Integrating Eq. (4.91) and ensuring the correct phasing of the sail pitch angle with the Sun-line, the pitch angle will be defined as

$$\alpha = \frac{\pi}{4} + \frac{f}{2} \tag{4.92}$$

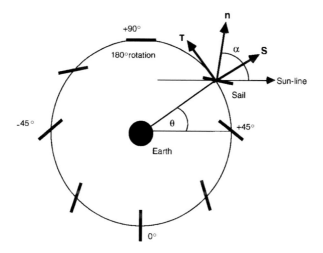

Fig. 4.28. Orbit rate steering law.

as shown in Fig. 4.28. It is found that the transverse solar radiation pressure acceleration may then be written as

$$T(f) = \kappa \left| \cos^3 \left(\frac{\pi}{4} + \frac{f}{2} \right) \right| \quad (0 \le f \le 2\pi) \tag{4.93}$$

Substituting the transverse acceleration in Eq. (4.89) and integrating, the change in semi-major axis per orbit is found to be

$$\Delta a = \frac{16\kappa}{3\mu} a^3 \tag{4.94}$$

which is clearly greater than that for the simple switching law.

An orbit-raising manoeuvre using this steering law is shown in Fig. 4.29. In this case a rapid 180° rotation is required once per orbit, again imposing demands on the sail attitude control actuators, particularly for low-altitude starting orbits. An alternative to the rapid rotation manoeuvres is to use a sail which is reflective on both sides. In this case the sail rotation rate is almost constant, slowly decreasing as the orbit period increases. While this alleviates the need for rapid rotations, additional mass is clearly added to the sail, thus reducing its performance. It can also be seen from Fig. 4.29 that the energy gain during the sunward arc of the orbit is still small. Fortunately, it is during this period of low energy gain that the 180° sail rotation must be performed. The main benefit of the orbit rate steering law is to align the solar radiation pressure force vector more closely to the solar sail velocity vector than is possible with the switching law. However, an even more efficient strategy is to continually maximise the rate of change of orbit energy.

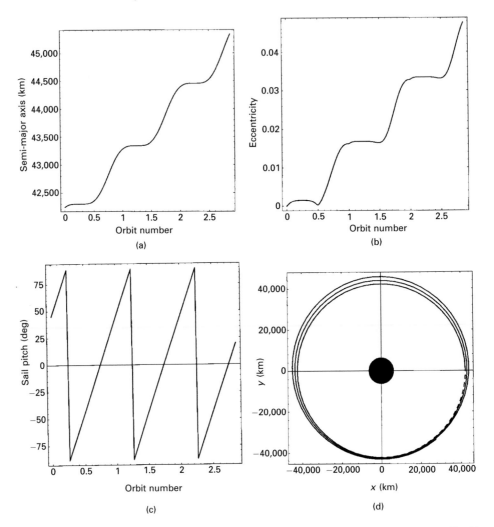

Fig. 4.29. Orbit rate spiral: (a) semi-major axis; (b) eccentricity; (c) sail pitch angle; (d) solar sail orbit.

4.4.2.3 Locally optimal steering

In order maximise the instantaneous rate of change of orbit energy, a locally optimal steering law will now be investigated. Firstly, taking the scalar product of the solar sail equation of motion (Eq. (4.86)), with the solar sail velocity vector **v**, it can be seen that

$$\frac{\mathrm{d}^2\mathbf{r}}{\mathrm{d}t^2}\cdot\mathbf{v}+\mu\frac{\mathbf{r}\cdot\mathbf{v}}{r^3}=\kappa(\boldsymbol{l}\cdot\mathbf{n})^2\mathbf{n}\cdot\mathbf{v} \tag{4.95}$$

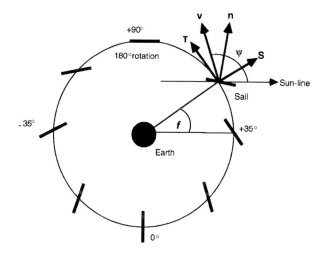

Fig. 4.30. Locally optimal steering law.

However the left side of Eq. (4.95) may be written as

$$\frac{d^2\mathbf{r}}{dt^2} \cdot \mathbf{v} + \mu \frac{\mathbf{r} \cdot \mathbf{v}}{r^3} = \frac{d}{dt}\left(\tfrac{1}{2}\mathbf{v} \cdot \mathbf{v} - \frac{\mu}{r}\right) \qquad (4.96)$$

which represents the rate of change of total orbit energy E. Therefore, the instantaneous rate of change of orbit energy is given by

$$\frac{dE}{dt} = \kappa(\boldsymbol{l} \cdot \mathbf{n})^2 \mathbf{n} \cdot \mathbf{v} \qquad (4.97)$$

In order to maximise the energy gain at all points along the trajectory, the sail pitch angle must now be chosen to maximise the component of solar radiation pressure force along the solar sail velocity vector. This is the same steering strategy used for Sun-centred orbits in section 4.3.4.1. Defining ψ as the angle of the solar sail velocity vector relative to the Sun-line, as shown in Fig. 4.30, it can be seen that

$$(\boldsymbol{l} \cdot \mathbf{n})^2 \mathbf{n} \cdot \mathbf{v} = \cos^2 \alpha \cos(\psi - \alpha) \qquad (4.98)$$

Therefore, to maximise the instantaneous rate of change of orbit energy the turning point of Eq. (4.98) must now be determined. This corresponds to finding the solar sail pitch angle which maximises the orbit energy gain for each value of ψ. The turning point is found from

$$\frac{d}{d\alpha}[\cos^2 \alpha \cos(\psi - \alpha)] = 0 \qquad (4.99)$$

which yields the optimal sail pitch angle as

$$\alpha^* = \frac{1}{2}\left[\psi - \sin^{-1}\left(\frac{\sin \psi}{3}\right)\right] \qquad (4.100)$$

Therefore, using this locally optimal steering law, the transverse acceleration of the solar sail is given by

$$T(f) = \kappa \cos^2 \alpha^* |\sin(\alpha^* - f)| \tag{4.101}$$

For a near circular orbit $\psi \sim f + \pi/2$ so that the transverse acceleration is a function of true anomaly only. Then, substituting Eq. (4.101) into Eq. (4.89) and integrating, the change in semi-major axis per orbit is obtained as

$$\Delta a = \frac{5.52 \, \kappa}{\mu} a^3 \tag{4.102}$$

An orbit-raising manoeuvre is shown in Fig. 4.31 using this locally optimal steering law. Again, in this case either a rapid 180° rotation is required once per orbit, or the sail must be reflective on both sides. As before, there is little energy gain during the sunward arc of each orbit which also coincides with the time for the sail rotation. It can be seen from Fig. 4.31 that the sail pitch angle profile is similar to that of the orbit rate steering law. However, as the eccentricity of the orbit grows, the pitch rate varies to track the rotations of the solar sail velocity vector. A longer 30-day spiral from a geostationary orbit is shown in Fig. 4.32 with a rotating Sun-line. It can be seen that as the orbit eccentricity grows, the major axis of the orbit remains normal to the Sun-line; then, as the Sun-line rotates, the argument of perigee of the orbit also increases causing a slow rotation of the entire orbit, within the orbit plane.

Although a geostationary orbit has been used as the starting point for these examples, for ease of illustration, other starting orbits are of course possible. One of the more likely starting orbits is GTO, the long transfer ellipse from low Earth orbit to geostationary orbit. Piggy-back rides on commercial communication satellite launches provide an attractive option for low-cost solar sail missions. However, the use of an elliptical starting orbit complicates the escape strategy. In particular, the angle of the major axis of the ellipse relative to the Sun-line has a significant effect on the efficiency of the escape manoeuvre. In addition, the initial orbit-raising scheme must be designed to quickly raise the perigee of the ellipse to avoid excessive air drag. Since the launch time and Sun aspect angle are chosen by the owner of the primary payload, and due to the high sail rotation rates required in elliptical orbits, the use of GTO is best avoided if possible.

4.4.2.4   *Polar orbit escape*

In order to avoid rapid sail rotations and to ensure that energy is gained at all points along the trajectory, polar orbit escape may be considered with the solar sail orbit plane normal to the Sun-line, as shown in Fig. 4.33. In this case the transverse acceleration is constant, viz.

$$T(f) = \kappa \cos^2 \alpha \sin \alpha \tag{4.103}$$

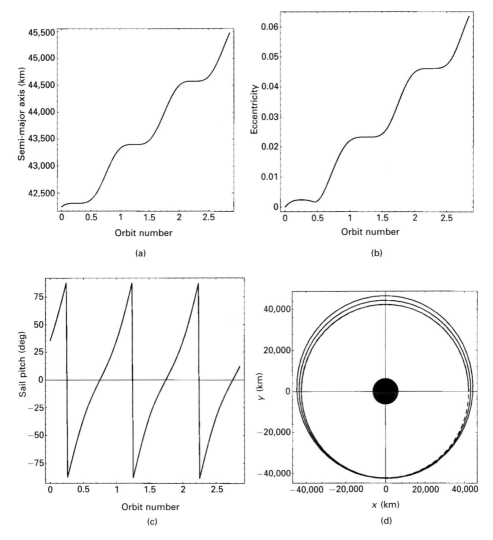

Fig. 4.31. Locally optimal spiral: (a) semi-major axis; (b) eccentricity; (c) sail pitch angle; (d) solar sail orbit.

For a near circular orbit the semi-major axis change per orbit is then obtained from Eq. (4.89) as

$$\Delta a = \frac{4\pi\kappa}{\mu} a^3 \cos^2\alpha \sin\alpha \qquad (4.104)$$

Clearly, the semi-major axis gain is maximised if the transverse acceleration is maximised using the usual sail pitch angle $\alpha = \tan^{-1}(1/\sqrt{2})$. Then, the change in

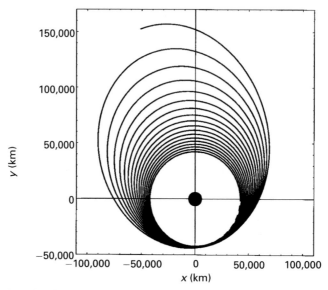

Fig. 4.32. 30-day spiral from geostationary orbit.

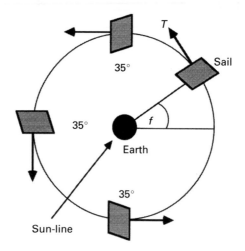

Fig. 4.33. Polar orbit steering law.

semi-major axis per orbit is given by

$$\Delta a = \frac{8\,\pi\kappa}{3\sqrt{3}\,\mu}a^3 \qquad (4.105)$$

The change in eccentricity during the spiral manoeuvre may also be obtained using

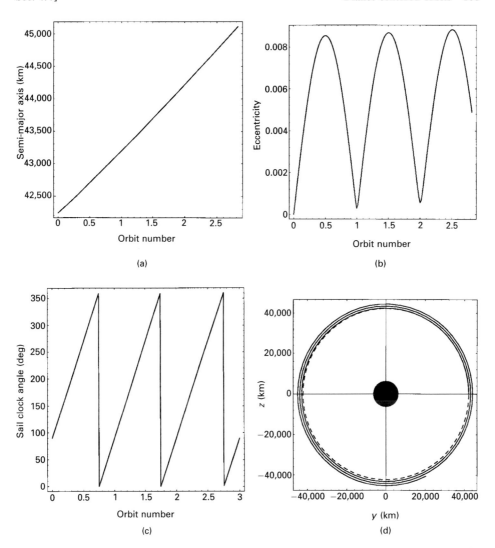

Fig. 4.34. Polar orbit spiral: (a) semi-major axis; (b) eccentricity; (c) sail clock angle; (d) solar sail orbit.

Eq. (4.14b) as

$$\Delta e = \frac{2a^2}{\mu} \int_0^{2\pi} T(f) \cos f \, df \qquad (4.106)$$

However, since the transverse force is constant, it is found that $\Delta e \sim 0$ so that the orbit remains quasi-circular as the solar sail spirals outwards. Although the sail pitch angle remains constant, the sail must still roll through $360°$ once per orbit to keep the transverse acceleration tangent to the trajectory. This is clearly less demanding than the rotation manoeuvres required for the other sail-steering laws.

An orbit-raising manoeuvre using polar orbit escape is shown in Fig. 4.34. Although an initial polar orbit at geostationary altitude is impractical, it will be used for comparison with the earlier solar sail-steering laws. It can be seen that the semi-major axis monotonically increases, demonstrating the constant orbit energy gain as the solar sail spirals outwards. It is also clear that the orbit remains quasi-circular, unlike other steering laws where the eccentricity increases due to the asymmetry of the transverse solar radiation pressure force experienced by the solar sail during each orbit. Since the orbit remains quasi-circular, the variational equation for the semi-major axis may be used to obtain a closed form solution for the outward spiral. The variational equation is separable and may be written as

$$\int_{a_0}^{a} \frac{da}{a^3} = \frac{2T}{\mu} \int_{f_0}^{f} df \tag{4.107}$$

Therefore, integrating Eq. (4.107) and rearranging, the orbit equation is obtained as

$$a(f) = a_0 \left[1 - \frac{4a_0^2}{\mu} \kappa \cos^2 \alpha \sin \alpha (f - f_0) \right]^{-1/2} \tag{4.108}$$

It is found that this relation gives good agreement with numerical simulation up to the point of escape where the orbit eccentricity rapidly increases.

Lastly, a comparison of the semi-major axis gain for each steering law is shown in Table. 4.3. It can be seen that although the transverse force is constant for polar orbit escape, the locally optimal tangential steering law clearly provides the greatest semi-major axis gain per orbit. The locally optimal law provides significantly better performance than the simple switching law. However, the benefit over the orbit rate law is more modest. Therefore, the simple orbit rate steering law has significant benefits due to its relatively good performance and simplicity of implementation. All that is required is an estimate of the orbit rate, knowledge of the Sun-line vector and the solar sail attitude. This is clearly less complex than the locally optimal steering law which also requires knowledge of the direction of the solar sail velocity vector. The velocity vector information could be estimated on-board or could be up-linked after ground-based orbit determination. However, for long duration spirals such ground-based processing represents an additional cost overhead for the mission. Therefore, it is desirable that sail steering during long escape spirals is both simple and autonomous.

Table 4.3. Planetary escape steering laws: $\Delta a = \varepsilon (\kappa a^3 / \mu)$

	On–off	Orbit rate	Optimal	Polar
ε	4.00	5.33	5.52	4.84

4.4.3 Minimum time escape trajectories

Time optimal escape trajectories have also been investigated using optimal control theory. By averaging the variational equations over one orbital period, short period terms may be eliminated, leading to a significant reduction in computational effort. Since the method of averaging is only valid for a small ratio of solar radiation pressure acceleration to local gravitational acceleration, such solutions are not uniformly valid up to the point of escape. They can, however, allow rapid generation of trajectories up to some sub-escape point, with the full set of variational equations used to generate the final few loops of the trajectory. For many optimal trajectories a rapid increase in eccentricity is found with a consequent lowering of the solar sail perigee radius. To avoid perigee radii less than the radius of the Earth, and to avoid the rapid attitude manoeuvres associated with high-eccentricity orbits, a penalty function may be added to bias against a low perigee. It has been found that a velocity-dependent steering law similar to that derived in section 4.4.4.3 produces solutions close to optimum without the complexity of optimal control theory. Furthermore, it is also found that optimal polar escape trajectories require less demanding attitude manoeuvres, but are somewhat slower than optimal equatorial escape trajectories, as expected from section 4.4.2.4.

In any choice of sail-steering law for planetary escape, whether optimal or suboptimal, the maximum sail turning rate will be the limiting factor in determining the efficiency of the trajectory. A solar sail with a low turning rate may not be able to reorient itself quickly enough during the sunward arc of each orbit in order to trim to the optimal pitch angle during the period of rapid energy gain while moving away from the Sun. As discussed, these rapid rotation manoeuvres can be eliminated using a sail film which is reflective on both sides. However, the mass saving achieved by allowing the use of smaller attitude control actuators must be traded-off against the additional mass which is added by the extra sail coating. As noted earlier, long escape spirals are best avoided by using a chemical kick-stage to inject the solar sail directly onto an interplanetary trajectory.

4.4.4 Approximate escape time

Although the escape time from some initial orbit can be readily obtained through numerical integration of the solar sail equations of motion with an appropriate steering law, it is useful to derive an approximate expression for the escape time. In order to do so it is necessary to assume that the outward spiral orbit remains quasi-circular up to the point of escape. As has been seen, this is not the case, so that the expression for the escape time is only approximate at best. Firstly, the average transverse acceleration obtained per orbit will be calculated using

$$\bar{T} = \frac{1}{2\pi} \int_0^{2\pi} T(f)\,\mathrm{d}f \tag{4.109}$$

Although any sail-steering law may be used, only the locally optimal law will be considered here. Again, assuming a quasi-circular orbit so that $\psi \sim f + \pi/2$ the optimal sail pitch angle can be obtained, as discussed in section 4.4.2.3. Then, substituting for the transverse acceleration from Eq. (4.101) it is found that $\bar{T} = 0.4393\,\kappa$. The variational equation for the semi-major axis can now be transformed from true anomaly as the independent variable to time using

$$\frac{da}{dt} = \frac{da}{df}\sqrt{\frac{\mu}{a^3}} \tag{4.110}$$

since the rate of change of true anomaly is just $\sqrt{\mu/a^3}$. Then from Eq. (4.87) the approximate rate of change of semi-major axis is obtained as

$$\frac{da}{dt} = \frac{2a^{3/2}}{\sqrt{\mu}}\,\bar{T} \tag{4.111}$$

This equation can now be separated and integrated from some initial semi-major axis a_0, up to escape as

$$\int_{a_0}^{\infty} \frac{da}{a^{3/2}} = \frac{2\bar{T}}{\sqrt{\mu}}\int_0^{\tau} dt \tag{4.112}$$

Then, evaluating the integral, the approximate escape time is obtained from

$$\tau = \frac{1}{\bar{T}}\sqrt{\frac{\mu}{a_0}} \tag{4.113}$$

Substituting for the appropriate physical constants, this approximate escape time can then be written conveniently as

$$\tau = \frac{2805}{\beta\sqrt{6371 + h(km)}}\ [\text{days}] \tag{4.114}$$

where h is the initial orbit altitude and β is the usual solar sail lightness number. The escape time for a range of starting orbits is shown in Fig. 4.35. It can be seen that the escape time is a strong function of the altitude of the starting orbit so that there is a strong motivation to begin missions directly from an Earth escape trajectory, or at least from geostationary orbit rather than GTO.

4.4.5   Solution by the Hamilton–Jacobi method

The dynamics of a planet-centred solar sail orbit will now be investigated through the use of Hamilton–Jacobi theory. It will be demonstrated that a general, closed-form analytic solution exists for a solar sail with a fixed attitude and an (assumed) fixed Sun-line. The analysis demonstrates yet another analytic trajectory representation and provides an effective means of rapidly generating preliminary orbits. The Hamilton–Jacobi equation for this problem is separable and may be solved using a set of parabolic co-ordinates. Hamilton–Jacobi theory is a powerful tool from classical mechanics used to investigate the dynamics of non-linear systems.

Firstly, a set of parabolic co-ordinates (ξ, η, θ) may be defined. These co-ordinates then transform the usual Cartesian co-ordinates, assuming the z axis is aligned with

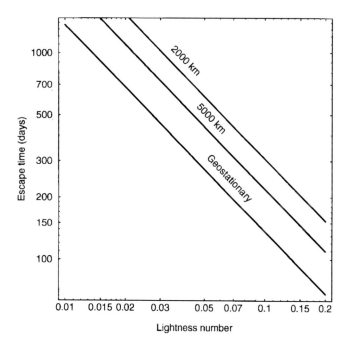

Fig. 4.35. Approximate Earth orbit escape time as a function of solar sail lightness number and starting orbit altitude.

the Sun-line. The co-ordinates will be defined as

$$x = \xi\eta\cos\theta \tag{4.115a}$$

$$y = \xi\eta\sin\theta \tag{4.115b}$$

$$z = \tfrac{1}{2}(\xi^2 - \eta^2) \tag{4.115c}$$

where constant ξ and η co-ordinates define diametrically oriented parabolae of revolution about the z axis, with θ the azimuthal angle as shown in Fig. 4.36. In Cartesian co-ordinates the kinetic energy and potential energy of the solar sail are given by

$$T = \frac{1}{2}\left[\left(\frac{dx}{dt}\right)^2 + \left(\frac{dy}{dt}\right)^2 + \left(\frac{dz}{dt}\right)^2\right] \tag{4.116a}$$

$$V = -\frac{\mu}{r} + \kappa z \tag{4.116b}$$

where the solar sail is assumed to be oriented normal to the Sun-line. Arbitrary fixed sail attitudes may be used by rotating the co-ordinate system so that the z axis lies along the sail normal. Then, using the new co-ordinate set, the kinetic and potential

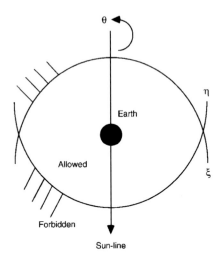

Fig. 4.36. Parabolic co-ordinates.

energy of the solar sail may be written as

$$T = \frac{1}{2} \left\{ \left[\left(\frac{d\xi}{dt} \right)^2 + \left(\frac{d\eta}{dt} \right)^2 \right] (\xi^2 + \eta^2) + \left(\frac{d\theta}{dt} \right)^2 \xi^2 \eta^2 \right\} \qquad (4.117a)$$

$$V = - \frac{2\mu}{\xi^2 + \eta^2} - \tfrac{1}{2} \kappa (\xi^2 - \eta^2) \qquad (4.117b)$$

The conjugate momenta corresponding to each position co-ordinate may also be obtained from

$$p_\xi = \frac{\partial T}{\partial \dot{\xi}} = \left[\frac{d\xi}{dt} \right] (\xi^2 + \eta^2) \qquad (4.118a)$$

$$p_\eta = \frac{\partial T}{\partial \dot{\eta}} = \left[\frac{d\eta}{dt} \right] (\xi^2 + \eta^2) \qquad (4.118b)$$

$$p_\theta = \frac{\partial T}{\partial \dot{\theta}} = \left[\frac{d\theta}{dt} \right] \xi^2 \eta^2 \qquad (4.118c)$$

where p_θ is an integral of the system since the potential V is independent of θ.

Equations (4.117) and (4.118) may now be used to construct the Hamiltonian of the system $H = T + V$. In terms of the parabolic co-ordinates and momenta it is found that

$$H = \frac{1}{2(\xi^2 + \eta^2)} \left\{ p_\xi^2 + p_\eta^2 + \left[\frac{1}{\xi^2} + \frac{1}{\eta^2} \right] p_\theta^2 \right\} - \tfrac{1}{2} \kappa (\xi^2 - \eta^2) - \frac{2\mu}{\xi^2 + \eta^2} \qquad (4.119)$$

Defining the Hamilton–Jacobi function as S, the Hamilton–Jacobi equation

$\partial S/\partial t + H = 0$ may now be written as

$$\frac{\partial S}{\partial t} + \frac{1}{2(\xi^2 + \eta^2)} \left\{ \left(\frac{\partial S}{\partial \xi}\right)^2 + \left(\frac{\partial S}{\partial \eta}\right)^2 + \left[\frac{1}{\xi^2} + \frac{1}{\eta^2}\right]\left(\frac{\partial S}{\partial \theta}\right)^2 \right\}$$

$$- \tfrac{1}{2}\kappa(\xi^2 - \eta^2) - \frac{2\mu}{\xi^2 + \eta^2} = 0 \qquad\qquad (4.120)$$

where $p_\xi = \partial S/\partial \xi$, $p_\eta = \partial S/\partial \eta$ and $p_\theta = \partial S/\partial \theta$. Since the Hamiltonian is independent of the azimuthal angle θ and time t, the Hamilton–Jacobi function becomes

$$S = \alpha_1 t + \alpha_3 \theta + \tilde{S}(\xi, \eta) \qquad\qquad (4.121)$$

where α_1 may be identified with the total energy of the system and α_3 with the azimuthal momentum p_θ. If a separability property is now used so that $\tilde{S}(\xi, \eta) = S_1(\xi) + S_2(\eta)$ the Hamilton–Jacobi equation may then be separated into

$$\left(\frac{\partial S_1}{\partial \xi}\right) + 2\alpha_1 \xi^2 + \alpha_3 \xi^{-2} - \kappa \xi^4 = -\alpha_2^2 \qquad\qquad (4.122a)$$

$$\left(\frac{\partial S_2}{\partial \eta}\right) + 2\alpha_1 \eta^2 + \alpha_3 \eta^{-2} - \kappa \eta^4 - 4\mu = \alpha_2^2 \qquad\qquad (4.122b)$$

where α_2 is the separation constant. Having obtained the partial derivatives of the Hamilton–Jacobi function, the momenta may now be written in terms of the parabolic co-ordinates and the constants of motion, viz.

$$p_\xi = F_1(\xi)^{1/2} \qquad\qquad (4.123a)$$

$$p_\eta = F_2(\eta)^{1/2} \qquad\qquad (4.123b)$$

where the functions F_1 and F_2 are defined as

$$F_1 = \frac{\kappa}{\xi}\left(\xi^6 - \frac{2\alpha_1}{\kappa}\xi^4 - \frac{\alpha_2^2}{\kappa}\xi^2 - \frac{\alpha_3^2}{\kappa}\right)^{1/2} \qquad\qquad (4.124a)$$

and

$$F_2 = \frac{\kappa}{\eta}\left(-\eta^6 - \frac{2\alpha_1}{\kappa}\eta^4 + \frac{\alpha_2^2 + 4\mu}{\kappa}\eta^2 - \frac{\alpha_3^2}{\kappa}\right)^{1/2} \qquad\qquad (1.24b)$$

The roots of these bicubic polynomials $\pm\xi_j$ $(j = 1, 2, 3)$ and $\pm\eta_j$ $(j = 1, 2, 3)$ then define the region of space accessible to the solar sail, as illustrated in Fig. 4.36.

The problem can now be fully solved by obtaining the parabolic co-ordinates through the inversion of elliptic quadratures. Defining a new time variable τ such that, $d\tau = (\xi + \eta)\,dt$, it may be shown that a closed solution is obtained in terms of Jacobian elliptic functions sn, cn as

$$\xi(\tau) = D_1 + D_2\,\text{sn}^2[D_3, D_4(\tau - \tau_0)] \qquad\qquad (4.125a)$$

$$\eta(\tau) = E_1 + E_2\,\text{cn}^2[E_3, E_4(\tau - \tau_0)] \qquad\qquad (4.125b)$$

where D_i, $E_i(i = 1\text{--}4)$ and τ_0 are constants of the motion defined through the

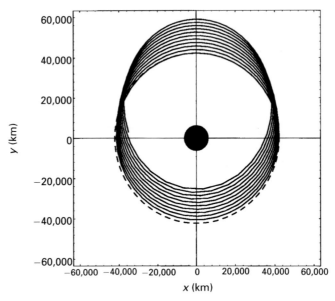

Fig. 4.37. Earth-centred orbit with the sail normal fixed along the Sun-line.

canonical constants α_j $(j = 1, 2, 3)$. An example orbit is shown in Fig. 4.37 using a solar sail in geostationary orbit with the sail normal directed along the Sun-line. It can be seen that the motion is clearly bounded by parabolae, as expected from the earlier analysis. Since the solutions are valid for a fixed sail attitude, multiple solutions may be patched together to generate more complex trajectories using a number of trajectory arcs with fixed sail attitudes. These patched arcs allow the rapid generation of preliminary orbits for the evaluation of manoeuvre strategies.

4.5   SUMMARY

In this chapter both Sun-centred and planet-centred solar sail orbits have been investigated. Firstly, it was shown that conic section orbits exist for solar sails with the effective solar gravity determined by the solar sail lightness number. These families of orbits were found to offer the possibility of simple transfer manoeuvres, and escape trajectories using high-performance sails. Logarithmic spiral trajectories were then considered as another simple means of interplanetary transfer. It was found, however, that logarithmic spiral trajectories did not satisfy the boundary conditions required for general interplanetary transfer. Locally optimal sail-steering laws were then obtained in closed, analytical form by maximising the rate of change of individual orbital elements, or combinations of elements. These locally optimal steering laws provide an effective means of manoeuvring, but do not in general allow arbitrary boundary conditions to be enforced, as required for interplanetary transfer.

To allow such transfers the calculation of globally optimal trajectories was outlined using optimal control theory.

After the discussion of Sun-centred orbits, planet-centred orbits were considered, in particular Earth escape trajectories. A variety of sail-steering laws were investigated and evaluated. It was found that rapid sail rotations are required for equatorial escape trajectories, while polar escape only requires a slow sail rotation once per orbit. However, polar escape was in general somewhat less efficient than near equatorial escape. The most effective means of escape was found to be a locally optimal steering law which maximised the component of the solar sail force vector along the solar sail velocity vector. The improvement over a simpler orbit rate steering law was, however, quite modest. Lastly, it was demonstrated that for a fixed solar sail attitude, and (assumed) fixed Sun-line, the solar sail equations of motion have a closed form solution obtained in terms of elliptic integrals.

4.6 FURTHER READING

Sun-centred trajectories

Bacon, R.H., 'Logarithmic Spiral – an Ideal Trajectory for an Interplanetary Vehicle with Engines of Low Sustained Thrust', *American Journal of Physics*, **27**, 12–18, 1959.

Tsu, T.C., 'Interplanetary Travel by Solar Sail', *American Rocket Society Journal*, **29**, 422–427, 1959.

London, H.S., 'Some Exact Solutions of the Equations of Motion of a Solar Sail With a Constant Setting', *American Rocket Society Journal*, **30**, 198–200, 1960.

Kiefer, J.W., 'Feasibility Considerations for a Solar-Powered Multi-Mission Probe', *Proceedings of the 15th International Astronautical Congress*, Warsaw, **1**, 383–416, 1965.

Sauer, C.G., 'A Comparison of Solar Sail and Ion Drive Trajectories for a Halley's Comet Rendezvous', AAS-77-4, *AAS/AIAA Astrodynamics Conference*, September 1977.

Van der Ha, J.C. & Modi, V.J., 'Long-term Evaluation of Three-Dimensional Heliocentric Solar Sail Trajectories with Arbitrary Fixed Sail Setting', *Celestial Mechanics*, **19**, 113–138, 1979.

Koblik, V.V. *et al.*, 'Controlled Solar Sailing Transfer Flights into Near-Sun Orbits Under Restrictions on Sail Temperature', *Cosmic Research*, **34**, 572–578, 1996.

Minimum time trajectories

Zhukov, A.N. & Lebedev, V.N., 'Variational Problem of Transfer Between Heliocentric Circular Orbits by Means of a Solar Sail', *Cosmic Research*, **2**, 41–44, 1964.

Sauer, C.G., 'Optimum Solar Sail Interplanetary Trajectories', AIAA-76-792, *AAS/AIAA Astrodynamics Conference*, August 1976.

Green, A.J., 'Optimal Escape Trajectory From a High Earth Orbit by Use of Solar Radiation Pressure', T-652, Master of Science Thesis, Massachusetts Institute of Technology, 1977.

Sackett, L.L. & Edelbaum, T.N., 'Optimal Solar Sail Spiral to Escape', *AAS/AIAA Astrodynamics Conference*, September 1977.

Sun, H. & Bryson, A.E., 'Minimum Time Solar Sailing from Geosynchronous Orbit to the Sun–Earth L_2 Point', AIAA-92-4657, *AAS/AIAA Astrodynamics Conference*, August 1992.

Simon, K. & Zakharov, Y., 'Optimisation of Interplanetary Trajectories with Solar Sail', IAF-95-A.2.08, *46th International Astronautical Federation Congress*, October 1995.

Planet-centred trajectories

Sands, N., 'Escape from Planetary Gravitational Fields by use of Solar Sails', *American Rocket Society Journal*, **31**, 527–531, 1961.

Fimple, W.R., 'Generalised Three-Dimensional Trajectory Analysis of Planetary Escape by Solar Sail', *American Rocket Society Journal*, **32**, 883–887, 1962.

Isayev, Y.N. & Kunitsyn, A.L., 'To the Problem of Satellite's Perturbed Motion Under the Influence of Solar Radiation Pressure, *Celestial Mechanics*, **6**, 44–51, 1972.

Van der Ha, J.C. & Modi, V.J., 'Solar Pressure Induced Orbital Perturbations and Control of a Satellite in an Arbitrary Orbit', AIAA-77-32, *AIAA 15th Aerospace Sciences Meeting*, January 1977.

Fekete, T.A. *et al.*, 'Trajectory Design for Solar Sailing from Low-Earth Orbit to the Moon', AAS-92-184, *AAS/AIAA Spaceflight Mechanics Meeting*, February 1992.

Miscellaneous

Roy, A.E., *Orbital Motion*, Adam Hilger, Bristol, 1982.

Battin, R.H., *An Introduction to the Methods and Mathematics of Astrodynamics*, AIAA Education Series, New York, 1987.

5

Non-Keplerian orbits

5.1 INTRODUCTION

When designing future missions, the engineer's imagination is limited by inevitable real-world constraints. For example, the propulsion system to be used will have a given specific impulse and so, for a finite propellant mass, will provide some total Δv. In addition, to achieve the desired mission goals a trajectory will be designed which fits within the envelope of the Δv available from the propulsion system. For solar sails, however, the propulsion system has in principle infinite specific impulse, freeing the engineer to consider new means of attaining mission goals. Furthermore, with high-performance solar sails, a potentially infinite specific impulse is combined with an acceleration of the same order as the local solar gravitational acceleration. With such tools at their disposal, engineers can consider novel forms of orbital acrobatics, again allowing new means of attaining mission goals.

As an example, consider the highly successful ESA/NASA Ulysses solar polar mission. From the outset the mission goal was clear: the payload principal investigators required observations of the solar magnetic field, the interplanetary dust complex and the cosmic-ray background from high above the ecliptic plane. How these goals were met by the engineers designing the mission was in principle of no concern to the investigators. To achieve the mission goals a spacecraft would be required to transport a suite of field and particle instruments into a polar orbit about the Sun. Clearly, a direct transfer into a solar polar orbit was impractical due to the huge Δv required from any standard form of chemical propulsion system. Therefore, a trajectory was designed which used a gravity assist at Jupiter to reach solar polar orbit with a reasonable spacecraft launch mass. Since the spacecraft orbit after the Jupiter encounter was a long ellipse almost perpendicular to the ecliptic plane, the trajectory provided only relatively short observation periods above the solar poles.

In contrast to the conventional mission designed for Ulysses, solar sails can offer mission planners the opportunity to consider some quite different approaches. For example, a solar sail may be used to transport a payload to a close polar orbit about the Sun using the cranking orbit manoeuvre described in Chapter 4. Such high-energy manoeuvres are possible using only moderate performance solar sails. Alternatively, if a high-performance solar sail is available the payload may be 'levitated' above the solar poles, providing continuous observations, rather than the relatively brief glimpse obtained with a conventional ballistic trajectory. Or, the solar sail may be manoeuvred into a circular orbit displaced high above the ecliptic plane, again providing continuous out-of-plane observations. It is this type of exotic, highly non-Keplerian orbit that will be investigated in this chapter. As will be seen in Chapter 6, these novel orbits provide significant leverage to enable new and exciting mission opportunities.

Firstly, a family of Sun-centred non-Keplerian orbits will be investigated. It will be demonstrated that circular, Sun-centred orbits can be displaced high above the ecliptic by directing a component of the solar radiation pressure force normal to the orbit plane. These families of Sun-centred non-Keplerian orbits are para-meterised solely by the solar sail orbit period. In particular, the orbit period may be chosen to minimise the required solar sail lightness number to obtain a given displacement distance, or chosen to synchronise the solar sail orbit with that of the Earth. Using a linear stability and control analysis, stable families of orbits will be identified and control schemes developed for the unstable families. It will be also be shown that Sun-centred non-Keplerian orbits can be patched to Keplerian orbits, or to other non-Keplerian orbits. Following the discussion of Sun-centred non-Keplerian orbits, planet-centred non-Keplerian orbits will be considered. Again, these are circular orbits but are displaced behind the planet along the Sun–planet line. Stable families of orbits are again identified and control schemes developed for the unstable families. Elaborate additional families of orbits are then obtained by patching individual orbits together. It will be shown that the solar sail lightness number requirements for planet-centred non-Keplerian orbits are not as demanding as for Sun-centred non-Keplerian orbits. Orbits about Mercury are particularly attractive since the ratio of the solar radiation pressure acceleration to the local gravitational acceleration is significantly higher than for orbits about the Earth.

As discussed earlier, one of the principal attractions of high-performance solar sails is that they can provide an acceleration which is comparable in magnitude with the local gravitational acceleration. However, at the Lagrange points in a Sun–planet three-body system this local acceleration is vanishingly small; therefore, for low-performance solar sails the regions of space near the Lagrange points provide near-term opportunities for exotic non-Keplerian orbits. It will be shown that the classical three-body problem is significantly modified for solar sails. Rather than singular equilibrium points, large surfaces are found to exist where solar sails will remain at rest. As will be seen in Chapter 6, these equilibrium surfaces provide near-term opportunities for novel missions using the unique properties of solar sails.

5.2 SUN-CENTRED NON-KEPLERIAN ORBITS

5.2.1   Introduction

In this section the dynamics of a new mode of operation for solar sails is discussed: Sun-centred non-Keplerian orbits. These are families of displaced circular solar sail orbits which are generated by orienting the solar sail so that a component of the solar radiation pressure force is directed normal to the orbit plane. A Sun-centred non-Keplerian orbit is unlike other solar sail orbits in that it is essentially an equilibrium solution to the equations of motion, unlike transfer trajectories which require a time-varying sail attitude to satisfy sets of boundary conditions. Since Sun-centred non-Keplerian orbits are the progenitor of planet-centred non-Keplerian orbits, their underlying dynamics will be discussed in some detail.

Firstly, it will be demonstrated that the solar sail orbit period T, orbit radius ρ and out-of-plane displacement distance z may be chosen independently with a suitable choice of solar sail pitch angle α and lightness number β. The solar sail orbit period may be chosen to be synchronous with a circular Keplerian orbit of the same radius or chosen to be some particular fixed value, such as a one-year Earth-synchronous orbit. Perhaps more importantly, the orbit period may also be chosen to minimise the solar sail lightness number required for a given orbit radius and displacement, thus generating a family of optimal non-Keplerian orbits.

Following the definition of non-Keplerian orbit types, their stability character-istics will be investigated and families of linearly stable orbits identified. It will be shown that the optimised orbits are stable for all orbit parameters, whereas the fixed period orbits have a finite region of stability near the orbital plane. The Keplerian synchronous orbits are always unstable. It will also be shown that the unstable families of orbits are in principle controllable using feedback to the sail pitch attitude alone. Linear control laws are then developed for the unstable orbit families which render them bound and stable. Lastly, by patching individual non-Keplerian orbits together, it will be shown that complex and elaborate new trajectories may be generated. For example, by patching four perpendicular orbits together the solar sail may be forced to follow the surface of a cube. Furthermore, by switching the solar sail attitude such that the sail is oriented edgewise to the Sun, the non-Keplerian orbit may be patched to a Sun-centred Keplerian ellipse.

5.2.2   Non-Keplerian orbit solutions

To investigate the dynamics of Sun-centred non-Keplerian orbits the equations of motion will now be considered in a Sun-centred rotating frame of reference. Equilibrium solutions to the equations of motion will then be found in this rotating frame. These equilibrium solutions correspond to displaced circular orbits when viewed from an inertial frame of reference. Since the orientation of the rotating frame is arbitrary, the axis of the orbit may in fact have any desired orientation with respect to the ecliptic plane.

An idealised, perfectly reflecting solar sail will now be considered at position $\mathbf{r}$ in a

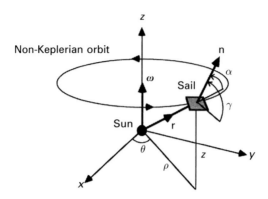

Fig. 5.1. Sun-centred non-Keplerian orbit frame of reference.

frame of reference rotating with angular velocity ω, as shown in Fig. 5.1. The sail orientation is again defined by a unit vector **n**, fixed in the rotating frame of reference, and the solar sail performance is characterised by the solar sail lightness number β. Since the sail orientation is fixed in the rotating frame, the solar sail must rotate once per orbit with respect to an inertial frame. The equation of motion for the solar sail in this rotating frame of reference may then be written as

$$\frac{\mathrm{d}^2\mathbf{r}}{\mathrm{d}t^2} + 2\boldsymbol{\omega} \times \frac{\mathrm{d}\mathbf{r}}{\mathrm{d}t} + \boldsymbol{\omega} \times (\boldsymbol{\omega} \times \mathbf{r}) = \mathbf{a} - \nabla V \tag{5.1}$$

where the terms on the left of Eq. (5.1) represent the kinematic, coriolis and centripetal accelerations respectively. These accelerations are equated to the solar radiation pressure acceleration **a** and solar gravitational acceleration $-\nabla V$ exerted on the solar sail. The solar radiation pressure acceleration and the two-body gravitational potential V are given by

$$\mathbf{a} = \beta \frac{\mu}{r^2} (\hat{\mathbf{r}} \cdot \mathbf{n})^2 \mathbf{n} \tag{5.2a}$$

$$V = -\frac{\mu}{r} \tag{5.2b}$$

as defined in section 2.3.3. Since the solar radiation pressure acceleration can never be directed sunward, the constraint $\hat{\mathbf{r}} \cdot \mathbf{n} \geq 0$ is imposed so that the sail pitch angle is always less than 90°. It is now noted that the centripetal term in Eq. (5.1) is conservative and so may be written in terms of a scalar potential Φ defined such that

$$\Phi = -\tfrac{1}{2}|\boldsymbol{\omega} \times \mathbf{r}|^2 \tag{5.3a}$$

$$\nabla\Phi = \boldsymbol{\omega} \times (\boldsymbol{\omega} \times \mathbf{r}) \tag{5.3b}$$

A new modified potential, $U = V + \Phi$, will now be defined so that a reduced equation of motion is obtained, viz.

$$\frac{d^2\mathbf{r}}{dt^2} + 2\boldsymbol{\omega} \times \frac{d\mathbf{r}}{dt} + \nabla U = \mathbf{a} \tag{5.4}$$

In the rotating frame of reference an equilibrium solution is required so that the first two terms of Eq. (5.4) must vanish. Since the vector $\mathbf{a}$ is oriented in direction $\mathbf{n}$, taking the vector product of $\mathbf{n}$ with Eq. (5.4) it is found that

$$\nabla U \times \mathbf{n} = 0 \Rightarrow \mathbf{n} = \varepsilon \nabla U \tag{5.5}$$

where ε is an arbitrary scalar multiplier. Using the normalisation condition $|\mathbf{n}| = 1$, ε is then identified as $|\nabla U|^{-1}$ so that the required sail attitude is defined simply by

$$\mathbf{n} = \frac{\nabla U}{|\nabla U|} \tag{5.6}$$

Since the solar sail is to have uniform azimuthal motion, there can be no component of the vector $\mathbf{n}$, and so of $\mathbf{a}$, in the azimuthal direction. Therefore $\mathbf{n}$ is contained in the plane spanned by $\hat{\mathbf{r}}$ and $\boldsymbol{\omega}$ so that the sail attitude may be described solely in terms of the pitch angle α between $\mathbf{n}$ and $\hat{\mathbf{r}}$. Taking vector and scalar products of Eq. (5.6) with $\hat{\mathbf{r}}$ it is found that

$$\tan \alpha = \frac{|\hat{\mathbf{r}} \times \nabla U|}{\hat{\mathbf{r}} \cdot \nabla U} \tag{5.7}$$

Similarly, the required solar sail lightness number is obtained by taking a scalar product of Eq. (5.4) with $\mathbf{n}$, again requiring an equilibrium solution, viz.

$$\beta = \frac{r^2 \, \nabla U \cdot \mathbf{n}}{\mu \, (\hat{\mathbf{r}} \cdot \mathbf{n})^2} \tag{5.8}$$

General functions for the solar sail attitude and lightness number have now been obtained in terms of the two-body rotating potential function U. For a given orbit period, orbit radius and displacement distance the solar sail pitch angle and lightness number can then be obtained.

If the set of cylindrical co-ordinates (ρ, θ, z) shown in Fig. 5.1 are now used, the rotating two-body potential function may be written as

$$U = -\left(\tfrac{1}{2}\rho^2\omega^2 + \frac{\mu}{r} \right) \tag{5.9}$$

where the angular velocity ω is related to the orbit period T by $\omega = 2\pi/T$. Evaluating the potential gradient in Eqs (5.7) and (5.8) it is found that

$$\tan \alpha = \frac{(z/\rho)(\omega/\tilde{\omega})^2}{(z/\rho)^2 + [1 - (\omega/\tilde{\omega})^2]}, \quad \tilde{\omega}^2 = \frac{\mu}{r^3} \tag{5.10a}$$

$$\beta = \left[1 + \left(\frac{z}{\rho}\right)^2\right]^{1/2} \frac{\{(z/\rho)^2 + [1 - (\omega/\tilde{\omega})^2]^2\}^{3/2}}{\{(z/\rho)^2 + [1 - (\omega/\tilde{\omega})^2]\}^2} \tag{5.10b}$$

where $\tilde{\omega}$ is the orbital angular velocity of a circular Keplerian orbit of radius r. For convenience, the parameter μ may now be chosen to be unity so that the unit of distance becomes the astronomical unit and the non-dimensional unit of time becomes $(2\pi)^{-1}$ years. These expressions for the solar sail pitch angle and lightness number may now be used to investigate various families of non-Keplerian orbits. It can be seen that Eqs (5.10) are a function of z/ρ and $\omega/\tilde{\omega}$ only so that the expressions have a scale invariance. In addition, the expressions are also independent of the azimuthal angle θ. Therefore, for a fixed solar sail lightness number, Eq. (5.10b) defines nested surfaces of revolution about the z axis.

5.2.2.1 Type I

For this family of orbits the solar sail orbit period will be chosen to be some fixed value for all values of orbit radius and displacement. For example, a family of Earth synchronous one-year orbits are generated by setting $\omega = 1$ in Eq. (5.10b). A section of the surfaces of equal solar sail lightness number generated by Eq. (5.10b) are shown in Fig. 5.2. In addition, the equivalent solar sail characteristic acceleration and loading values are listed in Table. 5.1. Each point on a surface of constant solar sail lightness number corresponds to a displaced orbit with a different orbit radius and displacement distance. The required direction of the sail normal vector **n** is also shown.

Topologically, the full three-dimensional surfaces form a family of nested tori for $\beta < 1$ and nested cylinders for $\beta > 1$, when the inner radius of the torus vanishes. When $\beta = 1$, the surface is an oblate sphere with an infinite line along the z axis representing non-orbiting equilibrium solutions along the axis of rotation of the

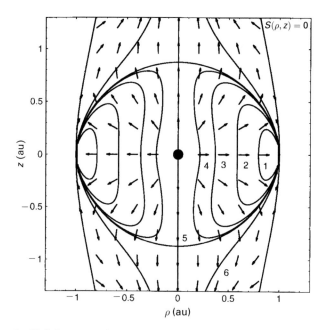

Fig. 5.2. Type I orbit lightness number contours (see Table 5.1 for values).

Table 5.1. Solar sail lightness number, characteristic acceleration and loading for type I, II and III Sun-centred non-Keplerian orbits

Type		Contour					
		1	2	3	4	5	6
I	β	0.5	0.8	0.95	0.99	1.0	1.3
	a_0 (mm s^{-2})	2.97	4.74	5.63	5.87	5.93	7.71
	σ (g m^{-2})	3.06	1.91	1.61	1.55	1.53	1.18
II	β	1	1.05	1.2	1.5	2	5
	a_0 (mm s^{-2})	5.93	6.23	7.12	8.90	11.86	29.66
	σ (g m^{-2})	1.52	1.46	1.28	1.02	0.77	0.31
III	β	0.2	0.5	0.8	0.95	0.99	1.0
	a_0 (mm s^{-2})	1.19	2.97	4.74	5.63	5.87	5.93
	σ (g m^{-2})	7.65	3.06	1.91	1.61	1.55	1.53

frame of reference. The nested tori intersect the ecliptic plane in a set of circular contours providing the required solar sail lightness number for one-year orbits in the ecliptic plane. From Eq. (5.10a) it can be seen that, in this case, a radial sail attitude with zero pitch angle is required. This decoupling of orbit period from orbit radius was discussed in section 4.3.2.2. In particular, it can be seen that Eq. (5.10b) reduces to Eq. (4.25) when $z = 0$. Furthermore, along the z axis it can be seen from Eqs. (5.10) that $\alpha = 0$ and $\beta = 1$, corresponding to an equilibrium solution at any distance along the axis of rotation of the frame of reference. In this simple equilibrium, the solar radiation pressure acceleration is exactly balanced by the solar gravitational acceleration. It is again noted that, owing to the symmetry of the problem, the axis of the orbits need not be normal to the ecliptic plane. A non-Keplerian orbit may be established about any axis passing through the origin so that orbits inclined to the ecliptic plane are possible.

The region of space in the rotating frame of reference in which non-Keplerian orbit solutions exist is bounded, being defined by the region interior to the surface $\hat{\mathbf{r}} \cdot \mathbf{n} = 0$. It can be seen from Eq. (5.8) that, as this limiting surface is approached, the solar sail lightness number $\beta \to \infty$. Using Eq. (5.6) it is found that the boundary is defined by

$$S(\rho, z) = \frac{\mu}{r} - \rho^2 \omega^2 = 0 \qquad (5.11)$$

where again $\omega = 1$ for the Earth synchronous case. Outside this surface $\hat{\mathbf{r}} \cdot \mathbf{n} < 0$, corresponding to a required sail pitch angle of greater than $90°$. In this case the total gravitational and centripetal force is outwards. The solar radiation pressure force will then only augment these forces and so equilibrium solutions in the rotating frame of reference will not be possible. Inside the boundary surface $\hat{\mathbf{r}} \cdot \mathbf{n} > 0$ and so equilibrium solutions are possible with the sum of the gravitational and centripetal

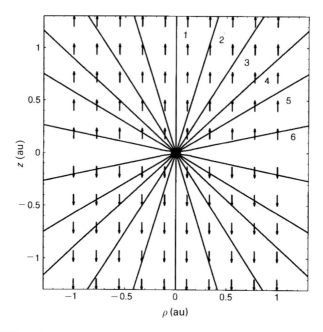

Fig. 5.3. Type II orbit lightness number contours (see Table 5.1 for values).

forces exactly balanced by the solar radiation pressure force. The surface $S = 0$ is shown as the outer contour in Fig. 5.2.

5.2.2.2   Type II

For this family of orbits the solar sail orbit period is chosen to be that of a Keplerian orbit of radius equal to the Sun–sail distance r, so that $\omega = \tilde{\omega}$. Surfaces of co-rotation (equal orbital period) are then defined by spheres of constant radius. Substituting for this functional form of ω, the required solar sail pitch angle and lightness number are obtained as

$$\tan \alpha = \frac{\rho}{z} \tag{5.12a}$$

$$\beta = \left[1 + \left(\frac{\rho}{z} \right)^2 \right]^{1/2} \tag{5.12b}$$

For a fixed solar sail lightness number, Eq. (5.12b) again defines nested surfaces of revolution about the z-axis. A section of these surfaces of constant solar sail lightness number is shown in Fig. 5.3 along with the required sail normal direction. It can be seen that the solar sail is always oriented perpendicular to the orbit plane. Therefore the constraint $\hat{\mathbf{r}} \cdot \mathbf{n} \geq 0$ is always satisfied and the family of orbits is not bounded. In addition, it can be seen that for any combination of orbit radius and displacement a solar sail lightness number greater than unity is always required. This is also evident from Eq. (5.12b).

5.2.2.3   Type III

To generate an optimal family of orbits the solar sail orbit period will now be treated as a free parameter with respect to which the solar sail lightness number may be minimised. Therefore, setting the derivative of β with respect to ω to zero in Eq. (5.10b), a quadratic in ω^2 is obtained, viz.

$$\omega^4 - \omega^2\tilde{\omega}^2\left[2 + 3\left(\frac{z}{\rho}\right)^2\right] + \tilde{\omega}^4\left[1 + \left(\frac{z}{\rho}\right)^2\right] = 0 \qquad (5.13)$$

This quadratic yields two solutions for ω^2, one of which fails to satisfy the condition $\hat{\mathbf{r}}\cdot\mathbf{n} \geq 0$ while the other solution always satisfies the condition. The necessary solar sail orbital angular velocity to minimise the required solar sail lightness number is then given by

$$\omega = \tilde{\omega}\left[1 + \frac{3}{2}\left(\frac{z}{\rho}\right)^2\right]^{1/2}\left[1 - \left(1 - \frac{1 + (z/\rho)^2}{[1 + (3/2)(z/\rho)^2]^2}\right)^{1/2}\right]^{1/2} \qquad (5.14)$$

Using this functional form for the orbital angular velocity, surfaces of constant solar sail lightness number may again be generated from Eq. (5.10b). A section of these surfaces is shown in Fig. 5.4 along with the required sail normal direction. It can be seen that this orbit family is also unbound, so that an optimal non-Keplerian orbit may be established at any point in the ρ–z plane. Most importantly, it can also be seen that the optimal family of orbits always requires a solar sail lightness number

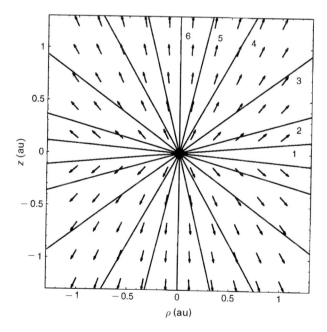

Fig. 5.4. Type III orbit lightness number contours (see Table 5.1 for values).

$0 < \beta < 1$. Therefore, by treating ω as a free parameter of the system, orbits with large out-of-plane displacements may be achieved with a solar sail lightness number lower than that required for the type I or type II families. This is clearly a significant advantage, although the orbit period can no longer be chosen arbitrarily.

5.2.2.4   Other orbit families

Other families of orbits may also be generated through an appropriate choice of ω. In particular, choosing $\omega = \rho^{-3/2}$ would generate a family of orbits which are synchronous with circular Keplerian orbits of the same orbit radius in the ecliptic plane. However, it is found that this potentially interesting family of orbits is forbidden as it violates the constraint $\hat{\mathbf{r}} \cdot \mathbf{n} \geq 0$. In general, to enforce this constraint it can be shown that ω must be chosen such that

$$\omega^2 \leq \frac{1}{\rho^3} \left[1 + \left(\frac{z}{\rho} \right)^2 \right]^{-1/2} \tag{5.15}$$

For a given orbit period this constraint then places bounds on the region of space in the ρ–z plane within which non-Keplerian orbit solutions can exist. Conversely, Eq. (5.15) provides the shortest orbit period possible at a given point in the ρ–z plane.

5.2.3   Sun-centred non-Keplerian orbit stability

Now that the three orbit families have been defined their linear stability character-istics will be investigated. This is achieved by linearising the equations of motion about the nominal orbit solution to obtain a linear variational equation. The linear variational equation then describes the motion of the solar sail in the neighbourhood of the nominal orbit. The stability of the resulting solar sail trajectory may be determined by examining the eigenvalues of the variational equation. If any of the eigenvalues are real and positive, the motion of the solar sail is exponentially divergent so that the orbit will be unstable. However, if all the eigenvalues are purely imaginary, the solar sail is expected to remain bound in a neighbourhood of the nominal orbit. It should be noted that only a linear analysis is performed, which in the present case provides only necessary conditions for stability but sufficient conditions for instability. The non-linear equations of motion will be linearised by perturbing the solar sail from its nominal orbit with the sail attitude fixed in the rotating frame of reference. The perturbation is therefore applied with the sail elevation angle γ fixed, as shown in Fig. 5.1, so that the sail pitch angle α varies during the perturbation.

5.2.3.1   Linearisation

The non-linear equation of motion will now be linearised using standard methods. Firstly, a perturbation δ will be added to the solar sail position vector at some point $\mathbf{r}_0 = (\rho_0, \theta_0, z_0)$ on the nominal orbit such that $\mathbf{r}_0 \rightarrow \mathbf{r}_0 + \delta$. Then, the variational equation is obtained from the non-linear equation of motion, Eq. (5.4), as

$$\frac{d^2\boldsymbol{\delta}}{dt^2} + 2\boldsymbol{\omega} \times \frac{d\boldsymbol{\delta}}{dt} + \nabla U(\mathbf{r}_0 + \boldsymbol{\delta}) - \mathbf{a}(\mathbf{r}_0 + \boldsymbol{\delta}, \mathbf{n}) = 0 \qquad (5.16)$$

where the vector $\boldsymbol{\delta} = (\xi, \psi, \eta)$ represents first-order displacements in the rotating frame of reference in the (ρ, θ, z) directions respectively. The potential gradient and the solar radiation pressure acceleration may be expanded in a Taylor series about the nominal orbit to first order as

$$\nabla U(\mathbf{r}_0 + \boldsymbol{\delta}) = \nabla U(\mathbf{r}_0) + \left[\frac{\partial}{\partial \mathbf{r}} \nabla U(\mathbf{r}) \right]_0 \boldsymbol{\delta} \qquad (5.17a)$$

$$\mathbf{a}(\mathbf{r}_0 + \boldsymbol{\delta}) = \mathbf{a}(\mathbf{r}_0) + \left[\frac{\partial}{\partial \mathbf{r}} \mathbf{a}(\mathbf{r}, \mathbf{n}) \right]_0 \boldsymbol{\delta} \qquad (5.17b)$$

where the subscript 0 indicates that the partial derivatives are evaluated along the nominal orbit. Then, since $\nabla U(\mathbf{r}_0) = \mathbf{a}(\mathbf{r}_0)$ along the nominal orbit, a linear variational equation with constant coefficients is obtained, viz.

$$\frac{d^2\boldsymbol{\delta}}{dt^2} + \mathbf{S}\frac{d\boldsymbol{\delta}}{dt} + (\mathbf{M} - \mathbf{N})\boldsymbol{\delta} = 0 \qquad (5.18)$$

where $\mathbf{M}$ and $\mathbf{N}$ are the gravity and radiation pressure gradient matrices and $\mathbf{S}$ is the skew-symmetric gyroscopic matrix given by

$$\mathbf{M} = \left[\frac{\partial}{\partial \mathbf{r}} \nabla U(\mathbf{r}) \right]_0 \qquad (5.19a)$$

$$\mathbf{N} = \left[\frac{\partial}{\partial \mathbf{r}} \mathbf{a}(\mathbf{r}, \mathbf{n}) \right]_0 \qquad (5.19b)$$

$$\mathbf{S} = \begin{bmatrix} 0 & -2 & 0 \\ 2 & 0 & 0 \\ 0 & 0 & 0 \end{bmatrix} \qquad (5.19c)$$

The components of the matrices $\mathbf{M}$ and $\mathbf{N}$ are then the partial derivatives of the solar sail acceleration components with respect to the position co-ordinates. In component form the variational equation then becomes

$$\frac{d^2\xi}{dt^2} - 2\omega\rho_0 \frac{d\psi}{dt} + L_{11}\xi + L_{13}\eta = 0 \qquad (5.20a)$$

$$\frac{d^2\psi}{dt^2} + \frac{2\omega}{\rho_0}\frac{d\xi}{dt} = 0 \qquad (5.20b)$$

$$\frac{d^2\eta}{dt^2} + L_{31}\xi + L_{33}\eta = 0 \qquad (5.20c)$$

where $L_{ij} = M_{ij} - N_{ij}$ $(i, j = 1\text{--}3)$. Owing to the azimuthal symmetry of the problem all derivatives with respect to θ in the matrices $\mathbf{M}$ and $\mathbf{N}$ will vanish. Therefore, the six terms $L_{2,j}$ and $L_{j,2}$ ($j = 1$–3) are zero. This set of three coupled, ordinary differential equations may be reduced to two by integrating Eq. (5.20b) to obtain

$$\frac{d\psi}{dt} = -\frac{2\omega}{\rho_0}(\xi - \xi_0) \tag{5.21}$$

This equation is in effect a linearised form of Kepler's third law, describing the orbital angular velocity of the solar sail relative to the nominal orbit due to the radial displacement ξ. This equation may then be substituted into Eq. (5.20a) to eliminate the azimuthal term. However, this substitution then leads to a constant term $4\omega^2\xi_0$ in Eq. (5.20a) so that the variational equation is no longer homogeneous. It can be shown that the non-homogeneity can be easily removed by rescaling the co-ordinates through a change of variable

$$\xi' = \xi - \frac{4\omega^2 L_{33}}{L_{11}L_{33} - L_{13}L_{31}}\xi_0 \tag{5.22a}$$

$$\eta' = \eta + \frac{4\omega^2 L_{13}}{L_{11}L_{33} - L_{13}L_{31}}\xi_0 \tag{5.22b}$$

Using this transformation a reduced variational system with a set of two coupled equations is then obtained, viz.

$$\frac{d^2}{dt^2}\begin{bmatrix} \xi' \\ \eta' \end{bmatrix} + \begin{bmatrix} L_{11} & L_{13} \\ L_{31} & L_{33} \end{bmatrix}\begin{bmatrix} \xi' \\ \eta' \end{bmatrix} = \begin{bmatrix} 0 \\ 0 \end{bmatrix} \tag{5.23}$$

Therefore, by eliminating the azimuthal co-ordinate the order of the variational system is reduced from three to two. However, by ignoring the azimuthal motion it must be remembered that the solar sail is then free to drift along the nominal orbit. Therefore, orbital stability will be decided by determining if the motion in the ρ–z plane is bound or unbound.

By evaluating the partial derivatives along the nominal orbit it is found that the coefficients of the matrix $\mathbf{L}$ are given by

$$L_{11} = 3\omega^2 + \tilde{\omega}^2\left[1 - 3\left(\frac{\rho}{r}\right)^2\right] + \frac{2\beta\tilde{\omega}^2}{r}\chi\left[\frac{2\rho\chi}{r^2} - \cos\gamma\right]\cos\gamma \tag{5.24a}$$

$$L_{13} = -3\tilde{\omega}^2\left[\frac{\rho z}{r^2}\right] + \frac{2\beta\tilde{\omega}^2}{r}\chi\left[\frac{2z\chi}{r^2} - \sin\gamma\right]\cos\gamma \tag{5.24b}$$

$$L_{31} = -3\tilde{\omega}^2\left[\frac{\rho z}{r^2}\right] + \frac{2\beta\tilde{\omega}^2}{r}\chi\left[\frac{2\rho\chi}{r^2} - \cos\gamma\right]\sin\gamma \tag{5.24c}$$

$$L_{33} = \tilde{\omega}^2\left[1 - 3\left(\frac{z}{r}\right)^2\right] + \frac{2\beta\tilde{\omega}^2}{r}\chi\left[\frac{2z\chi}{r^2} - \sin\gamma\right]\sin\gamma \tag{5.24d}$$

where the auxiliary coefficient χ is defined by

$$\chi = \rho\cos\gamma + z\sin\gamma \tag{5.24e}$$

The stability characteristics of the non-Keplerian orbit families may now be investigated by calculating the eigenvalues of the variational equation. The eigenvalues may be obtained by substituting an exponential solution of the form

$$\begin{bmatrix} \xi' \\ \eta' \end{bmatrix} = \begin{bmatrix} \xi_0 \\ \eta_0 \end{bmatrix} \exp(\lambda t) \tag{5.25}$$

where in general λ is a complex quantity. Substituting this solution into Eq. (5.23) yields a matrix equation of the form

$$\begin{bmatrix} \lambda^2 + L_{11} & L_{13} \\ L_{31} & \lambda^2 + L_{33} \end{bmatrix} \begin{bmatrix} \xi_0 \\ \eta_0 \end{bmatrix} = \begin{bmatrix} 0 \\ 0 \end{bmatrix} \tag{5.26}$$

For non-trivial solutions a vanishing determinant of the matrix equation is required. This then provides the characteristic polynomial of the system as

$$f(\lambda^2) = \lambda^4 + \text{tr}(\mathbf{L})\lambda^2 + \det(\mathbf{L}) = 0 \tag{5.27}$$

with the trace of the matrix $\text{tr}(\mathbf{L})$ given by $L_{11} + L_{33}$ and its determinant $\det(\mathbf{L})$ given by $L_{11}L_{33} - L_{13}L_{31}$. The characteristic polynomial has four complex roots λ_j ($j = 1$–4), the four frequencies of the eigenmodes of the system. Formally these eigenvalues may be written as

$$\lambda_j = \frac{\pm 1}{\sqrt{2}} \{-\text{tr}(\mathbf{L}) \pm [\text{tr}(\mathbf{L})^2 - 4\det(\mathbf{L})]^{1/2}\}^{1/2} \tag{5.28}$$

where the negative root gives rise to a long period response and the positive root a short period response. The solar sail motion in the neighbourhood of the nominal orbit is then defined by a superposition of the four eigenmodes as

$$\begin{bmatrix} \xi' \\ \eta' \end{bmatrix} = \sum_{j=1}^{4} \begin{bmatrix} \xi_{0j} \\ \eta_{0j} \end{bmatrix} \exp(\lambda_j t) \tag{5.29}$$

The stability characteristics of the families of non-Keplerian orbits may now be investigated by numerically searching for regions with purely imaginary eigenvalues, $\lambda_j^2 < 0$ ($j = 1$–4), giving stable, bound oscillations in the ρ–z plane. For purely imaginary eigenvalues it is required that both $\det(\mathbf{L}) > 0$ and $\text{tr}(\mathbf{L}) > 0$, as shown in Fig. 5.5. It can be seen that with these two conditions all roots of Eq. (5.27) will lie in the left hand complex plane.

5.2.3.2 Stability of orbit types

Firstly, the stable regions of the type I synchronous orbits will be mapped with $\omega = 1$. Then, using the scale invariance of the problem, the stability of the type II Keplerian synchronous and type III optimal orbits will also be determined. Setting $\omega = 1$ in Eq. (5.27) it is found that there exists a stable family of orbits with $\lambda_j^2 < 0$ ($j = 1$–4), near the ecliptic plane. This region is bounded by the surface C_1, the section of which is shown in Fig. 5.6. The surface intersects the z axis at 0.794 au, corresponding to the maximum out-of-plane displacement for a stable orbit. It

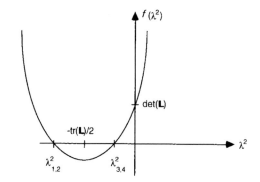

Fig. 5.5. Roots of the characteristic polynomial.

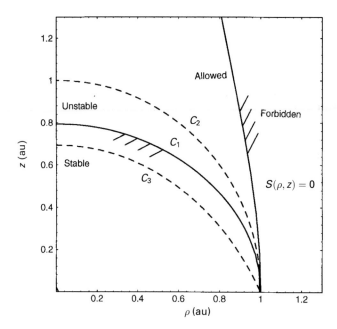

Fig. 5.6. Stable and unstable regions of the ρ–z plane.

should be noted that the z axis is not included in the set of stable orbits. Solutions along the z axis correspond to non-orbiting equilibrium solutions which are marginally unstable since the solar radiation pressure force and the solar gravitational force balance at all points along the axis. The stability of such orbits will be considered in the next section. For the type II Keplerian synchronous case the surface $\omega = 1$ is a unit sphere, the section of which is shown in Fig. 5.6 by the curve C_2. It can be seen that this surface lies outside the stable region of the map so that the $\omega = 1$ Keplerian synchronous orbits will be unstable. Furthermore, the surface defined by type III orbits with $\omega = 1$, shown as C_3 in Fig. 5.6, may also be generated and is found to lie within the stable region. From Eqs (5.10) it can be seen that there exists a scaling law of the form

$$\frac{z_1}{\rho_1} \to \frac{z_2}{\rho_2}, \qquad \frac{\omega_1}{\tilde{\omega}} \to \frac{\omega_2}{\tilde{\omega}} \tag{5.30}$$

as $\omega_1 \to \omega_2$. Therefore, the hierarchy of surfaces shown in Fig. 5.6 is found to hold for any value of ω. The type II Keplerian synchronous mode is therefore always unstable and the type III optimal mode is always stable. The instability of the type II Keplerian synchronous mode can also be verified by noting that with $\omega = \tilde{\omega}$ Eqs (5.24) reduce to

$$\mathrm{tr}(\mathbf{L}) = 4\tilde{\omega}^2 \left(\frac{z}{r}\right)^2 \tag{5.31a}$$

$$\det(\mathbf{L}) = -\tilde{\omega}^4 \left(\frac{\rho}{r}\right)^2 \tag{5.31b}$$

Therefore, since $\det(\mathbf{L}) < 0$ and $\mathrm{tr}(\mathbf{L}) > 0$ two of the eigenvalues are positive, as can be seen from Fig. 5.5. As an example, a typical unstable response for a one-year type I orbit is shown in Fig. 5.7 using numerical integration of the full non-linear equations of motion. This unstable orbit has a radius of 0.5 au and a displacement distance of 0.8 au which lies in the region of unstable type I orbits. It can be seen that the orbit has a rather long instability timescale, with the instability manifesting itself in the sail falling sunward.

Lastly, the coupling of the orbit perturbations in the ρ–z plane to the azimuthal motion may be found by integrating Eq. (5.21), viz.

$$\psi(t) = \psi_0 + \left(\frac{2\omega\xi_0}{\rho_0}\right)t - \frac{2\omega}{\rho_0} \int_0^t \xi(\tau)\,\mathrm{d}\tau \tag{5.32}$$

where $\xi(t)$ is the response defined by Eq. (5.29). The first-order drift in azimuthal position can then be obtained as

$$\psi(t) = \psi_0 + \frac{2\omega\xi_0}{\rho_0}\left(1 - \frac{4\omega^2 L_{33}}{L_{11}L_{33} - L_{13}L_{31}}\right)t - \frac{2\omega}{\rho_0}\sum_{j=1}^{4}\frac{\xi_{0j}}{\lambda_j}\exp(\lambda_j t) \tag{5.33}$$

For the stable orbit families the azimuthal perturbations are of the form of periodic oscillations with a secular drift. This drift is due to the solar sail having an initial perturbation along the ρ axis so that it has a first-order difference in orbit period from the nominal orbit. The solar sail motion is therefore constrained to a torus around the nominal non-Keplerian orbit.

5.2.3.3   Levitated solar sail stability

Having established the stability characteristics of the three families of non-Keplerian orbits, the stability of the one-dimensional equilibrium solutions along the z axis will now be considered. This case is investigated separately to illustrate an effect of the exact solar radiation pressure model derived in Chapter 2. It will be shown that the exact solar radiation pressure model leads to exponential divergence and so

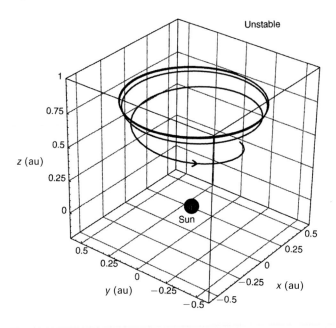

Fig. 5.7. Unstable one-year type I orbit ($\rho_0 = 0.5\,\text{au}$, $z_0 = 0.8\,\text{au}$, $\xi_0 = 10^{-2}\rho_0$, $\eta_0 = 10^{-2}z_0$).

instability. For a radially oriented solar sail the equation of motion may be written as

$$\frac{d^2\mathbf{r}}{dt} = -(1 - \beta)\frac{\mu}{r^2}\hat{\mathbf{r}} \tag{5.34}$$

as defined in section 4.3.2.1. It can be seen from Eq. (5.34) that a solar sail lightness number of unity allows equilibrium solutions. If the solar sail is initially at rest it will then 'levitate' above the Sun, at any desired distance. This is, of course, due to solar gravity and solar radiation pressure both having an inverse square variation with heliocentric distance. If the solar sail is initially at rest at some position $\mathbf{r}_0$ and a perturbation $\boldsymbol{\delta} = (\xi, \eta)$ is applied, the resulting solar sail motion is obtained from Eq. (5.34) as

$$\frac{d^2\xi}{dt^2} = -\frac{\mu}{r_0^3}\xi \tag{5.35a}$$

$$\frac{d^2\eta}{dt^2} = 0 \tag{5.35b}$$

It can be seen from Eqs (5.35) that while the transverse motion is stable, the vertical motion of the solar sail will be of the form

$$\eta(t) = \eta_1 + \eta_2 t \tag{5.36}$$

where η_1 and η_2 are constants. Therefore, for an inverse square variation of solar radiation pressure, a levitating solar sail has marginal instability. That is, if perturbed the sail will move from its equilibrium point, but with linear rather than exponential growth.

This stability analysis will now be reconsidered with the modified solar radiation pressure from a uniformly bright solar disc, defined by Eq. (2.44) from section 2.5.2. The equation of motion now becomes

$$\frac{d^2\mathbf{r}}{dt^2} = -[1 - \beta F(r)]\frac{\mu}{r^2}\hat{\mathbf{r}} \qquad (5.37)$$

where the solar sail lightness number is now defined as the ratio of the inverse square solar radiation pressure force to the solar gravitational force exerted on the solar sail. Therefore, the solar sail lightness number is not, in this case, the ratio of the actual forces acting on the sail. Clearly then, there is no longer a unique solar sail lightness number providing equilibrium at all distances. The solar sail lightness number required for equilibrium will now be a function of r, given by

$$\beta^*(r) = \frac{1}{F(r)} \qquad (5.38)$$

where $F(r) \rightarrow 1$ as $r \rightarrow \infty$. The solar sail will again be considered to be stationary at some distance r_0 with $\beta = \beta^*(r_0)$ and a perturbation δ applied along the z axis such that $r_0 \rightarrow r_0 + \delta$. Then, expanding Eq. (5.37) in a Taylor series to first order a variational equation is obtained, viz.

$$\frac{d^2\delta}{dt^2} - M(r_0)\delta = 0 \qquad (5.39a)$$

$$M(r_0) = \frac{\mu}{r_0^3}\left(\frac{r}{F(r)}\frac{dF}{dr}\right)_{r=r_0} \qquad (5.39b)$$

The stability characteristics of the equilibrium solution may now be investigated by determining the sign of the function $M(r_0)$. If $M(r_0) > 0$ then Eq. (5.39a) has an unbound exponential solution leading to instability, whereas if $M(r_0) < 0$ then the solution is bound. Evaluating the derivative of $F(r)$ and making the substitution $v = (R_S/r_0)^2$ the function $M(r_0)$ reduces to

$$M(r_0) = \frac{2\mu}{r_0^3}K(v) \qquad (5.40a)$$

$$K(v) = 1 - \frac{3}{2}v\frac{(1-v)^{1/2}}{1-(1-v)^{3/2}} \qquad (5.40b)$$

where the sign of the function K now determines whether the eigenvalues of the variational equation are real or purely imaginary. The asymptotic behaviour of the function K will now be examined to determine the sign of K in the range $0 < v < 1$. Firstly, as $v \rightarrow 1$ it can be seen that $K \rightarrow 1$. Then, as $v \rightarrow 0$ it is found that K may be expanded as

$$K = 1 - \frac{1 - \frac{1}{2}v + O(v^2)}{1 - \frac{1}{4}v + O(v^2)} \rightarrow 0 \qquad (5.41)$$

so that $K \to 0$ as $v \to 0$. Since it may also be shown that K has only one real root at $v = 0$, and so does not change sign for $0 < v < 1$, it is concluded that $K > 0$ for $0 < v < 1$. Therefore, there are two real eigenvalues of opposite sign yielding exponential divergence and so instability. The instability timescale is found to be

$$\omega = \tilde{\omega}\sqrt{2K} \tag{5.42}$$

It has been shown then that by considering the Sun as a uniformly bright, extended source of radiation that a solar sail of a given lightness number has in principle only one possible equilibrium point and that this point is exponentially unstable. The instability timescale is long (e.g. 0.96 years at 0.1 au), but the existence of the instability necessitates the need for active station-keeping. Physically the instability may be understood by considering the ratio of forces exerted on the solar sail as it is perturbed. It can be shown that if the solar sail is perturbed outwards, then the solar radiation pressure force falls more slowly than the solar gravitational force. Therefore, the solar sail is accelerated outwards. Conversely, if the solar sail is perturbed sunward, the solar gravitational force increases more quickly than the solar radiation pressure force so that the solar sail is accelerated sunward. All such simple equilibria are therefore exponentially unstable.

5.2.4   Sun-centred non-Keplerian orbit control

Since the stability analysis has shown that some non-Keplerian orbit families have regions of instability, simple closed loop control schemes that ensure asymptotic stability will now be investigated. If the solar sail azimuthal position is unimportant then the reduced two-dimensional variational equation may be used to investigate control laws which stabilise the motion in the ρ–z plane. It will be shown that the two-dimensional system is in principle controllable using state feedback to the sail pitch attitude alone. In this context a controllable system is one in which there always exists an input such that the system can be transferred between two arbitrary states within a finite time.

5.2.4.1   Orbit controllability

The first type of control to be investigated is proportional and derivative feedback to the sail elevation angle γ. However, it must firstly be established that the orbit families are in fact controllable using the sail pitch attitude alone. If the equation of motion is again linearised by allowing first-order changes in the solar sail position $\mathbf{r}_0 \to \mathbf{r}_0 + \boldsymbol{\delta}$ and also the sail attitude $\mathbf{n}_0 \to \mathbf{n}_0 + \delta\mathbf{n}$, such that $|\mathbf{n}_0 + \delta\mathbf{n}| = 1$, a modified variational equation is then obtained, viz.

$$\frac{d^2\boldsymbol{\delta}}{dt^2} + \mathbf{S}\frac{d\boldsymbol{\delta}}{dt} + (\mathbf{M} - \mathbf{N})\boldsymbol{\delta} = \mathbf{K}\,\delta\mathbf{n} \tag{5.43a}$$

$$\mathbf{K} = \left[\frac{\partial}{\partial\mathbf{n}}\mathbf{a}(\mathbf{r}, \mathbf{n})\right]_0 \tag{5.43b}$$

where the matrix $\mathbf{K}$ gives the first-order variation in the solar radiation pressure acceleration due to changes in the sail attitude. The azimuthal co-ordinate may again be eliminated using Eq. (5.21) and the co-ordinate transformations defined by Eqs (5.22) then used to reduce the modified variational equation to the variables $\delta' = (\xi', \eta')$. The variational equation then becomes

$$\frac{d^2\delta'}{dt^2} + \mathbf{L}\delta' = \mathbf{K}\,\delta\gamma \tag{5.44a}$$

$$\mathbf{K} = \begin{bmatrix} \dfrac{\partial a_\rho}{\partial \gamma} \\[2mm] \dfrac{\partial a_z}{\partial \gamma} \end{bmatrix} \tag{5.44b}$$

where the first-order attitude change $\delta\mathbf{n}$ now becomes the change in the sail elevation angle $\delta\gamma$. The solar radiation pressure acceleration partial derivatives, given by the components of $\mathbf{K} = (K_1, K_2)^{\mathrm{T}}$, are found to be

$$K_1 = \frac{\beta z}{2r^4}(\rho\cos\gamma + z\sin\gamma)\left[3\cos 2\gamma\left(1 - \frac{\rho}{z}\tan 2\gamma\right) + 1\right] \tag{5.45a}$$

$$K_2 = \frac{\beta\rho}{2r^4}(\rho\cos\gamma + z\sin\gamma)\left[3\cos 2\gamma\left(1 + \frac{z}{\rho}\tan 2\gamma\right) - 1\right] \tag{5.45b}$$

The variational equation is now reduced further by transforming to a set of four first-order equations in the state variable $\mathbf{x} = (\delta', d\delta'/dt)$. The variational equation may then be written in standard state space form as

$$\frac{d\mathbf{x}}{dt} = \mathbf{L}^*\mathbf{x} + \mathbf{K}^*\,\delta\gamma \tag{5.46a}$$

$$\mathbf{L}^* = \begin{bmatrix} \mathbf{0} & \mathbf{I} \\ -\mathbf{L} & \mathbf{0} \end{bmatrix} \tag{5.46b}$$

$$\mathbf{K}^* = \begin{bmatrix} \mathbf{0} \\ \mathbf{K} \end{bmatrix} \tag{5.46c}$$

To determine the controllability of the orbits, the rank of the 4×4 controllability matrix $\mathbf{C} = [\mathbf{K}^*, \mathbf{L}^*\mathbf{K}, \mathbf{L}^{*2}\mathbf{K}^*, \mathbf{L}^{*3}\mathbf{K}^*]$ is calculated from the system matrix $\mathbf{L}^*$ and the input matrix $\mathbf{K}^*$. For the orbits to be controllable the matrix $\mathbf{C}$ must have full rank and so must have a non-zero determinant. Evaluating the controllability matrix it is found that

$$\mathbf{C} = \begin{bmatrix} 0 & K_1 & 0 & -L_{11}K_1 - L_{13}K_2 \\ 0 & K_2 & 0 & -L_{31}K_1 - L_{33}K_2 \\ K_1 & 0 & -L_{11}K_1 - L_{13}K_2 & 0 \\ K_2 & 0 & -L_{31}K_1 - L_{33}K_2 & 0 \end{bmatrix} \tag{5.47}$$

Therefore, after some reduction the determinant is found to be

$$\det(\mathbf{C}) = -[K_1(L_{31}K_1 + L_{33}K_2) - K_2(L_{11}K_1 + L_{13}K_2)]^2 \qquad (5.48)$$

It is found that $\det(\mathbf{C})$ is non-zero for all the non-Keplerian orbit families so that state feedback to the sail elevation angle can ensure asymptotic stability in the ρ–z plane. Such controllability is to be expected as the motion in the ρ and z axes is strongly coupled through the matrix $\mathbf{L}$.

5.2.4.2   Control by variable sail elevation

Now that the controllability of the non-Keplerian orbit families has been established, a closed loop control law using the sail elevation angle will be considered. The sail elevation angle will be related to the state variables through a general feedback law of the form

$$\delta\gamma = \sum_{j=1}^{4} g_j x_j \qquad (5.49)$$

where x_j ($j = 1$–4) are the four components of the state vector $\mathbf{x}$. The four feedback gains g_j ($j = 1$–4) are then chosen to ensure that all four of the eigenvalues of the system lie in the left-hand complex plane so that the orbit has asymptotic stability. Since the reference state of the orbit is $\mathbf{x} = 0$ the controller is a state regulator. Equation (5.49) can then be substituted into Eq. (5.46a) and a new characteristic polynomial obtained. The Routh–Hurwitz criterion then defines limits to the range of values the gains may take so that all of the eigenvalues are in the left-hand complex plane. In addition, the gains may be selected by plotting the positions of the closed loop eigenvalues in the complex plane using the root locus technique. The closed loop response of the controller is shown in Fig. 5.8 for the unstable type I orbit considered in section 5.2.3.2. The gains have been chosen to minimise the damping time while avoiding excessive overshoot. It can be seen from Fig. 5.9 that only small trims to the sail attitude are required to ensure stability. However, the damping timescale is rather long owing to the coupling between the magnitude and direction of the solar radiation pressure force. It is also found that the initial errors damp asymptotically to non-zero values. This is due to the variable transformations, defined by Eqs (5.22), used to obtain the reduced variational equation. That is, the control law ensures that both ξ' and $\eta' \to 0$, leaving the constant terms in Eqs (5.22) as residual errors. Physically these residual errors arise from the solar sail having excess orbital angular momentum due to the initial error along the ρ axis. This excess angular momentum can in fact be removed through the addition of an integral term to the control law and allowing trims to the solar sail yaw attitude.

5.2.4.3   Control by fixed sail pitch

An extremely simple closed loop scheme will now be investigated which ensures stability for all of the non-Keplerian orbit families. The control scheme does not damp errors, but is used only to stabilise unstable orbits, such as one-year type I orbits high above the ecliptic plane. The control law only requires that the sail pitch

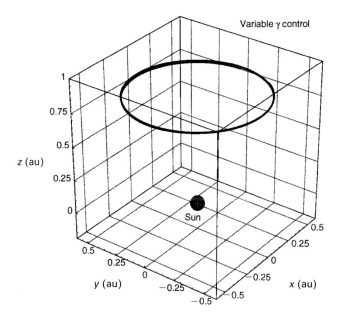

Fig. 5.8. Variable γ control for an unstable type I orbit.

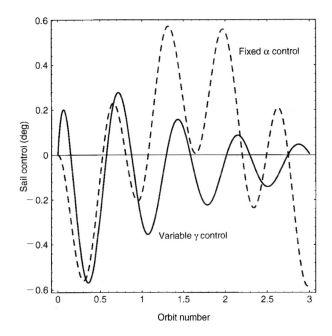

Fig. 5.9. Sail elevation angle trim $\delta\gamma$ for orbit control.

angle α remains fixed in the rotating frame of reference. Therefore, as the solar sail moves away from the nominal orbit, the sail elevation angle γ must vary. However, as will be seen, maintaining a fixed sail pitch attitude results in a rather simple implementation for the scheme.

For a fixed sail pitch angle the components of the radiation pressure acceleration (a_ρ, a_z) may be written as

$$a_\rho = \frac{\beta}{r^2}\cos^2\alpha\left(\cos\alpha\frac{\rho}{r} - \sin\alpha\frac{z}{r}\right) \tag{5.50a}$$

$$a_z = \frac{\beta}{r^2}\cos^2\alpha\left(\sin\alpha\frac{\rho}{r} - \cos\alpha\frac{z}{r}\right) \tag{5.50b}$$

These expressions may then be expanded to first order in a Taylor series, using a similar procedure to section 5.2.3.1. Therefore, a new system matrix may be formed and its stability characteristics examined. It is found that the reduced variational equation now becomes

$$\frac{d^2}{dt^2}\begin{bmatrix}\xi'\\\eta'\end{bmatrix} + \begin{bmatrix}P_{11} & P_{13}\\P_{31} & P_{33}\end{bmatrix}\begin{bmatrix}\xi'\\\eta'\end{bmatrix} = \begin{bmatrix}0\\0\end{bmatrix} \tag{5.51}$$

where the coefficients of the matrix $\mathbf{P}$ are given by

$$P_{11} = 3\tilde{\omega}^2 + \tilde{\omega}^2\left[1 - 3\left(\frac{\rho}{r}\right)^2\right] + 3\beta\tilde{\omega}^2\cos^2\alpha\frac{\rho}{r}\chi_1 - \beta\tilde{\omega}^2\cos^3\alpha \tag{5.52a}$$

$$P_{13} = -3\tilde{\omega}^2\left(\frac{\rho z}{r^2}\right) + 3\beta\tilde{\omega}^2\cos^2\alpha\frac{z}{r}\chi_1 + \beta\tilde{\omega}^2\cos^2\alpha\sin\alpha \tag{5.52b}$$

$$P_{31} = -3\tilde{\omega}^2\left(\frac{\rho z}{r^2}\right) + 3\beta\tilde{\omega}^2\cos^2\alpha\frac{\rho}{r}\chi_2 - \beta\tilde{\omega}^2\cos^2\alpha\sin\alpha \tag{5.52c}$$

$$P_{33} = \tilde{\omega}^2\left[1 - 3\left(\frac{z}{r}\right)^2\right] + 3\beta\tilde{\omega}^2\cos^2\alpha\frac{z}{r}\chi_2 - \beta\tilde{\omega}^2\cos^3\alpha \tag{5.52d}$$

with the auxiliary coefficients χ_1 and χ_2 defined by

$$\chi_1 = \frac{1}{r}(\rho\cos\alpha - z\sin\alpha) \tag{5.53a}$$

$$\chi_2 = \frac{1}{r}(\rho\sin\alpha + z\cos\alpha) \tag{5.53b}$$

To determine the stability of this system a new characteristic polynomial is formed and its eigenvalues found, viz.

$$\omega^4 + \text{tr}(\mathbf{P})\omega^2 + \det(\mathbf{P}) = 0 \tag{5.54}$$

The conditions for purely imaginary eigenvalues, and hence stability, are again that $\text{tr}(\mathbf{P}) > 0$ and $\det(\mathbf{P}) > 0$. However, owing to the scale invariance of the problem, as discussed in section 5.2.3.2, only the stability characteristics of the type II orbit family need be examined. Substituting for $\omega = \tilde{\omega}$ it is found that

$$\text{tr}(\mathbf{P}) = \tilde{\omega}^2 \left[2 + \left(\frac{z}{r} \right)^2 \right] \tag{5.55a}$$

$$\det(\mathbf{P}) = \tilde{\omega}^4 \left(\frac{\rho}{r} \right)^2 \tag{5.55b}$$

both of which are clearly positive. It can be seen then that the type II orbit family is rendered stable using the simple fixed sail pitch attitude control law. Therefore, due to the scale invariance of the problem, all other orbit families are rendered stable. There will, however, still be an azimuthal drift along the nominal orbit due to the excess orbital angular momentum of the solar sail owing to the initial error along the ρ-axis.

The required change in the sail elevation angle $\delta\gamma$ can be written as a feedback law by evaluating the change in sail pitch angle with first-order changes in ρ and z. The pitch angle α may be written as

$$\alpha = \gamma - \tan^{-1} \left(\frac{z}{\rho} \right) \tag{5.56}$$

so that $\delta\alpha$ may be formed from $\partial\alpha/\partial\rho$ and $\partial\alpha/\partial z$. The required control law to maintain a fixed sail pitch angle is then $\delta\gamma = -\delta\alpha$, viz.

$$\delta\gamma = \frac{1}{\rho} \frac{1}{1 + (z/\rho)^2} \left[\eta - \left(\frac{z}{\rho} \right) \xi \right] \tag{5.57}$$

This control scheme is clearly appealing since in practice no state variable information is required. A Sun sensor would measure the change in sail pitch angle $\delta\alpha$ which would then be used directly to command an opposite change in sail elevation angle $\delta\gamma$. It is this simplicity which makes the control law so attractive. The performance of the control law is illustrated in Fig. 5.10. In this case the unstable type I orbit considered in section 5.2.3.2 is stabilised using small trims to the sail elevation angle to maintain a fixed sail pitch angle. As can be seen, the orbit is now stable although there is no active damping of the initial errors. Again, only small trims to the sail orientation are required, as shown in Fig. 5.9.

5.2.5 Patched orbits

In this section it will be demonstrated that Sun-centred non-Keplerian orbits may be patched together, providing large families of new orbits. Firstly, non-Keplerian orbits will be patched to Keplerian orbits by orienting the solar sail edgewise to the Sun. Then, the conditions required to transfer between Sun-centred non-Keplerian orbits will be investigated. It will be shown that transfer between non-Keplerian orbits is only possible if the orbits are identical, but have different orientations. In particular, a series of non-Keplerian orbits may be used to transfer the solar sail from a displaced orbit above the ecliptic plane to an identical but retrograde orbit

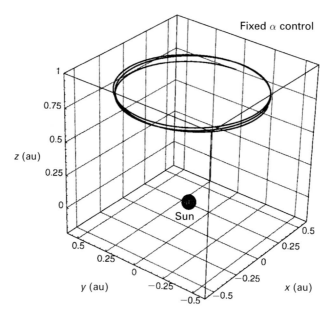

Fig. 5.10. Fixed α control for an unstable type I orbit.

below. In addition to generating new families of orbits, the use of patched orbits enlarges the scope for novel mission applications, as will be discussed in Chapter 6.

5.2.5.1   Patch to a Keplerian orbit

Firstly, a solar sail will be considered on a Sun-centred non-Keplerian orbit with radius ρ and displacement z. Then, at some point on the orbit, the solar sail will be rotated so that it is oriented edgewise to the Sun. With the solar radiation pressure force removed, the non-Keplerian orbit radius and solar sail velocity determine the size and shape of the resulting Keplerian ellipse, as shown in Fig. 5.11. The orientation of the Keplerian orbit is also determined by non-Keplerian orbit geometry and the time at which the orbits are patched. The orbit inclination is obtained from $\tan^{-1}(z/\rho)$ while the ascending node angle is determined by the angular distance around the non-Keplerian orbit at which the patch takes place. All of the resulting Keplerian orbits will have an argument of pericentre of $270°$, since the patch always occurs at aphelion.

On any non-Keplerian orbit the solar radiation pressure force exerted on the solar sail is always normal to the solar sail velocity vector so that the total energy E is constant, viz.

$$E = \tfrac{1}{2}\omega^2\rho^2 - \frac{\mu}{\sqrt{\rho^2 + z^2}} \tag{5.58}$$

Therefore, given that the total orbit energy is $-\mu/2a$, the semi-major axis of the Keplerian orbit obtained by orienting the sail edgewise to the Sun is given by

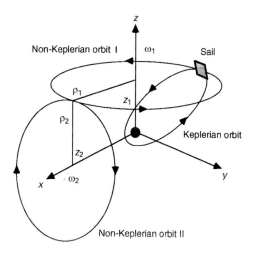

Fig. 5.11. Patched Keplerian and non-Keplerian orbits.

$$a = \left[\frac{2}{\sqrt{\rho^2 + z^2}} - \frac{\omega^2 \rho^2}{\mu}\right]^{-1} \tag{5.59}$$

Similarly, noting that the patch point corresponds to the aphelion of the Keplerian orbit, viz.

$$a(1 + e) = \sqrt{\rho^2 + z^2} \tag{5.60}$$

the eccentricity of the resulting ellipse may also be obtained as

$$e = 1 - \frac{\omega^2 \rho^2}{\mu}\sqrt{\rho^2 + z^2} \tag{5.61}$$

Therefore, for a given non-Keplerian orbit, the size, shape and orientation of the Keplerian orbit are obtained. It should be noted that the range of eccentricities available is limited by the constraint on the solar sail orbital angular velocity, defined by Eq. (5.15). This constraint limits the eccentricity of the Keplerian orbit such that $0 < e \leq 1$. Therefore, it is not possible to transfer to a precisely circular orbit. It can also been seen from Eq. (5.61) that $e \to 1$ as $\rho \to 0$. In this case the solar sail is in equilibrium so that the resulting Keplerian orbit is rectilinear.

5.2.5.2   Patch to a non-Keplerian orbit

Now that patching to Keplerian orbits has been considered, the possibility of patching to other non-Keplerian orbits will be investigated. For energy continuity between the orbits it is required that

$$\rho_1 \omega_1 = \rho_2 \omega_2 \tag{5.62}$$

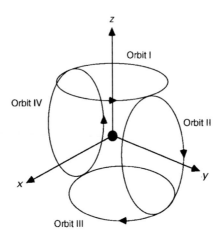

Fig. 5.12. 'Cubic' patched non-Keplerian orbit.

as shown in Fig. 5.11. Similarly, from Eq. (5.10b), it can be shown that the condition for continuity of the solar sail lightness number results in the requirement that

$$z_1^2\tilde{\omega}^4 + \rho_1^2\omega_1^2(\omega_1^2 - 2\tilde{\omega}^2) = z_2^2\tilde{\omega}^4 + \rho_2^2\omega_2^2(\omega_2^2 - 2\tilde{\omega}^2) \qquad (5.63)$$

However, using Eq. (5.62) it is found from Eq. (5.63) that the condition $\omega_1 = \omega_2$, and so $\rho_1 = \rho_2$ is required. Therefore, the two orbits must be identical, but with different orientations. Using this patching procedure, four non-Keplerian orbits may be patched together to form an elaborate new trajectory in which the solar sail orbits over the surface of a cube, as shown in Fig. 5.12. It should be noted that orbit III is retrograde with respect to orbit I. Furthermore, the entire patched trajectory is symmetric to rotations and so may be oriented in any way with respect to the ecliptic plane.

5.3   PLANET-CENTRED NON-KEPLERIAN ORBITS

5.3.1   Introduction

Following the development of the families of Sun-centred non-Keplerian orbits, planet-centred non-Keplerian orbits will now be investigated. These orbits are again generated by orienting the solar sail such that a component of the solar radiation pressure force is directed out of the orbit plane. Therefore, planet-centred non-Keplerian orbits are again circular orbits, but are displaced behind the planet in the anti-Sun direction. As before, the solar sail orbit period, orbit radius and out-of-plane displacement distance may be chosen independently with a suitable choice of solar sail pitch angle and acceleration.

It will be assumed that the radiation field is uniform over the scale of the problem (tens of planetary radii) so that the ratio of the solar radiation pressure force to the local gravitational force increases with increasing distance from the planetary centre. It is this relation which leads to interesting new orbits. In addition, during the initial analysis it will be assumed that the planet is the only gravitating body. The region in which this assumption remains valid will clearly be limited owing to solar and, if applicable, lunar gravitational perturbations. For example, given that the lunar sphere of influence has a radius of order 10 Earth radii, Earth-centred non-Keplerian orbits with a displacement of up to 40–50 Earth radii may be considered. Over this distance the solar radiation pressure varies by only 4×10^{-3} from its value at 1 au. For other planets different constraints are imposed upon the orbits. For example, Mercury provides large displaced orbits due to its small mass and close proximity to the Sun; however, for Mercury the rapid annual rotation of the Sun-line and solar gravitational perturbations are both important.

After discussing the various types of planet-centred non-Keplerian orbits their stability characteristics will be investigated and linearly stable families of orbits identified. For the unstable families simple control laws are again developed. As will be seen, these control laws are suitable for stabilising the unstable orbit families against perturbations, such as the Sun-line rotation and solar gravity. Lastly, it will be demonstrated that by patching individual Keplerian and non-Keplerian orbits, complex new families of trajectories may again be generated.

5.3.2   Non-Keplerian orbit solutions

Following the analysis of section 5.2.2, the equations of motion will again be considered for a perfectly reflecting solar sail in a rotating frame of reference, with the origin of the frame centred on a point mass planet. The axis of rotation $\boldsymbol{\omega}$ will be directed along the Sun–planet line, as shown in Fig. 5.13. The sail attitude is again defined by a unit vector $\mathbf{n}$ fixed in the rotating frame of reference with the magnitude

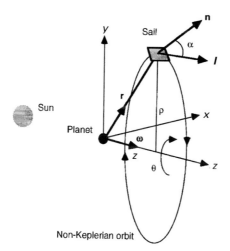

Fig. 5.13. Planet-centred non-Keplerian orbit frame of reference.

of the solar radiation pressure acceleration defined by κ. For Earth-centred orbits this is just the solar sail characteristic acceleration. Since the sail attitude is fixed in the rotating frame of reference, the solar sail must rotate once per orbit with respect to an inertial frame of reference.

The equation of motion for a solar sail in this rotating frame of reference under the action a gravitational potential and uniform radiation field is given by

$$\frac{d^2\mathbf{r}}{dt^2} + 2\boldsymbol{\omega} \times \frac{d\mathbf{r}}{dt} + \boldsymbol{\omega} \times (\boldsymbol{\omega} \times \mathbf{r}) = \mathbf{a} - \nabla V \qquad (5.64)$$

The solar radiation pressure acceleration $\mathbf{a}$ and the two-body gravitational potential are given by

$$\mathbf{a} = \kappa(\mathbf{l} \cdot \mathbf{n})^2 \mathbf{n} \qquad (5.65a)$$

$$V = -\frac{\mu}{r} \qquad (5.65b)$$

where the unit vector $\mathbf{l}$ is directed along the Sun-line. The requirement $\mathbf{l} \cdot \mathbf{n} > 0$ is imposed to ensure that the sail normal is always directed away from the Sun. This requirement consequently constrains the solar sail motion to the planetary night-side. Equation (5.64) may be simplified by again introducing the scalar potential Φ to represent the conservative centripetal acceleration, viz.

$$\Phi = -\tfrac{1}{2} |\boldsymbol{\omega} \times \mathbf{r}|^2 \qquad (5.66a)$$

$$\nabla\Phi = \boldsymbol{\omega} \times (\boldsymbol{\omega} \times \mathbf{r}) \qquad (5.66b)$$

Similarly, the solar radiation pressure acceleration is, in this case, conservative, and so may be written in terms of a scalar potential Γ if required, viz.

$$\Gamma = \kappa(\mathbf{l} \cdot \mathbf{n})^2 (\mathbf{r} \cdot \mathbf{n}) \qquad (5.67a)$$

$$\nabla\Gamma = \kappa(\mathbf{l} \cdot \mathbf{n})^2 \mathbf{n} \qquad (5.67b)$$

Defining a new composite potential function $U = V + \Phi$, the equation of motion is then reduced to

$$\frac{d^2\mathbf{r}}{dt^2} + 2\boldsymbol{\omega} \times \frac{d\mathbf{r}}{dt} + \nabla U = \mathbf{a} \qquad (5.68)$$

In the rotating frame of reference equilibrium solutions are again required so that the first two terms of Eq. (5.68) must vanish. The solar sail attitude required for an equilibrium solution in the rotating frame is therefore defined by

$$\mathbf{n} = \frac{\nabla U}{|\nabla U|} \qquad (5.69)$$

Since the solar sail is to have uniform azimuthal motion, there can be no transverse component of the solar radiation pressure acceleration. Therefore, the sail attitude may again be defined by the pitch angle α between $\mathbf{l}$ and $\mathbf{n}$, given by

$$\tan \alpha = \frac{|\mathbf{l} \times \nabla U|}{\mathbf{l} \cdot \nabla U} \qquad (5.70)$$

Similarly, the required solar radiation pressure acceleration may be obtained by taking a scalar product of $\mathbf{n}$ with Eq. (5.68). Again, requiring an equilibrium solution in the rotating frame of reference, it is found that

$$\kappa = \frac{\nabla U \cdot \mathbf{n}}{(\mathbf{l} \cdot \mathbf{n})^2} \tag{5.71}$$

Using planet-centred cylindrical polar co-ordinates (ρ, θ, z) defined in Fig. 5.13, the rotating two-body potential function may be written as

$$U = -\left[\tfrac{1}{2}\rho^2 \omega^2 + \frac{\mu}{r}\right] \tag{5.72}$$

where the angular velocity ω is again related to the orbit period T by $\omega = 2\pi/T$. Therefore, evaluating the potential gradient it is found that the required solar sail pitch angle and acceleration are given by

$$\tan \alpha = \frac{\rho}{z}\left[1 - \left(\frac{\omega}{\tilde{\omega}}\right)^2\right], \qquad \tilde{\omega}^2 = \frac{\mu}{r^3} \tag{5.73a}$$

$$\kappa = \tilde{\omega}^2 \left\{1 + \left(\frac{\rho}{z}\right)^2 \left[1 - \left(\frac{\omega}{\tilde{\omega}}\right)^2\right]^2\right\}^{3/2} z \tag{5.73b}$$

If the unit of length is now chosen to be the planetary radius and μ is chosen to be unity, κ will then be the solar radiation pressure acceleration in units of the gravitational acceleration at 1 planetary radius. This measure will be used so that the analysis of the resulting orbit families is generic. The required solar sail acceleration can also be transformed into the characteristic acceleration and lightness number by appropriate rescaling.

5.3.2.1   Type I

For this family of orbits the solar sail orbit period will be fixed at some particular value for all values of orbit radius and displacement distance. This is equivalent to choosing the orbits to be synchronous with some particular Keplerian orbit with orbit radius r_0, such that $\omega = r_0^{-3/2}$. From Eq. (5.73b) surfaces of constant solar sail acceleration may be generated, with ω chosen to be some fixed value. A section of these surfaces can be seen in Fig. 5.14 with the solar sail orbit period chosen to be synchronous with a Keplerian orbit of radius 30 planetary radii (period of 9.6 days in the geocentric case). The solar sail characteristic acceleration is listed in Table. 5.2 for equivalent orbits at Mercury, Venus, Earth and Mars. Again, each point on a surface of constant solar sail acceleration corresponds to a displaced orbit with a different orbit radius and displacement distance. It can be seen that for low values of solar sail acceleration there are two disconnected surfaces. These surfaces correspond to large orbits with small displacements near 30 planetary radii, and small orbits with large displacements away from the planetary centre. As the solar sail acceleration increases, these surfaces then expand and connect. The required direction of the sail normal vector $\mathbf{n}$ is also shown. It can be seen that along the z axis of the system the

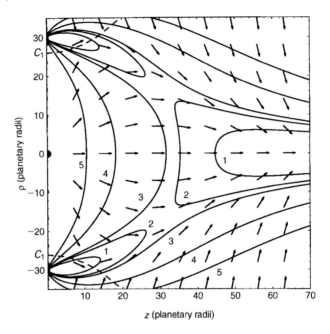

Fig. 5.14. Type I orbit acceleration contours (see Table 5.2 for values).

Table 5.2. Solar sail characteristic acceleration for Mercury, Venus, Earth and Mars planet-centred non-Keplerian orbits

Type		Contour				
		1	2	3	4	5
I	Mercury	0.28	0.45	0.56	1.69	5.07
a_0 (mm s^{-2})	Venus	2.32	3.71	4.64	13.92	41.76
	Earth	4.91	7.86	9.82	29.46	88.38
	Mars	4.29	6.87	8.59	25.76	77.29
II	Mercury	0.11	0.45	1.13	2.25	5.07
a_0 (mm s^{-2})	Venus	0.93	3.71	9.28	18.56	41.76
	Earth	1.96	7.86	19.64	39.28	88.38
	Mars	1.72	6.87	17.17	34.35	77.29
III	Mercury	0.17	0.23	0.34	0.56	1.13
a_0 (mm s^{-2})	Venus	1.39	1.86	2.78	4.64	9.28
	Earth	2.94	3.93	5.89	9.82	19.64
	Mars	2.58	3.43	5.15	8.59	17.17

sail pitch angle is zero, corresponding to non-orbiting equilibrium solutions with the solar radiation pressure acceleration balancing the local gravitational acceleration.

5.3.2.2 Type II

For this family the solar sail orbit period will be chosen such that $\omega = \rho^{-3/2}$. This corresponds to the orbits being synchronous with Keplerian orbits above the planetary terminator with an orbit radius equal to the non-Keplerian orbit radius ρ. Therefore, cylindrical surfaces of co-rotation will extend in the anti-Sun direction. Solar sails with the same radius ρ will then orbit synchronously with each other at differing displacement distances z. From Eqs (5.73) the required solar sail pitch angle and acceleration are obtained as

$$\tan \alpha = \frac{\rho}{z} \left\{ 1 - \left[1 + \left(\frac{z}{\rho} \right)^2 \right]^{3/2} \right\} \tag{5.74a}$$

$$\kappa = \tilde{\omega}^2 \left[1 + \left(\frac{\rho}{z} \right)^2 \left\{ 1 - \left[1 + \left(\frac{z}{\rho} \right)^2 \right]^{3/2} \right\}^2 \right]^{3/2} z \tag{5.74b}$$

A section of the surfaces of constant solar sail acceleration generated by Eq. (5.74b) is shown in Fig. 5.15. It can be seen that the surfaces of constant solar sail acceleration approach the $z = 0$ plane as $\kappa \to 0$ corresponding to Keplerian orbits

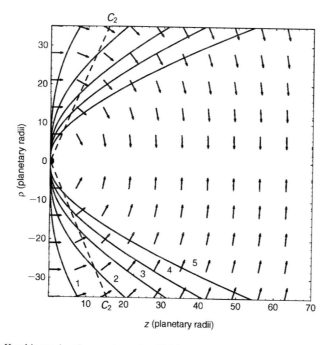

Fig. 5.15. Type II orbit acceleration contours (see Table 5.2 for values).

with $\omega = \rho^{-3/2}$ and $z = 0$. Since the required sail pitch angle quickly increases for large displacements, the required solar sail acceleration also increases rapidly.

5.3.2.3   Type III

The solar sail orbit period will now be treated as a free parameter with which to minimise the required solar sail acceleration and so generate a family of optimal non-Keplerian orbits. Therefore, setting the derivative of κ with respect to ω to zero it is found that orbits with

$$\omega = \tilde{\omega} \tag{5.75}$$

will minimise the required solar sail acceleration. For an optimal orbit the solar sail orbit period is therefore equal to the orbit period of a Keplerian orbit of radius r. With this orbit period the required solar sail pitch angle and acceleration are obtained from Eqs (5.73) as

$$\tan \alpha = 0 \tag{5.76a}$$

$$\kappa = \tilde{\omega}^2 z \tag{5.76b}$$

so that the sail normal is directed along the Sun-line. In this case the general three-dimensional motion of the solar sail can be also obtained in closed form using the Hamilton–Jacobi method, as discussed in section 4.4.5. The non-Keplerian orbits then correspond to the intersection of the parabolic co-ordinates.

Surfaces of constant solar sail acceleration may now be generated by fixing κ at some value κ_0. Then, Eq. (5.76b) can be inverted to yield

$$\rho(z) = \left[\frac{z^{2/3}}{\kappa_0^{2/3}} - z^2 \right]^{1/2} \tag{5.77}$$

which defines a surface of revolution about the z axis. It can also be shown that the locus of orbits of maximum radius is defined by the cone

$$\rho = \sqrt{2}\, z \tag{5.78}$$

Sections of these surfaces of constant solar sail acceleration are shown in Fig. 5.16. The line C_3 shows the locus of orbits of maximum radius, as defined by Eq. (5.78). Non-orbiting equilibrium solutions are again shown along the z axis. It can be seen from Table 5.2 that extremely large solar sail accelerations are required for close Earth-centred orbits. However, for orbits about other bodies the requirements on the solar sail are not as demanding, especially for Mercury which is closer to the Sun and has a lower planetary mass. Furthermore, for this optimal family of orbits, the axis of symmetry of the system need not in fact lie along the Sun-line. Any axis k passing through the origin, such that $(l \cdot k) > 0$, will generate new off-axis orbits, as shown in Fig. 5.17. However, since the sail normal is directed along the new axis $\mathbf{n} = k$, the required solar sail acceleration is increased by a factor $(l \cdot k)^{-2}$ owing to the oblique incidence of photons on the sail. Using such off-axis orbits the solar sail can move, for a period, along the $+z$ axis over the planetary day side.

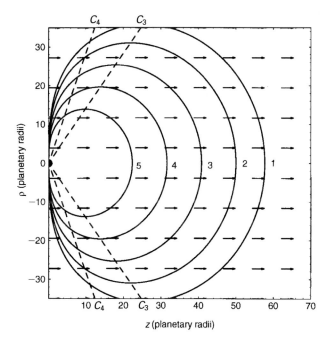

Fig. 5.16. Type III orbit acceleration contours (see Table 5.2 for values).

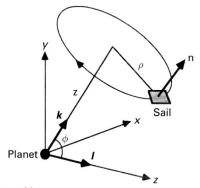

Fig. 5.17. Off-axis non-Keplerian orbit.

5.3.3   Planet-centred non-Keplerian orbit stability

The stability of the families of planet-centred non-Keplerian orbits will now be investigated through a linear perturbation analysis similar to that of section 5.2.3.1. However, it should be noted that since the radiation field is uniform, the solar radiation pressure acceleration remains constant during the perturbation. Therefore, applying a perturbation $\boldsymbol{\delta}$ such that $\mathbf{r}_0 \rightarrow \mathbf{r}_0 + \boldsymbol{\delta}$ a linear variational equation may again be obtained, viz.

$$\frac{d^2\boldsymbol{\delta}}{dt^2} + 2\boldsymbol{\omega} \times \frac{d\boldsymbol{\delta}}{dt} + \nabla U(\mathbf{r}_0 + \boldsymbol{\delta}) - \mathbf{a}(\mathbf{r}_0, \mathbf{n}) = 0 \tag{5.79}$$

The potential gradient is now expanded about the nominal orbit using a Taylor series and written in component form, again using the same analysis of section 5.2.3.1. Again, the azimuthal terms are removed to reduce the stability problem to the ρ–z plane. It is then found that the variational equation reduces to a matrix equation of the form

$$\frac{d^2}{dt^2} \begin{bmatrix} \xi' \\ \eta' \end{bmatrix} + \begin{bmatrix} M_{11} & M_{13} \\ M_{31} & M_{33} \end{bmatrix} \begin{bmatrix} \xi' \\ \eta' \end{bmatrix} = \begin{bmatrix} 0 \\ 0 \end{bmatrix} \tag{5.80}$$

As with Sun-centred non-Keplerian orbits, the solar sail is free to drift along the nominal orbit in the azimuthal direction. Evaluating the partial derivatives of the potential gradient it is found that the coefficients of the matrix $\mathbf{M}$ may be written as

$$M_{11} = 3\omega^2 + \tilde{\omega}^2 \left[1 - 3\left(\frac{\rho}{r}\right)^2 \right] \tag{5.81a}$$

$$M_{13} = M_{31} = -3\tilde{\omega}^2 \left(\frac{\rho z}{r^2}\right) \tag{5.81b}$$

$$M_{33} = \tilde{\omega}^2 \left[1 - 3\left(\frac{z}{r}\right)^2 \right] \tag{5.81c}$$

The stability of the orbit families may now be investigated by again calculating the eigenvalues of the system. The eigenvalues may be obtained by substituting an exponential solution of the form

$$\begin{bmatrix} \xi' \\ \eta' \end{bmatrix} = \begin{bmatrix} \xi_0 \\ \eta_0 \end{bmatrix} \exp(\lambda t) \tag{5.82}$$

Substituting this solution into Eq. (5.80) yields a matrix equation of the form

$$\begin{bmatrix} \lambda^2 + M_{11} & M_{13} \\ M_{31} & \lambda^2 + M_{33} \end{bmatrix} \begin{bmatrix} \xi_0 \\ \eta_0 \end{bmatrix} = \begin{bmatrix} 0 \\ 0 \end{bmatrix} \tag{5.83}$$

For non-trivial solutions it is required that the determinant of this matrix equation vanishes. The characteristic polynomial of the variational equation is then found to be

$$\lambda^4 + \text{tr}(\mathbf{M})\lambda^2 + \det(\mathbf{M}) = 0 \tag{5.84}$$

The stability characteristics of each orbit type may now be investigated by substituting for the appropriate functional form of ω. Then, regions in the ρ–z plane may be identified where the roots of the characteristic polynomial are purely imaginary with $\omega_j^2 < 0\,(j = 1\text{–}4)$ indicating stable, bound oscillations. Such purely imaginary eigenvalues are obtained if $\text{tr}(\mathbf{M}) > 0$ and $\det(\mathbf{M}) > 0$. Again, only a linear analysis is presented which provides only necessary conditions for stability, but sufficient conditions for instability.

5.3.3.1   *Type I stability*

For this family of orbits the orbit period is fixed for all values of orbit radius and displacement. Then, the non-Keplerian orbits are synchronous with some particular Keplerian orbit with an orbit radius r_0 such that $\omega = r_0^{-3/2}$. From Eqs (5.81) it is found that

$$\text{tr}(\mathbf{M}) = \tilde{\omega}^2 \left[3 - \left(\frac{r_0}{r} \right)^3 \right] \tag{5.85a}$$

$$\det(\mathbf{M}) = \tilde{\omega}^4 \left[3 \left(\frac{r}{r_0} \right)^3 - 9 \left(\frac{r}{r_0} \right)^3 \left(\frac{z}{r} \right)^2 - 2 \right] \tag{5.85b}$$

Clearly, for $\text{tr}(\mathbf{M})$ to be strictly positive it is required that

$$r > \left(\tfrac{1}{3} \right)^{1/3} r_0 \tag{5.86}$$

so that the region defined by $r < \left(\tfrac{1}{3} \right)^{1/3} r_0$ will necessarily be unstable. A stricter condition for stability can be obtained from the determination of the overall stability map which requires a numerical solution of $\det(\mathbf{M}) > 0$. It can be shown that this condition reduces to

$$\frac{2}{3} \left(\frac{r_0}{r} \right)^3 + 3 \left(\frac{z}{r} \right)^2 - 1 < 0 \tag{5.87}$$

Substituting an equality in Eq. (5.87) yields the section of the resulting boundary surface, shown as C_1 in Fig. 5.14. It can be seen that only orbits with a large radius and small displacement distance will be stable. The intersection of the boundary surface with the ρ axis occurs at the point $\rho = \left(\tfrac{2}{3} \right)^{1/3} r_0$.

5.3.3.2   *Type II stability*

For this family of orbits the solar sail orbit period will be chosen such that $\omega = \rho^{-3/2}$. Again, this corresponds to the orbits being synchronous with Keplerian orbits above the planetary terminator with an orbit radius equal to the non-Keplerian orbit radius ρ. Then, from Eqs (5.81) it is found that

$$\text{tr}(\mathbf{M}) = \tilde{\omega}^2 \left[3 \left(\frac{r}{\rho} \right)^2 - 1 \right] \tag{5.88a}$$

$$\det(\mathbf{M}) = -\tilde{\omega}^4 \left[2 - 3 \left(\frac{r}{\rho} \right)^3 + 9 \left(\frac{r z^2}{\rho^3} \right) \right] \tag{5.88b}$$

Firstly, it can be seen that $\text{tr}(\mathbf{M})$ is always strictly positive while $\det(\mathbf{M})$ is strictly positive if

$$\frac{2}{3} \left(\frac{\rho}{r} \right)^3 + 3 \left(\frac{z}{r} \right)^2 - 1 < 0 \tag{5.89}$$

The boundary between stable and unstable orbits can be obtained by substituting an equality in Eq. (5.89) and rewriting as

$$\frac{2}{3}\left[1 + \left(\frac{z}{\rho}\right)^2\right]^{-3/2} + 3\left(\frac{z}{\rho}\right)^2\left[1 + \left(\frac{z}{\rho}\right)^2\right]^{-1} - 1 = 0 \tag{5.90}$$

Since Eq. (5.90) is a function of z/ρ only, the boundary between stable and unstable orbits is of the form $\rho = \varepsilon z$, where ε is a constant. This constant may be found from Eq. (5.90) using

$$\tfrac{2}{3}(1 + \varepsilon^{-2})^{-3/2} + 3\varepsilon^{-2}(1 + \varepsilon^{-2})^{-1} - 1 = 0 \tag{5.91}$$

This equation is further simplified by making the substitution $\gamma = (1 + \varepsilon^{-2})^{1/2}$ to clear the radical and obtain

$$\gamma^3 - \tfrac{3}{2}\gamma + \tfrac{1}{3} = 0 \tag{5.92}$$

The solution to this cubic yields $\varepsilon = 2.264$, which therefore defines the boundary between stable and unstable orbits. The condition for stability is therefore

$$\rho > 2.264z \tag{5.93}$$

which defines a cone separating the stable and unstable families of type II orbits. The section of the partitioning cone is shown in Fig. 5.15 as the line C_2.

5.3.3.3   *Type III stability*

For this family of orbits the required solar sail acceleration is minimised by choosing the orbit angular velocity as $\tilde{\omega}$, the orbital angular velocity of a Keplerian orbit of radius r. Therefore, from Eqs (5.81) it is found that

$$\mathrm{tr}(\mathbf{M}) = 2\tilde{\omega}^2 \tag{5.94a}$$

$$\det(\mathbf{M}) = \tilde{\omega}^4\left[4 - 3\left(\frac{\rho}{r}\right)^2 - 12\left(\frac{z}{r}\right)^2\right] \tag{5.94b}$$

It can be seen that $\mathrm{tr}(\mathbf{M})$ is clearly positive while $\det(\mathbf{M})$ is also positive provided that

$$\rho > 2\sqrt{2}\,z \tag{5.95}$$

so that there are again two distinct regions of stability and instability. The partitioning is again defined by a cone, whose section is shown in Fig. 5.16 as the line C_4. A typical unstable response for a 15.51-day Earth-centred optimal orbit is shown in Fig. 5.18 using a numerical integration of the full non-linear equations of motion with Sun-line rotation and solar gravitational perturbations. It can be seen that the orbit is unbound with the solar radiation pressure acceleration driving the spacecraft out of the gravitational potential well. However, as will be seen, such orbits can be stabilised through the use of simple feedback control laws.

5.3.4   **Planet-centred non-Keplerian orbit control**

The stability analysis of section 5.3.3 has shown that unstable families of planet-

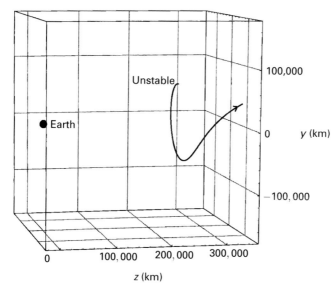

Fig. 5.18. Unstable type III orbit ($\rho_0 = 10$ Earth radii, $z_0 = 40$ Earth radii, $\xi_0 = 10^{-2}\rho_0$, $\eta_0 = 10^{-2}z_0$).

centred non-Keplerian orbits exist. Furthermore, for the stable families, it is found that Sun-line rotation and solar gravity induce large amplitude oscillations about the nominal orbit. It is therefore necessary to develop simple control schemes to stabilise the unstable orbit families and to damp the solar sail response to perturbations. Following section 5.2.4, the reduced two-dimensional system will be used so that the azimuthal motion of the solar sail will again be ignored. It will be shown that the orbit families are controllable using state feedback to the sail pitch attitude or feedback using small trims to the sail area.

5.3.4.1   Orbit controllability

It will now be demonstrated that the reduced two-dimensional problem is controllable using either trims of the solar sail pitch attitude $\delta\alpha$ or allowing small trims of the sail area. These trims to the sail area then result in small variations in the solar sail acceleration $\delta\kappa$. Making use of Eq. (5.80) and allowing first-order changes in the solar sail attitude or acceleration, a modified variational equation is obtained. The variational equation may again be reduced to the variables $\boldsymbol{\delta}' = (\xi', \eta')$ by eliminating the azimuthal co-ordinate, viz.

$$\frac{d^2\boldsymbol{\delta}'}{dt^2} + \mathbf{M}\boldsymbol{\delta}' = \mathbf{K}\mathbf{u} \tag{5.96a}$$

$$\mathbf{M} = \begin{bmatrix} M_{11} & M_{13} \\ M_{31} & M_{33} \end{bmatrix} \tag{5.96b}$$

where the control $\mathbf{u}$ is actuated through either the solar sail attitude or the sail area.

For pitch control and area control, respectively, the matrix $\mathbf{K}$ is defined as

$$\mathbf{K} = \begin{bmatrix} \dfrac{\partial a_\rho}{\partial \alpha} \\[2mm] \dfrac{\partial a_z}{\partial \alpha} \end{bmatrix} \tag{5.97a}$$

$$\mathbf{K} = \begin{bmatrix} \dfrac{\partial a_\rho}{\partial \kappa} \\[2mm] \dfrac{\partial a_z}{\partial \kappa} \end{bmatrix} \tag{5.97b}$$

Then, calculating the partial derivatives of solar radiation pressure acceleration, it is found that for pitch control the components of $\mathbf{K}$ are given by

$$K_1 = \kappa \cos^3 \alpha (1 - 2\tan^2 \alpha) \tag{5.98a}$$

$$K_2 = -3\,\kappa \cos^2 \alpha \sin \alpha \tag{5.98b}$$

and for area control the components of $\mathbf{K}$ are given by

$$K_1 = \cos^2 \alpha \sin \alpha \tag{5.99a}$$

$$K_2 = \cos^3 \alpha \tag{5.99b}$$

Writing the variational equation in standard state variable form, a set of four first-order equations are again obtained which may be written in matrix form as

$$\frac{d\mathbf{x}}{dt} = \mathbf{M}^*\mathbf{x} + \mathbf{K}^*\mathbf{u} \tag{5.100a}$$

$$\mathbf{M}^* = \begin{bmatrix} \mathbf{0} & \mathbf{I} \\ -\mathbf{M} & \mathbf{0} \end{bmatrix} \tag{5.100b}$$

$$\mathbf{K}^* = \begin{bmatrix} \mathbf{0} \\ \mathbf{K} \end{bmatrix} \tag{5.100c}$$

where the state vector $\mathbf{x} = (\delta', d\delta'/dt)$. To determine the controllability of the families of orbits the rank of the 4×4 controllability matrix

$$\mathbf{C} = [\mathbf{K}^*, \mathbf{M}^*\mathbf{K}^*, \mathbf{M}^{*2}\mathbf{K}^*, \mathbf{M}^{*3}\mathbf{K}^*]$$

is again formed from the system matrix $\mathbf{M}^*$ and the input matrix $\mathbf{K}^*$. For the system to be fully controllable the matrix $\mathbf{C}$ must have full rank and so non-zero determinant. Evaluating the controllability matrix it is found that

$$\mathbf{C} = \begin{bmatrix} 0 & K_1 & 0 & -M_{11}K_1 - M_{13}K_2 \\ 0 & K_2 & 0 & -M_{31}K_1 - M_{33}K_2 \\ K_1 & 0 & -M_{11}K_1 - M_{13}K_2 & 0 \\ K_2 & 0 & -M_{31}K_1 - M_{33}K_2 & 0 \end{bmatrix} \tag{5.101}$$

Therefore, the matrix determinant is then found to be

$$\det(\mathbf{C}) = -[K_1(M_{31}K_1 + M_{33}K_2) - K_2(M_{11}K_1 + M_{13}K_2)]^2 \qquad (5.102)$$

For type III optimal orbits with pitch control, Eq. (5.102) reduces to

$$\det(\mathbf{C}) = -\frac{9\,\kappa^6\rho^2}{r^4} \qquad (5.103)$$

while for type III optimal orbits with area control, Eq. (5.102) reduces to

$$\det(\mathbf{C}) = -\frac{9\,\kappa^2\rho^2}{r^4} \qquad (5.104)$$

so that type III orbits are fully controllable using either sail pitch or sail area, ensuring asymptotic stability in the ρ–z plane. In addition, it is found that all other orbit types are in principle controllable. Again, such controllability is to be expected as the motion in the ρ and z axes is strongly coupled through the matrix $\mathbf{M}$.

5.3.4.2 Control by variable sail area

Since it has been demonstrated that the planet-centred non-Keplerian orbit families are controllable, closed loop feedback control will now be investigated. Whereas sail pitch control was utilised for the Sun-centred non-Keplerian orbit families, sail area control will now be considered for optimal type III planet-centred non-Keplerian orbits. Such control may be appropriate for this family of orbits since the nominal sail pitch angle is zero. Therefore, the solar sail may be oriented in a fixed attitude with the sail normal directed along the Sun-line and with control actuated through small trims to the sail area. Such a Sun-facing attitude can in principle be obtained by passive means using a slightly conical sail with the apex of the cone directed along the Sun-line. Full state feedback will now be used to generate small variations in the sail acceleration, viz.

$$\delta\kappa = \sum_{j=1}^{4} g_j x_j \qquad (5.105)$$

where x_j $(j = 1\text{–}4)$ are the four components of the state vector. The four gains g_j $(j = 1\text{–}4)$ are chosen to ensure that all four of the eigenvalues of the system are in the left-hand complex plane ensuring asymptotic stability. Constraints on the gains can again be found using the Routh–Hurwitz criterion and the gains selected using the standard root locus method for linear systems. The closed loop response of the unstable 15.51-day type III optimal orbit considered in section 5.3.3.3 is shown in Fig. 5.19. The rotating Sun-line perturbations induce a periodic forcing term in the dynamics of the orbit so that a small periodic response in the ρ–z plane is obtained. However, it can be seen in Fig. 5.20 that only small trims of the sail area are required to stabilise the orbit. Similar stabilisation can also be obtained using feedback control to the sail pitch.

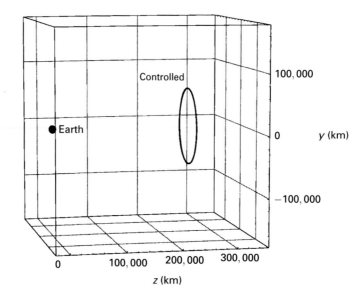

Fig. 5.19. Variable area control for an unstable type III orbit.

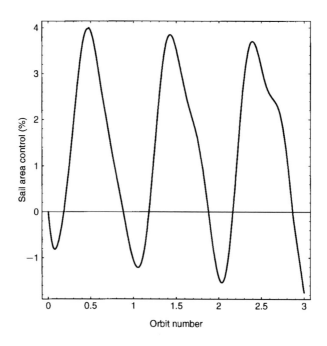

Fig. 5.20. Sail area trim $\delta\kappa$ for orbit control.

5.3.5 Patched orbits

It will now be shown that planet-centred non-Keplerian orbits may be patched together in a similar manner to Sun-centred non-Keplerian orbits. Firstly, it will be shown that planet-centred orbits may be patched to Keplerian orbits by orienting the solar sail edgewise to the Sun. Then, the conditions required for patching on-axis and off-axis orbits will be determined. It will be shown that such transfers are indeed possible, but that the range of initial and final orbits is somewhat limited. In addition to patching off-axis and on-axis orbits, it will also be shown that multiple off-axis orbits may be patched together. By an appropriate choice of orbit geometry, large families of interesting new periodic trajectories can be generated.

5.3.5.1   Patch to a Keplerian orbit

As with Sun-centred non-Keplerian orbits, the sail may be oriented edgewise to the Sun so that the solar sail is transferred onto a Keplerian orbit. This Keplerian orbit may be circular or elliptical depending on the geometry of the initial non-Keplerian orbit. Using the same analysis as in section 5.2.5.1, it is found that the semi-major axis of the Keplerian orbit is given by

$$a = \left[\frac{2}{\sqrt{\rho^2 + z^2}} - \frac{\omega^2 \rho^2}{\mu} \right]^{-1} \tag{5.106}$$

Similarly, since the patch point corresponds to the apocentre of the Keplerian orbit, viz.

$$a(1 + e) = \sqrt{\rho^2 + z^2} \tag{5.107}$$

the eccentricity of the resulting ellipse may also be obtained as

$$e = 1 - \frac{\omega^2 \rho^2}{\mu} \sqrt{\rho^2 + z^2} \tag{5.108}$$

As before, for a given non-Keplerian orbit, the size and shape of the Keplerian orbit can be obtained. In particular, the solar sail will transfer to a circular orbit if

$$\omega = \sqrt{\frac{\mu}{\rho^3}} \left[1 + \left(\frac{z}{\rho} \right)^2 \right]^{-1/4} \tag{5.109}$$

In addition to transfers between on-axis orbits and Keplerian orbits, transfers from off-axis orbits are also possible. However, in this case the non-Keplerian orbit angular velocity must be chosen as $\tilde{\omega}$, and so is no longer a free parameter. Other interesting manoeuvres include transfer from an off-axis orbit to a near polar elliptical orbit along the planetary terminator.

5.3.5.2   Patch to a non-Keplerian orbit

Now that patching to Keplerian orbits has been considered, a patch between an off-axis and an on-axis non-Keplerian orbit will be investigated, as shown in Fig. 5.21.

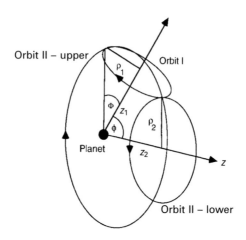

Fig. 5.21. Upper and lower patched non-Keplerian orbits.

The off-axis orbit is an optimal type III orbit with $\omega = \tilde{\omega}$, and is elevated at an angle ϕ relative to the Sun-line. The solar sail acceleration required for each orbit is given by Eqs (5.76b) and (5.73b) as

$$\kappa_1 = \frac{z_1}{r^3}\cos^{-2}\phi \tag{5.110a}$$

$$\kappa_2 = \frac{z_2}{r^3}\left\{1 + \left(\frac{\rho_2}{z_2}\right)^2\left[1 - \left(\frac{\omega_2}{\tilde{\omega}}\right)^2\right]^2\right\}^{3/2} \tag{5.110b}$$

where the sail pitch angle on orbit II is defined by Eq. (5.73a). Then, for energy continuity between the orbits it is required that

$$\rho_1\tilde{\omega} = \rho_2\omega_2 \tag{5.111}$$

The required solar sail accelerations may now be equated to obtain a function $\kappa_1 - \kappa_2 = 0$. This function can then be reduced to two variables by eliminating the radii and displacements of the two orbits using

$$\tan\Phi = \frac{\rho_1}{z_1} \tag{5.112a}$$

$$\tan(\phi - \Phi) = \frac{\rho_2}{z_2} \tag{5.112b}$$

where Φ is defined on the cone formed by orbit I and $\phi - \Phi$ is defined on the cone formed by orbit II (lower). The resulting condition for a patched orbit with equal energy and equal solar sail acceleration then becomes $F(\phi, \Phi) = 0$, where

$$F(\phi, \Phi) = \left\{1 + \tan^2(\phi - \Phi)\left[1 - \left(\frac{\sin\Phi}{\sin(\phi - \Phi)}\right)^2\right]^2\right\}^{3/2} - \frac{\cos\Phi}{\cos(\phi - \Phi)} \tag{5.113}$$

The numerical solution to this equation is shown in Fig. 5.22 as the curve T_1. This

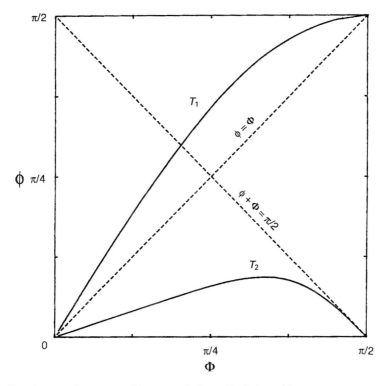

Fig. 5.22. Requirements for upper and lower patched non-Keplerian orbits.

curve represents the constraints imposed upon transfers between orbit I and orbit II (lower). It can be seen that $\phi \geq \Phi$, as required for a patched orbit to exist. As the elevation ϕ of orbit I increases, the angular size Φ of orbit I also increases so that the two orbits still intersect. It should also be noted that the system is rotationally symmetric about the z-axis so that the solar sail may transfer back from orbit II (lower) to orbit I at any azimuthal position.

In addition to transfers from the lower arc of orbit I, a transfer can be made from the upper arc, as shown in Fig. 5.21. The requirements for such transfers are obtained from $F(\phi, -\Phi) = 0$, shown in Fig. 5.22 as curve T_2. This curve represents the constraints imposed upon transfers between orbit I and orbit II (upper). It can be seen that a much smaller range of orbit geometries are available using this form of transfer. Since the transfer is from the upper arc, smaller values of elevation angle ϕ must be used to keep orbit II on the planetary night-side. Lastly, along with patching off-axis and on-axis orbits, off-axis orbits may also be patched together. For continuity of energy and solar sail acceleration, each orbit must have the same orbit radius and displacement distance, but will have a different azimuthal position about to the Sun-line. Therefore, the solar sail can orbit on a large patched orbit composed of small off-axis orbits spaced uniformly around the Sun-line.

5.4   SOLAR SAILS IN RESTRICTED THREE-BODY SYSTEMS

5.4.1   Introduction

The classical circular restricted three-body problem has five well-known equilibrium solutions L_j ($j = 1$–5) where an infinitesimal mass will remain at rest with respect to two primary masses in orbit about their common centre-of-mass. These are the well-known Lagrange points. The classical restricted problem has also been extended to include radiation pressure exerted on the infinitesimal mass from either or both of the primary masses. This formulation generates four new additional equilibria with interesting stability characteristics. However, for a point mass the radiation pressure force vector is constrained to lie along the line connecting the source of radiation pressure and the mass. For the planet–Sun–sail three-body problem however, the solar sail attitude may be freely oriented so that the direction of the solar radiation pressure force vector is not constrained in this way. Furthermore, the magnitude of the solar radiation pressure force may be chosen through the solar sail lightness number. Therefore, since these parameters can be selected, it is clear that rich new possibilities will arise for artificial equilibrium points. In fact, it will be demonstrated that there is a continuum of new equilibrium solutions parameterised by the solar sail attitude and lightness number. The stability of these new equilibrium solutions will also be investigated and their instability established. Lastly, by linearising the three-body equations of motion, it will be shown that a displaced out-of-plane orbit is possible about the lunar L_2 Lagrange point. These new equilibrium solutions to the three-body problem have many useful applications, as will be discussed in Chapter 6.

5.4.2   The classical restricted three-body problem

The classical Lagrange equilibrium solutions to the restricted three-body problem L_j ($j = 1$–5) have been studied in great depth since their discovery in 1772. More recently, attention has been given to the modified photo-gravitational problem, where one or both of the primary masses is a source of radiation pressure. It has been shown that the five in-plane equilibrium solutions are modified by the existence of radiation pressure and that four new out-of-plane solutions L_j ($j = 6$–9) are generated with L_8 and L_9 only existing when both of the primary masses are luminous. Other studies have included the relativistic Poynting–Robertson effect in the equations of motion which has a bearing on the stability of the solutions.

The geometry of the circular restricted three-body problem is shown in Fig. 5.23, with the five classical Lagrange points. The problem is characterised by the mass ratio of the system defined as $\mu = m_2/(m_1 + m_2)$. The motion of m_1 and m_2 about their common centre-of-mass then defines a frame of reference (x, y, z) rotating with angular velocity $\boldsymbol{\omega}$. Dimensionless units are used so that the distance between the primary masses, the sum of the primary masses and the gravitational constant are all chosen to be unity. The classical equilibrium solutions are found at the points L_j ($j = 4$–5) which are located at vertices of the equilateral triangle defined by

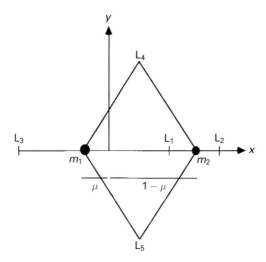

Fig. 5.23. Classical circular restricted three-body problem.

$r_1 = r_2 = 1$ and the collinear points L_j $(j = 1\text{–}3)$ which are obtained from the solution to a quintic polynomial. It may be shown that the equilateral points are linearly stable provided $\mu < 0.0385$ (Routh's value) and that the collinear points are always unstable.

With the addition of radiation pressure from m_1, the gravitational potential of m_1 is modified by the ratio of the radiation pressure force to the gravitational force β. If the third infinitesimal body is a point mass, this modification results in the classical Lagrange points moving to new positions as β increases. In particular, it can be shown that the triangular points $L_{4,5}$ are now defined by the constraints $r_1 = (1 - \beta)^{1/3}$, $r_2 = 1$. Therefore, as $\beta \to 1$ the triangular points will coalesce onto m_1. It is also found that L_1 and L_3 coalesce onto m_1 while L_2 moves onto m_2. Furthermore, two new equilibrium solutions $L_{6,7}$ appear out of the ecliptic plane, moving asymptotically towards the z axis with $z \to \infty$ as $\beta \to 1$.

5.4.3   Equilibrium solutions

An idealised, perfectly reflecting solar sail will now be considered in a frame of reference rotating with angular velocity ω with two point primary masses m_1 and m_2, as shown in Fig. 5.24. The angular velocity of the frame of reference is defined as $\omega = 1$ so that the frame co-rotates with the primary masses. The sail attitude is defined by a unit vector $\mathbf{n}$, fixed in the rotating frame of reference, and the ratio of the solar radiation pressure force to the solar gravitational force exerted on the solar sail is again defined by the solar sail lightness number β. Since the sail attitude is to be fixed in the rotating frame of reference the solar sail must rotate about the normal to the plane of the system once with respect to a fixed inertial frame in time $2\pi/\omega$. The units of the system will be chosen such that the gravitational constant, the

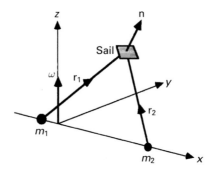

Fig. 5.24. Solar sail circular restricted three-body problem.

distance between the two primary masses, the sum of the primary masses, and so the angular velocity of co-rotation, are all taken to be unity. The vector equation of motion for a solar sail in this rotating frame of reference may be written in familiar form as

$$\frac{d^2\mathbf{r}}{dt^2} + 2\boldsymbol{\omega} \times \frac{d\mathbf{r}}{dt} + \boldsymbol{\omega} \times (\boldsymbol{\omega} \times \mathbf{r}) = \mathbf{a} - \nabla V \tag{5.114}$$

where the solar radiation pressure acceleration $\mathbf{a}$ and the three-body gravitational potential V are defined by

$$\mathbf{a} = \beta \frac{1-\mu}{r_1^2} (\hat{\mathbf{r}}_1 \cdot \mathbf{n})^2 \mathbf{n} \tag{5.115a}$$

$$V = -\left(\frac{1-\mu}{r_1} + \frac{\mu}{r_2} \right) \tag{5.115b}$$

where $\hat{\mathbf{r}}_1$ is directed along the Sun-line. Since the solar radiation pressure force can never be directed sunward, the sail attitude is constrained such that $\hat{\mathbf{r}}_1 \cdot \mathbf{n} \geq 0$. The solar sail position vectors are defined with respect to the rotating frame of reference as

$$\mathbf{r}_1 = (x + \mu, y, z) \tag{5.116a}$$

$$\mathbf{r}_2 = [x - (1-\mu), y, z] \tag{5.116b}$$

where $\mu = m_2(m_1 + m_2)$ is the mass ratio of the system. Since the centripetal term in Eq. (5.114) is conservative it may again be written as a scalar potential Φ such that

$$\Phi = -\tfrac{1}{2}|\boldsymbol{\omega} \times \mathbf{r}|^2 \tag{5.117a}$$

$$\nabla\Phi = \boldsymbol{\omega} \times (\boldsymbol{\omega} \times \mathbf{r}) \tag{5.117b}$$

A new modified potential, $U = V + \Phi$, can then be defined so that a reduced equation of motion is obtained, viz.

$$\frac{d^2\mathbf{r}}{dt^2} + 2\boldsymbol{\omega} \times \frac{d\mathbf{r}}{dt} + \nabla U = \mathbf{a} \tag{5.118}$$

In the rotating frame of reference equilibrium solutions are required so that the first two terms of Eq. (5.118) must vanish. The five classical equilibrium points are then given as the solutions to $\nabla U = 0$. However, for the solar sail three-body problem there exists an additional acceleration $\mathbf{a}$ which is a function of the solar sail lightness number β and attitude $\mathbf{n}$ so that new artificial equilibrium solutions may be generated.

Since the vector $\mathbf{a}$ is oriented in direction $\mathbf{n}$, then taking the vector product of $\mathbf{n}$ with Eq. (5.118) it follows that

$$\nabla U \times \mathbf{n} = 0 \Rightarrow \mathbf{n} = \varepsilon \nabla U \qquad (5.119)$$

where ε is an arbitrary scalar multiplier. Using the normalisation condition $|\mathbf{n}| = 1$, ε is identified as $|\nabla U|^{-1}$ so that the required sail attitude is again defined by

$$\mathbf{n} = \frac{\nabla U}{|\nabla U|} \qquad (5.120)$$

The sail attitude may then be expressed in terms of the sail cone and clock angles (α, δ), defined with respect to the co-ordinate triad $(\hat{\mathbf{r}}_1, \boldsymbol{\omega} \times \hat{\mathbf{r}}_1, \hat{\mathbf{r}}_1 \times (\boldsymbol{\omega} \times \hat{\mathbf{r}}_1))$ centred on the solar sail. As before, the cone angle α is defined as the angle between $\mathbf{n}$ and $\hat{\mathbf{r}}_1$ and the clock angle δ is defined as the angle of the projection of $\mathbf{n}$ in the plane normal to $\hat{\mathbf{r}}_1$ with respect to a reference direction $\hat{\mathbf{r}}_1 \times (\boldsymbol{\omega} \times \hat{\mathbf{r}}_1)$. Therefore, taking vector and scalar products of Eq. (5.120) with $\hat{\mathbf{r}}_1$, these angles may be written as

$$\tan \alpha = \frac{|\hat{\mathbf{r}}_1 \times \nabla U|}{\hat{\mathbf{r}}_1 \cdot \nabla U} \qquad (5.121\mathrm{a})$$

$$\tan \delta = \frac{|\hat{\mathbf{r}}_1 \times (\boldsymbol{\omega} \times \hat{\mathbf{r}}_1) \times (\hat{\mathbf{r}}_1 \times \nabla U(\mathbf{r}))|}{\hat{\mathbf{r}}_1 \times (\boldsymbol{\omega} \times \hat{\mathbf{r}}_1) \cdot (\hat{\mathbf{r}}_1 \times \nabla U(\mathbf{r}))} \qquad (5.121\mathrm{b})$$

The solar sail lightness number required may also be obtained by taking a scalar product of Eq. (5.118) with $\mathbf{n}$. Again requiring an equilibrium solution, it is found that

$$\beta = \frac{r_1^2}{1 - \mu} \frac{\nabla U \cdot \mathbf{n}}{(\hat{\mathbf{r}}_1 \cdot \mathbf{n})^2} \qquad (5.122)$$

Functions have now been obtained for the solar sail attitude and lightness number required for equilibrium solutions in the rotating frame of reference. Since the solar sail lightness number and attitude can in principle be chosen at will, the set of five classical equilibrium solutions will be replaced by an infinite set of artificially generated equilibrium solutions. The classical solutions then correspond to the subset $\beta = 0$. This infinite set of solutions is parameterised into surfaces by the solar sail lightness number. A particular equilibrium solution on a given surface is then defined by the sail cone and clock attitude angles.

5.4.4   Regions of existence of equilibrium solutions

Now that the existence of new equilibrium solutions has been established, the regions in which these solutions may exist will be investigated. These regions are defined by

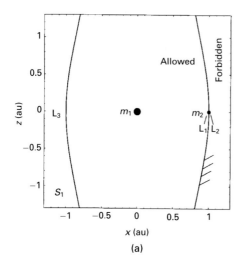

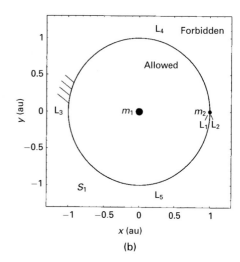

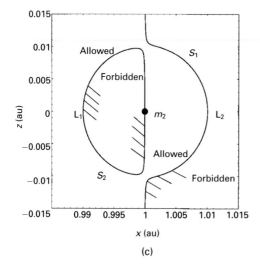

Fig. 5.25. Region of existence of equilibrium solutions: (a) x–z plane; (b) x–y plane; (c) x–z plane near m_2.

the constraint

$$\hat{\mathbf{r}}_1 \cdot \nabla U \geq 0 \qquad (5.123)$$

with the boundary surface defined by an equality in Eq. (5.123). This constraint may be understood physically since the solar radiation pressure acceleration vector $\mathbf{a}$, and so the sail attitude vector $\mathbf{n}$, can never be directed sunward. In scalar form the three-body potential U may be written as

$$U = -\left[\tfrac{1}{2}(x^2 + y^2) + \frac{1 - \mu}{r_1} + \frac{\mu}{r_2} \right] \qquad (5.124)$$

Therefore, evaluating the gradient of the potential U in Eq. (5.123), a function $S = 0$ is obtained, viz.

$$S = x(x + \mu) + y^2 - \frac{1 - \mu}{r_1} - \mu \frac{\mathbf{r}_1 \cdot \mathbf{r}_2}{r_2^3} \qquad (5.125)$$

which defines the required boundary. The function S has two topologically disconnected surfaces S_1 and S_2 which define the boundary to the region of existence of equilibrium solutions, as shown in Fig. 5.25. The outer surface S_1 possesses cylindrical topology and excludes solutions along the x-axis from $-\infty < x < L_3$ and $L_2 < x < +\infty$ while the inner surface S_2 excludes solutions along $L_1 < x < 1 - \mu$. The region of existence of equilibrium solutions lies between these two surfaces in the region exterior to S_2 but interior to S_1 defined by $S_1 \cap S_2'$. All of the five classical equilibrium solutions lie on $S_1 \cup S_2$ since they are the solutions to $\nabla U = 0$ in Eq. (5.123). In general, the surfaces of constant solar sail lightness number will approach the boundary surface asymptotically with $\beta \to \infty$ in the limit as $\hat{\mathbf{r}}_1 \cdot \nabla U \to 0$. This limiting behaviour is clear from Eq. (5.122).

5.4.5 Equilibrium solutions in the Earth–Sun system

Surfaces of constant solar sail lightness number may now be generated from Eq. (5.122) for the Earth–Sun system which has a mass ratio of $\mu = 3.036 \times 10^{-6}$. For ease of illustration, sections of these surfaces through the x–y and x–z planes will be shown. Then, only a single sail attitude angle will be required to describe completely the sail orientation required for an equilibrium solution. In general, however, the cone and clock angles are both required to describe the sail attitude for an equilibrium solution at some arbitrary location.

Sections of the surfaces of equal solar sail lightness number generated by Eq. (5.122) are shown in Fig. 5.26. The sections of these surfaces define families of curves of equal solar sail lightness number representing subsets of the continuum of new artificial equilibrium solutions. The required direction of the sail normal is also shown. Far from the Earth it can be seen that topologically the surfaces form a family of nested tori with the inner radius of the torus vanishing as $\beta \to 1$. For $\beta > 1$ the surfaces form a family of nested cylinders. In the x–y plane the curves are nearly circular, representing the solar sail lightness number required for a circular one-year orbit. Similarly, in the x–z plane it can be seen that the solutions are essentially one-year Sun-centred non-Keplerian orbits. Along the z axis of the system there are equilibrium solutions above the poles of the Sun with $\beta = 1$. Lastly, a detailed plot of the section of the surfaces near the Earth is shown in Fig. 5.26(c) – see also Table 5.3. It can be seen that their are surfaces around the classical L_1 and L_2 points which expand with increased solar sail lightness number, but always contain a classical Lagrange point as this corresponds to the solution $\beta \neq 0$ and $\hat{\mathbf{r}}_1 \cdot \mathbf{n} = 0$. These interesting solutions close to the Earth have near-term applications since only low solar sail lightness numbers are required. Such applications will be discussed in Chapter 6.

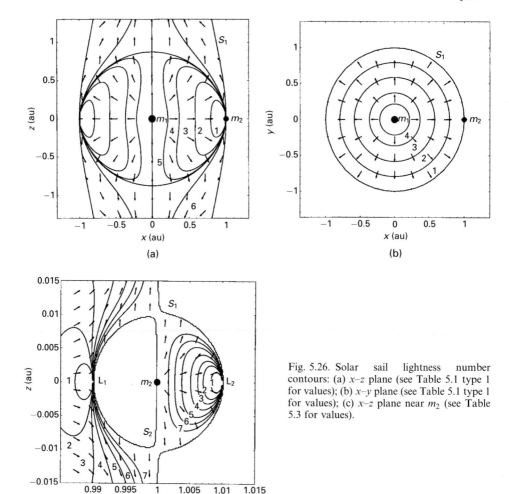

Fig. 5.26. Solar sail lightness number contours: (a) x–z plane (see Table 5.1 type 1 for values); (b) x–y plane (see Table 5.1 type 1 for values); (c) x–z plane near m_2 (see Table 5.3 for values).

Table 5.3. Solar sail lightness number, characteristic acceleration and loading for Lagrange point equilibrium solutions in the vicinity of the Earth

Contour	1	2	3	4	5	6	7
β	0.02	0.04	0.06	0.1	0.2	0.4	1.0
a_0 (mm s^{-2})	0.12	0.24	0.36	0.59	1.19	2.37	5.93
σ (g m^{-2})	76.50	38.25	25.50	15.30	7.65	3.83	1.53

5.4.6   Stability of equilibrium solutions

Now that the existence of the artificial equilibrium solutions has been established it is again necessary to examine their stability. It will be assumed that the solar sail is in equilibrium at some arbitrary position $\mathbf{r}_0$. Then, the equation of motion in a local neighbourhood of $\mathbf{r}_0$ is obtained in the usual manner by applying a perturbation $\boldsymbol{\delta}$, such that $\mathbf{r}_0 \to \mathbf{r}_0 + \boldsymbol{\delta}$. Since $\mathbf{r}_0$ is an equilibrium solution, a variational equation is obtained as

$$\frac{d^2\boldsymbol{\delta}}{dt^2} + 2\boldsymbol{\omega} \times \frac{d\boldsymbol{\delta}}{dt} + \nabla U(\mathbf{r}_0 + \boldsymbol{\delta}) - \mathbf{a}(\mathbf{r}_0 + \boldsymbol{\delta}, \mathbf{n}) = 0 \qquad (5.126)$$

The potential gradient and radiation pressure acceleration may then be expanded in a Taylor series about the nominal equilibrium point $\mathbf{r}_0$ to first order as

$$\nabla U(\mathbf{r}_0 + \boldsymbol{\delta}) = \nabla U(\mathbf{r}_0) + \left[\frac{\partial}{\partial \mathbf{r}} \nabla U(\mathbf{r}) \right]_0 \boldsymbol{\delta} \qquad (5.127a)$$

$$\mathbf{a}(\mathbf{r}_0 + \boldsymbol{\delta}) = \mathbf{a}(\mathbf{r}_0) + \left[\frac{\partial}{\partial \mathbf{r}} \mathbf{a}(\mathbf{r}, \mathbf{n}) \right]_0 \boldsymbol{\delta} \qquad (5.127b)$$

Then, since $\nabla U(\mathbf{r}_0) = \mathbf{a}(\mathbf{r}_0)$ at the nominal equilibrium point, a linear variational system with constant coefficients is obtained, viz.

$$\frac{d^2\boldsymbol{\delta}}{dt^2} + \mathbf{S}\frac{d\boldsymbol{\delta}}{dt} + (\mathbf{M} - \mathbf{N})\boldsymbol{\delta} = 0 \qquad (5.128)$$

where $\mathbf{M}$ and $\mathbf{N}$ are the gravity and radiation gradient matrices and $\mathbf{S}$ is the skew symmetric gyroscopic matrix. The stability of the equilibrium solutions may be investigated in the usual manner by examining the eigenvalues of the resulting characteristic polynomial. The eigenvalues may be obtained by substituting an exponential solution of the form

$$\boldsymbol{\delta} = \boldsymbol{\delta}_0 \exp(st) \qquad (5.129)$$

where $s = \sigma + i\rho$ is some complex number with real constants σ and ρ, while $i = \sqrt{-1}$. Substituting this solution into Eq. (5.128) yields a matrix equation of the form

$$(s^2\mathbf{I} + s\mathbf{S} + \mathbf{T})\boldsymbol{\delta}_0 = 0 \qquad (5.130)$$

where $\mathbf{T} = \mathbf{M} - \mathbf{N}$. For non-trivial solutions a vanishing determinant is required, which then yields the characteristic polynomial of the system $P(s) = 0$, viz.

$$P(s) = \sum_{j=0}^{6} a_{6-j} s^j \qquad (5.131)$$

The fundamental theorem of algebra demands that $P(s)$ has roots $\sigma_j + i\rho_j$ $(j = 1\text{–}6)$. For asymptotic stability it is required that all of the resulting eigenvalues are in the left-hand complex plane so that $\sigma_j < 0$ $(j = 1\text{–}6)$. However, to ensure that the solutions are bound, the weaker condition (i.e. that all the roots of $P(s)$ are at least purely imaginary) is required. This condition then constrains the motion to a local neighbourhood of the nominal equilibrium solution. Again, linear analysis provides the necessary conditions for stability and the sufficient conditions for instability.

Evaluating the characteristic polynomial it is found that the coefficients of the polynomial $P(s)$ are given by

$$a_0 = 1 \tag{5.132a}$$

$$a_1 = 0 \tag{5.132b}$$

$$a_2 = T_{11} + T_{22} + T_{33} + 4 \tag{5.132c}$$

$$a_3 = 2(T_{21} - T_{12}) \tag{5.132d}$$

$$a_4 = T_{11}T_{22} + T_{11}T_{33} + T_{22}T_{33} - T_{23}T_{32} - T_{13}T_{31} - T_{12}T_{21} + 4T_{33} \tag{5.132e}$$

$$a_5 = 2T_{33}(T_{21} - T_{12}) + 2(T_{32}T_{13} - T_{23}T_{31}) \tag{5.132f}$$

$$a_6 = T_{11}(T_{22}T_{33} - T_{23}T_{32}) - T_{12}(T_{33}T_{21} - T_{23}T_{31})$$
$$\quad - T_{13}(T_{22}T_{31} - T_{21}T_{32}) \tag{5.132g}$$

Since $a_1 = 0$, an application of the Routh–Hurwitz criterion implies that at least one eigenvalue will not lie in the left-hand complex plane. Therefore, the system does not naturally possess asymptotic stability. Given this fact, the condition for linear stability with purely imaginary eigenvalues will now be established. Substituting for purely imaginary eigenvalues $s = i\sigma$, the characteristic polynomial becomes

$$P(i\sigma) = -\sigma^6 + a_2\sigma^4 - ia_3\sigma^3 - a_4\sigma^2 + ia_5\sigma + a_6 \tag{5.133}$$

For the condition $P(s) = 0$ to hold it is required that both the real and purely imaginary parts of the polynomial are identically zero, viz.

$$-\sigma^6 + a_2\sigma^4 - a_4\sigma^2 + a_6 = 0 \tag{5.134a}$$

$$i\sigma(a_5 - \sigma^2 a_3) = 0 \tag{5.134b}$$

Six consistent solutions of Eqs (5.134) are now required with $a_j^2 > 0$ ($j = 1$–6). From Eq. (5.134b) it can be seen that $\sigma_1 = 0$ and $\sigma_{2,3} = \pm\sqrt{a_5/a_3}$. However, the solution $\sigma_1 = 0$ is obviously inconsistent with Eq. (5.134a). It is also found that the remaining solutions σ_2 and σ_3 are not generally consistent with Eq. (5.134a). It is therefore concluded that the equilibrium solutions are in general unstable. Consistency can, however, be achieved if both $a_5 = 0$ and $a_3 = 0$. Taken together these conditions require that the solar sail is oriented along the Sun-line with $\mathbf{n} = \hat{\mathbf{r}}$. This then leads to a modified version of the classical restricted three-body problem where the equilibrium points L_4 and L_5 are linearly stable for small mass ratios. Although the equilibrium solutions are in general unstable, it is also found that they are controllable using either feedback to the sail attitude, or trims of the sail area.

5.4.7   Lunar Lagrange point orbits

In this section the dynamics of a solar sail in the neighbourhood of a co-linear lunar Lagrange point will be investigated. In a frame of reference rotating with the Earth–Moon system, the Sun-line will appear to rotate once per synodic lunar month at an angular rate w_s. However, if the solar sail is always oriented along the Sun-line, a slow, forced periodic trajectory about a co-linear Lagrange point is possible. Furthermore, if the sail is pitched such that a component of the solar radiation pressure force is directed out of the orbit plane, the forced trajectory is displaced above the Earth–Moon plane. It will be assumed that the orbit of the Moon about the Earth is circular. In addition, the inclination of the lunar orbit relative to the ecliptic plane will be ignored in this first-order analysis.

The equation of motion of a solar sail in the neighbourhood of a co-linear Lagrange point can be obtained from the linearisation of section 5.4.6 as

$$\frac{d^2\boldsymbol{\delta}}{dt^2} + \mathbf{S}\frac{d\boldsymbol{\delta}}{dt} + \mathbf{M}\boldsymbol{\delta} = \kappa(\boldsymbol{l}\cdot\mathbf{n})^2\mathbf{n} \tag{5.135}$$

where $\boldsymbol{\delta}$ is the position vector of the solar sail relative to the Lagrange point and the matrix $\mathbf{M}$ is evaluated at the Lagrange point. The term on the right of Eq. (5.135) is the solar radiation pressure experienced by a solar sail in the vicinity of the Lagrange point while κ is the acceleration experienced by the solar sail when the sail normal $\mathbf{n}$ is directed along the Sun-line $\boldsymbol{l}$. In the rotating frame of reference the Sun-line will rotate once per synodic lunar month so that

$$\boldsymbol{l} = [\cos(w_s t), -\sin(w_s t), 0] \tag{5.136}$$

Then, writing Eq. (5.135) in component form it is found that

$$\frac{d^2\xi}{dt^2} - 2\frac{d\eta}{dt} + U^0_{xx}\xi = \kappa\cos(w_s t)\cos^3\alpha \tag{5.137a}$$

$$\frac{d^2\eta}{dt^2} + 2\frac{d\xi}{dt} + U^0_{yy}\eta = -\kappa\sin(w_s t)\cos^3\alpha \tag{5.137b}$$

$$\frac{d^2\zeta}{dt^2} + U^0_{zz}\zeta = \kappa\cos^2\alpha\sin\alpha \tag{5.137c}$$

where ξ, η and ζ represent small displacements along the x, y and z axes, as shown in Fig. 5.27, and U^0_{ij} are the (i, j) partial derivatives of the three-body potential evaluated at the Lagrange point. The sail attitude is now fixed such that the sail normal points along the Sun-line, but is pitched at an angle α to the plane of the plane of the Earth–Moon system. It is found that the general solution to Eqs (5.137) have divergent modes with unbound motion. However, these modes can be suppressed through an appropriate choice of initial conditions. A particular periodic in-plane solution of the following form will now be required, viz.

$$\begin{bmatrix} \xi(t) \\ \eta(t) \end{bmatrix} = \begin{bmatrix} \xi_0\cos(w_s t) \\ \eta_0\sin(w_s t) \end{bmatrix} \tag{5.138}$$

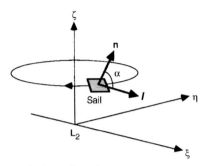

Fig. 5.27. Displaced orbit at the lunar L_2 point.

This periodic solution is now substituted in Eqs (5.137) to yield a relationship between the amplitude of motion along the x–y axes as

$$\frac{\xi_0}{\eta_0} = -\frac{\omega_s^2 + 2\omega_s - U_{yy}^0}{\omega_s^2 + 2\omega_s - U_{xx}^0} \tag{5.139}$$

Therefore, since $U_{xx}^0 \neq U_{yy}^0$ the in-plane trajectory will be an ellipse centred on the Lagrange point. The uncoupled out-of-plane motion defined by Eq. (5.137c) may now be obtained. Choosing the initial conditions such that $(d\zeta/dt)_0 = 0$, the out-of-plane solution can be conveniently expressed as

$$\zeta(t) = \frac{\kappa}{U_{zz}^0} \cos^2 \alpha \sin \alpha + \left[\zeta_0 - \frac{\kappa}{U_{zz}^0} \cos^2 \alpha \sin \alpha \right] \cos(\omega_s t) \tag{5.140}$$

It can be seen that once the solar sail is pitched, the resulting out-of-plane motion is of the form of periodic oscillations displaced to an out-of-plane distance $\kappa \cos^2 \alpha \sin \alpha / U_{zz}^0$. However, by choosing the initial out-of-plane distance ζ_0 as $\kappa \cos^2 \alpha \sin \alpha / U_{zz}^0$ the solar sail remains fixed at this distance. The z component of the solar radiation pressure acceleration, and so the out-of-plane distance, may then be maximised by an optimal choice of sail pitch angle of 35.26°. Using this optimal pitch angle the solar sail will then execute a displaced out-of-plane elliptical orbit centred at the Lagrange point. Potential applications for this orbit include lunar far-side communications using a displaced orbit at the lunar L_2 point. For example, at an out-of-plane distance of 3.5×10^3 km above the L_2 point both the lunar far-side and the equatorial regions of the Earth are visible, allowing continuous lunar far-side communications. It is found that the solar sail acceleration required for this orbit is only of order $0.2 \, \text{mm s}^{-2}$. The in-plane trajectory is a narrow ellipse normal to the Earth–Moon line with a period of 29.53 days (synodic lunar month). However, since the initial conditions are chosen to suppress the natural divergent modes of the orbit, active control is required.

5.5   EFFECT OF A REAL SOLAR SAIL MODEL

In the previous sections an ideal solar sail has been considered for ease of illustration. However, a real solar sail model which includes absorption will have the effect of reducing the range of non-Keplerian orbits available. In particular, the fact that the solar sail force vector is no longer normal to the sail surface leads, in some instances, to regions of space in which non-Keplerian orbits can no longer be generated. To capture these effects the solar sail model developed in section 2.6.1 will now be used. To allow closed-form solutions the solar sail will be assumed to have perfect specular reflectivity although the sail will have an overall reflectivity $\tilde{r}$ less than unity. Then, using Eqs (2.57) the solar sail acceleration in direction $\mathbf{m}$ may be written as the sum of components normal $\mathbf{n}$ and transverse $\mathbf{t}$ to the sail surface, viz.

$$a\mathbf{m} = \tfrac{1}{2}\beta\tfrac{\mu}{r^2}(1+\tilde{r})(\hat{\mathbf{r}}\cdot\mathbf{n})^2\mathbf{n} + \tfrac{1}{2}\beta\tfrac{\mu}{r^2}(1-\tilde{r})(\hat{\mathbf{r}}\cdot\mathbf{n})(\hat{\mathbf{r}}\cdot\mathbf{t})\mathbf{t} \qquad (5.141)$$

It can be seen that the main effect of the non-perfect reflectivity of the sail is to reduce the acceleration magnitude and to introduce an offset in the direction of the solar radiation pressure acceleration. The acceleration now acts in direction $\mathbf{m}$ rather than normal to the sail surface in direction $\mathbf{n}$. This offset is the centre-line angle ϕ, as discussed in section 2.6.1.

The analysis presented in the previous sections can be repeated using the solar sail acceleration vector defined by Eq. (5.141). Again, a rotating frame of reference is used with equilibrium solutions in the rotating frame corresponding to non-Keplerian orbits when viewed from an inertial frame of reference. The equation of motion may again be written as

$$\frac{d^2\mathbf{r}}{dt^2} + 2\boldsymbol{\omega} \times \frac{d\mathbf{r}}{dt} + \nabla U = a\mathbf{m} \qquad (5.142)$$

where the new solar radiation pressure acceleration model has been used. For an equilibrium solution the first two terms of Eq. (5.142) will vanish. Therefore, the solar sail attitude must be chosen such that the solar radiation pressure acceleration vector, acting in direction $\mathbf{m}$, balances the sum of the gravitational and centripetal accelerations. The required force direction is then obtained as

$$\mathbf{m} = \frac{\nabla U}{|\nabla U|} \qquad (5.143)$$

The unit vector $\mathbf{m}$ can now be defined by the cone angle θ between the radial direction $\hat{\mathbf{r}}$ and $\mathbf{m}$. Taking vector and scalar products of Eq. (5.143) with $\hat{\mathbf{r}}$ it is found that

$$\tan\theta = \frac{|\hat{\mathbf{r}} \times \nabla U|}{\hat{\mathbf{r}}\cdot\nabla U} \qquad (5.144)$$

In addition, using Eq. (2.59) the centre-line angle can be obtained from the ratio of the transverse and normal accelerations as

$$\tan\phi = \frac{1-\tilde{r}}{1+\tilde{r}}\tan\alpha \qquad (5.145)$$

Noting that $\mathbf{n}\cdot\mathbf{t} = 0$ and taking a scalar product of Eq. (5.142) with the unit vector $\mathbf{n}$

also gives the required solar sail lightness number as

$$\tilde{\beta} = \frac{2r^2}{\mu} \frac{\nabla U \cdot \mathbf{n}}{(1+\tilde{r})(\hat{\mathbf{r}} \cdot \mathbf{n})^2} \tag{5.146}$$

These expressions are quite general and can be applied to any potential function U. For illustration, the family of type I Sun-centred non-Keplerian orbits, as discussed in section 5.2.2.1, will now be considered. Substituting for the potential gradient in Eq. (5.144) yields the required sail cone angle as a function of the orbit radius, displacement distance and orbital angular velocity as

$$\tan \theta = \frac{(z/\rho)(\omega/\tilde{\omega})^2}{(z/\rho)^2 + [1 - (\omega/\tilde{\omega})^2]} \tag{5.147}$$

Then, after some reduction, Eq. (5.145) may be written as

$$\tan \phi = \frac{\tilde{r}}{(1+\tilde{r})\tan \theta} \left[1 - \left(1 - \frac{1-\tilde{r}^2}{\tilde{r}^2} \tan^2 \theta \right)^{1/2} \right] \tag{5.148}$$

so that the centre-line angle can be obtained immediately from the cone angle. Lastly, using Eq. (5.146) it is found that the required solar sail lightness number $\tilde{\beta}$ may be obtained in terms of the lightness number for a perfect solar sail β as

$$\tilde{\beta} = \frac{2}{(1+\tilde{r})} \frac{\sqrt{1 + \tan^2 \phi}}{(1 - \tan \theta \tan \phi)^2} \beta \tag{5.149}$$

where β is defined by Eq. (5.10b). Therefore, the required solar sail pitch angle and lightness number can now be obtained.

The effect of a non-perfect solar sail is shown in Fig. 5.28 for a one-year type I orbit with a sail reflectivity $\tilde{r}$ of 0.9. It can be seen that the topology of the surfaces of constant solar sail lightness number is unchanged by the non-perfect solar sail model. However, the main effect of the non-perfect sail is to reduce the out-of-plane displacement distance which may be achieved for a given solar sail lightness number. In addition, the transition from surfaces of nested tori to surfaces of nested cylinders occurs at a light number of $1/(1+\tilde{r})$, rather than unity for a perfect solar sail. It is also found that the family of type II orbits have regions close to the ecliptic plane where displaced orbits cannot be generated. This exclusion region only appears when sail absorption is considered. Similar exclusion regions are also found for the families of planet-centred non-Keplerian orbits.

5.6   SUMMARY

In this chapter it has been shown that solar sails may be used to establish large new families of displaced non-Keplerian orbits. These families are found to exist for either Sun-centred or planet-centred orbits. By choosing the orbit period to be fixed, or to be some function of the orbit radius and displacement distance, large families of orbits may be generated. Using a linear stability and control analysis, unstable

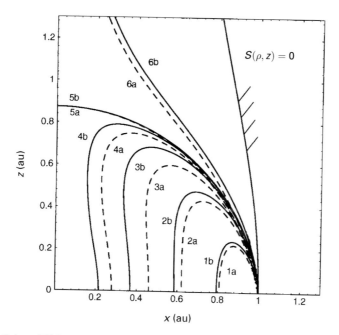

Fig. 5.28. Solar sail lightness number contours for a non-perfect sail: ([1a] 0.5, [2a] 0.8, [3a] 0.95, [4a] 1.03, [5a] 1.05, [6a] 1.3) and a perfect sail ([1b] 0.5, [2b] 0.8, [3b] 0.95, [4b] 0.99, [5b] 1.0, [6b] 1.3).

families of orbits have been identified and control schemes developed for these unstable families. In addition, it has been shown that new families of orbits may be generated by patching non-Keplerian orbits to Keplerian orbits, or to other non-Keplerian orbits. As an extension to the families of displaced non-Keplerian orbits, new equilibrium solutions for solar sails in three-body systems have been presented. Near the classical Lagrange points it has been found that these new equilibrium solutions require only relatively modest solar sail lightness numbers owing to the small local gravitational acceleration which the solar sail must balance in these regions. Although this chapter has only discussed the dynamics, stability and control of the families of non-Keplerian orbits, mission applications will be considered in Chapter 6.

5.7 FURTHER READING

Sun-centred non-Keplerian orbits

McInnes, C.R. & Simmons, J.F.L., 'Halo Orbits for Solar Sails – Dynamics and Applications', *European Space Agency Journal*, **13**, 3, 229–234, 1989.

McInnes, C.R., 'Advanced Trajectories for Solar Sail Spacecraft', PhD Thesis, Department of Physics and Astronomy, University of Glasgow, October 1991.

McInnes, C.R. & Simmons, J.F.L., 'Halo Orbits for Solar Sails I – Heliocentric Case', *Journal of Spacecraft and Rockets*, **29**, 4, 466–471, 1992.
Molostov, A.A. & Shvartsburg, A.A., 'Heliocentric Halos for a Solar Sail with Absorption', *Soviet Physics Doklady*, **37**, 3, 149-152, 1992.
Molostov, A.A. & Shvartsburg, A.A., 'Heliocentric Synchronous Halos for a Solar Sail with Absorption', *Soviet Physics Doklady*, **37**, 4, 195–197, 1992.
Mashkevich, S.V. & Shvartsburg, A.A., ' "Best" Solar Sail for Heliocentric Halos', *Soviet Physics Doklady*, **37**, 6, 290–293, 1992.

Planet-centred non-Keplerian orbits

McInnes, C.R. & Simmons, J.F.L., 'Halo Orbits for Solar Sails II – Geocentric Case', *Journal of Spacecraft and Rockets*, **29**, 4, 472–479, 1992.
Shvartsburg, A.A., 'Geocentric Halos for a Solar Sail with Absorption', *Soviet Physics Doklady*, **38**, 2, 85–88, 1993.
Glotova, M.Y. & Shvartsburg, A.A., 'Geocentric Synchronous Halos for a Solar Sail', *Soviet Physics Doklady*, **38**, 12, 449–501, 1993.

Artificial Lagrange points

McInnes, C.R., 'Solar Sail Halo Trajectories: Dynamics and Applications', IAF-91-334, *42nd International Astronautical Congress*, Montreal, October, 1991.
McInnes, C.R., 'Solar Sail Trajectories at the Lunar L_2 Lagrange Point', *Journal of Spacecraft and Rockets*, **30**, 6, 782–784, 1993.
McInnes, C.R., McDonald, A.J.C., Simmons, J.F.L. & MacDonald, E.W., 'Solar Sail Parking in Restricted Three-Body Systems', *Journal of Guidance, Control and Dynamics*, **17**, 2, 399–406, 1994.

6

Mission application case studies

6.1 INTRODUCTION

Conventional spacecraft have long been utilised for a diverse range of Earth observation, planetary science and space physics mission applications. While these missions have revolutionised our knowledge and understanding of the solar system, they have inevitably been limited by the use of chemical propulsion, both alone and in combination with gravity assist manoeuvres. Most noticeably, almost all interplanetary missions to-date have been constrained to the ecliptic plane, in part due to the energy requirements for injecting spacecraft into out-of-ecliptic trajectories. Likewise, to-date there has not been a comet rendezvous mission, a planetary sample return mission or a dedicated fast mission to the outer solar system beyond 100 au. All of these missions are either greatly enhanced or indeed enabled by solar sail propulsion. While attitudes to new propulsion technologies are now changing, as evidenced by flight tests of solar electric propulsion, solar sailing has yet to catch up with its long-term rival. As will be seen, however, some mission applications are absolutely unique to solar sailing and so represent a trump card for proponents of solar sailing to use to advance their views. Therefore, if missions with unrivalled scientific or operational value in which solar sailing truly enables the mission can be identified, then solar sailing will at last be demonstrated in orbit.

In this chapter a selection of recent concepts for solar sail missions will be discussed in some detail. These missions represent current thinking on the range of applications which are either enabled or greatly enhanced by solar sail technology. The applications described range from near-term Lagrange point missions using moderate performance solar sails to future outer solar systems missions requiring advanced, high-performance solar sails. What is evident from these descriptions is the diverse range of novel missions now being considered. In addition to viewing solar sailing as a form of high specific impulse propulsion for interplanetary transfer, some missions studies are now utilising the almost unlimited Δv capacity of solar sails to deliver payloads to exotic non-Keplerian orbits.

One such application which satisfies the earlier mission-enabling criteria and utilises a unique non-Keplerian orbit is the Geostorm mission. The mission requires only a moderate performance solar sail to orbit at an artificial Lagrange point sunward of the L_1 point, while remaining close to the Sun–Earth line. At this location the solar sail payload can detect passing coronal mass ejections, which can induce geomagnetic storms at Earth, before they would reach conventional spacecraft at the natural L_1 point. This particular mission brings together solutions to a number of difficulties which solar sailing has faced in the past. Firstly, the mission orbit is quite unique to solar sailing. While solar-electric propulsion can in principle be used to achieve a similar orbit, the orbit can only be maintained until the spacecraft propellant load has been depleted. In addition, the mission application is of high value to an influential group of end users. Enhanced warning of geomagnetic activity will allow terrestrial power distribution networks and both military and civil satellite communication links to be protected. Lastly, the risk of using solar sail technology will be minimised by deploying the sail only when a ballistic transfer to an L_1 halo orbit has been completed. If the sail fails to deploy, the mission can continue at the natural L_1 point so that the mission is not lost, merely degraded. All of these aspects of the mission lead to a quite compelling concept.

In addition to the Geostorm mission other recent concepts will be discussed: for example, a Mercury orbiter, small body sample return and fast missions to the outer solar system. It will be seen that, in general, the optimum use of solar sailing is for high-energy, relatively long-duration missions which are either impossible or extremely difficult to achieve using other propulsion options. It will also be seen that solar sail performance is greatly enhanced in the inner solar system where the solar radiation pressure force experienced by the solar sail is greater. Conversely, solar sailing is not a particularly competitive propulsion option for low-energy missions such as lunar or Mars orbiters. The effective Δv for a transfer to Mars is not quite high enough to offset significantly the sail mass compared to the mass of a chemical propulsion system. While solar sailing shows benefits for such a mission, it is not as compelling as other mission concepts, unless a sample return is required. Similarly, solar sails are not suited to extended operation in Earth orbit. Atmospheric drag limits the minimum altitude at which the sail can be deployed (typically above 600–900 km, depending on solar activity) and the high local gravitational acceleration leads to long spiral times. For these reasons all of the missions described in this chapter rely on a chemical stage to deliver the stowed solar sail to an Earth escape trajectory prior to deployment. The Earth escape capacity of current expendable launch vehicles is therefore a key input to the design of solar sail mission concepts.

In keeping with current requirements for low-cost missions, only small expendable launch vehicles will be considered. For example, the Orbital Sciences Corporation air-launched Pegasus vehicle can deliver just over 100 kg to an Earth escape trajectory. Owing to advances in micro-spacecraft technologies, it will be seen that a high-energy interplanetary solar sail mission is now feasible from the small, low-cost Pegasus vehicle. The more capable Taurus launch vehicle can deliver payloads between 330 and 475 kg to an Earth escape trajectory, depending on the particular vehicle configuration selected. The more recent Lockheed-Martin Athena family of launch vehicles can also deliver between 300 and 450 kg to an Earth escape trajectory. In addition, the Ariane V auxiliary payload ASAP ring can provide low-cost launch opportunities to GTO, if a chemical kick-stage can be incorporated for Earth escape. While each of the eight slots in the ring can accommodate a single 100 kg payload, adjacent slots may be used as attachment points for a single 200 kg payload. This provides sufficient mass for both a small solar sail with a modest payload and an Earth escape kick-stage. Particularly large solar sails may, however, require the more expensive Boeing Delta II family of launch vehicles, which can deliver between 750 and 1300 kg to an Earth escape trajectory depending on the particular vehicle configuration selected.

6.2 GEOSTORM MISSION

6.2.1 Background

On 13 March 1988, six million people in Quebec province were left without electrical power due to the temporary loss of 9450 MW of capacity from the Hydro-Quebec power company grid. This massive power black-out was due to a geomagnetic storm inducing currents in power transmission lines which led to switchgear and transformer failures. Other recent geomagnetic storm events have resulted in the partial or total loss of communication satellites, disruption to satellite navigation systems and interference to terrestrial radar networks. As reliance on networked global communication systems grows, disruption due to geomagnetic storms is likely to increase. The primary goal of the Geostorm mission is to provide enhanced warning of such storms to allow preventative action to be taken to protect vulnerable systems.

Geomagnetic storms are principally the result of coronal mass ejections (CMEs), the violent release of large volumes of plasma from the solar corona. These high-speed bubbles of plasma are transported outwards through the solar system, often driving shock waves into the slower solar wind. If the geometry of a CME release is correct, the CME trajectory may lead to the plasma impinging on the Earth's magnetosphere. If the magnetic field lines of a CME are anti-parallel to those of the Earth, magnetic reconnection can occur allowing the CME to deposit energy into the magnetosphere. The resulting perturbation to the Earth's magnetic field can then ultimately lead to currents being induced in terrestrial conductors. Most susceptible to these induced currents are long electricity power transmission lines and long pipelines. Areas with igneous rock are particularly susceptible since this rock type has low electrical conductivity. Rather than seeking paths through the ground, the induced currents will flow through man-made conductors. The surge in current can easy overload transformers and capacitors in power transmission lines and also leads to accelerated corrosion in pipelines. The Alaska oil pipeline has, for example, sustained currents of up to 1000 A during storm conditions, leading to accelerated corrosion and erroneous data from flowmeters.

Under benign conditions the solar wind forms a bow shock with the Earth's magnetic field, resulting in most energetic particles being deflected at the magnetopause, some 10 Earth radii above the day side of the Earth. During storms, however, the magnetosphere can be compressed with the magnetopause lowered to only 4 Earth radii. Therefore, geostationary satellites located at 6.6 Earth radii are no longer protected by the magnetosphere and are subject to a flux of energetic particles which can physically damage the satellite electronics and can lead to differential charge accumulation on the satellite body. Eventually discharges can result in arcing, thus damaging or failing satellite components. In addition to damaging geostationary spacecraft, solar events can also lead to heating of the upper atmosphere causing it to expand. The resulting increased density in the upper atmosphere then accelerates the orbit decay of satellites in low Earth orbit due to air drag. This increased orbit decay then requires thruster activity and propellant depletion for orbit maintenance, or indeed directly shortens the life of uncontrolled satellites.

Historically, the prediction of geomagnetic storms has been an inexact affair owing to the complexity of the physical process involved. However, as a result of the increasing civil and military reliance on satellite communications and the economic consequences of major power black-outs, the accurate prediction of 'space weather' is now of considerable importance. Currently, predictions of future activity are made by the National Oceanic and Atmospheric Administration (NOAA) Space Environment Centre in Colorado using terrestrial data, such as images of the solar disc, and more recently using real-time solar wind data obtained for the Advanced Composition Explorer (ACE) spacecraft. The ACE spacecraft is a 785 kg spin-stabilised platform launched on 25 August 1997. The spacecraft is stationed on a halo orbit about the interior L_1 Lagrange balance point some 1.5 million km sunward of the Earth, as shown in Fig. 6.1. From this vantage point the spacecraft has a continuous view of the Sun and its instruments are free from any disturbances by the Earth's magnetic field. A halo orbit is required to ensure that when the spacecraft is viewed from Earth for data returns it is sufficiently far from the solar radio disc to avoid interference from solar radio noise. Importantly, since the spacecraft is located sunward of the Earth, solar wind disturbances sensed by the suite of instruments on-

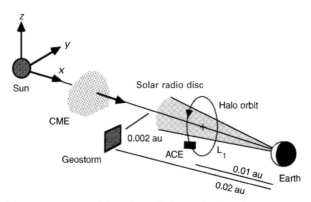

Fig. 6.1. ACE and Geostorm spacecraft locations relative to the L_1 point.

board the ACE spacecraft can be used to provide early warning of impending geomagnetic storms. In particular, the spacecraft magnetometers can detect the polarity of a CME which determines if the CME will deposit energy in the magnetosphere, as discussed earlier. Typically a prediction of order 1 hour can be made from the Lagrange point orbit, enhancing the quality of forecasts and alerts to user groups.

6.2.2   Mission concept

The enhanced storm warning provided by the ACE mission is limited by the need to orbit the L_1 point on the Sun–Earth line. Since L_1 is a natural equilibrium point, only a modest Δv budget is required for station-keeping (due to the inherent instability of L_1 halo orbits). Moving closer to the Sun would enable even greater enhancements of warning times but would require an artificially generated Lagrange point. For a conventional spacecraft, a large Δv budget would be required to artificially orbit sunward of the L_1 point, while remaining close to the Sun–Earth line. For example, to achieve the Geostorm mission orbit described below, a conventional spacecraft would required a Δv budget of order $9\,\mathrm{km\,s^{-1}}$ per year of operation. However, since solar sails have in principle unlimited Δv, such artificial Lagrange points can be sustained indefinitely and in fact require only relatively modest solar sails. The underpinning analysis of such artificial Lagrange points was considered in some detail in section 5.4.

The concept of enhancing warning times by orbiting sunward of L_1 was proposed to NOAA by the NASA Goddard Space Flight Centre in the early 1990s. It was not until June 1996, however, that NOAA requested that the NASA Jet Propulsion Laboratory develop a mission concept for an operational storm warning mission. The principal motivation for this request was to explore mission opportunities to ensure continuity of real-time solar wind data following the end of the ACE mission, projected for 2000–2002. In addition, developments in microspacecraft technologies and the successful demonstration of the IAE space inflatable structure in May 1996 (described in sections 1.2.5 and 3.4) prompted serious consideration of solar sail technology. Ultimately, the goal is to station a solar sail twice as far from the Earth as L_1 while remaining close to the Sun–Earth line, as shown in Fig. 6.1. Since CMEs will be detected earlier than at the natural L_1 point, warning times for operational forecasts and alerts will be at least doubled. As will be seen, this significant enhancement of warning time only requires a modest solar sail with a total loading of order $29\,\mathrm{g\,m^{-2}}$. This high-design loading then allows the use of commercially available $7.6\,\mu\mathrm{m}$ Kapton film for the sail substrate, and does not require the manufacture of specialised thin films.

It is clear then that Geostorm represents an unique mission application which is truly enabled by solar sailing. In addition, this key operational use of solar sail technology only requires a modest solar sail, which is important for a first mission. In earlier development plans such a modest solar sail would otherwise have served only as a flight test for more ambitious future interplanetary missions. However, with missions such as Geostorm, investments in solar sail development can now be amortised much earlier in the development of the technology. It is also somewhat ironic to note that although more advanced mission applications for higher performance solar sails have been considered for some time, these unique applications for modest solar sails have been recognised only recently. The Geostorm mission concept is now an inter-agency partnership between NOAA, the US Air Force and the US Department of Energy and has been proposed to the NASA New Millennium programme as a candidate for the fifth Deep Space mission (DS-5). If selected, the mission will be launched in the 2002–2003 timeframe and is likely to be the first operational solar sail mission.

6.2.3   Mission orbit

The baseline mission concept requires that the solar sail is firstly transferred to a conventional halo orbit at L_1. The launch to L_1 is likely to be a piggy-back to GTO or a dedicated launch using a Taurus class vehicle. For a launch to GTO an upper-stage will provide the necessary Δv of order $740 \, \mathrm{m \, s}^{-1}$ to transfer to the Lagrange point halo orbit at a heliocentric distance of 0.99 au. At L_1 the sail will be deployed and the solar sail will then spiral inwards to the operating station at a heliocentric distance of 0.98 au. In the event of a failure in the deployment sequence of the sail, the sail package can be jettisoned and the mission will continue at the initial L_1 halo orbit. In this eventuality real-time storm warning data is still provided, although the enhancement in storm warning time is not. The back-up mission at L_1 is of particular importance as it insures the end users of the mission data against the use of an untested technology. To avoid down-link interference from the solar radio disc, the solar sail will also be displaced away from the Sun–Earth line. For example, a displacement of 0.002 au normal to the Sun–Earth line corresponds to the solar sail being stationed on the edge of the 5° solar radio disc, as viewed from Earth.

In order to remain in equilibrium at this point, 0.98 au from the Sun and 0.002 au from the Sun–Earth line, the solar sail loading and orientation must be chosen appropriately. The analysis of section 5.4 provides the required parameters for equilibrium of an ideal, perfectly reflecting solar sail in a restricted three-body system with potential U. Recalling the key results from this section, the sail pitch angle α and sail lightness number β are given by

$$\tan \alpha = \frac{|\hat{\mathbf{r}} \times \nabla U|}{\hat{\mathbf{r}}_1 \cdot \nabla U} \tag{6.1a}$$

$$\beta = \frac{r_1^2}{1 - \mu} \frac{\nabla U \cdot \mathbf{n}}{(\hat{\mathbf{r}}_1 \cdot \mathbf{n})^2} \tag{6.1b}$$

where Eq. (6.1b) generates nested surfaces upon which the solar sail will remain in equilibrium. For a real solar sail, the sail film will have non-perfect reflectivity so that a more accurate analysis is required. This analysis, which has been presented in section 5.5, provides the required cone angle θ and the sail centre-line angle ϕ

as

$$\tan \theta = \frac{|\hat{\mathbf{r}}_1 \times \nabla U|}{\hat{\mathbf{r}}_1 \cdot \nabla U} \tag{6.2a}$$

$$\tan \phi = \frac{\tilde{r}}{(1 + \tilde{r}) \tan \theta} \left[1 - \left(1 - \frac{1 - \tilde{r}^2}{\tilde{r}^2} \tan^2 \theta \right)^{1/2} \right] \tag{6.2b}$$

where $\tilde{r}$ is the sail reflectivity. Recall that the cone angle defines the required orientation of the solar radiation pressure force vector while the centre-line angle defines the angle between the force vector and the normal to the sail surface. The centre-line angle arises from the non-perfect reflectivity of the sail film. From section 5.5 the required solar sail pitch angle and lightness number $\tilde{\beta}$ are now found to be

$$\alpha = \theta + \phi \tag{6.3a}$$

$$\tilde{\beta} = \frac{2}{1 + \tilde{r}} \frac{\sqrt{1 + \tan^2 \phi}}{(1 - \tan \theta \tan \phi)^2} \beta \tag{6.3b}$$

where β is the lightness number required for an ideal solar sail, defined by Eq. (6.1b). These results now provide the solar sail lightness number and pitch angle for a solar sail with reflectivity $\tilde{r}$ to remain in equilibrium at some desired operating point. Although the sail reflectivity can be estimated pre-flight, it is only through on-orbit parameter estimation that the true reflectivity of the entire sail can be determined. While the mission orbit is designed for an estimated reflectivity, the actual orbit used can be modified to account for variations from the pre-flight estimate. Therefore, if the actual sail reflectivity is less than expected, the artificial equilibrium point will not be displaced as far sunward as planned and so will only lead to a modest degradation of the expected warning enhancement.

For a nominal mission orbit 0.98 au from the Sun and 0.002 au from the Sun–Earth line, Eqs (6.3) can be used to determine the required solar sail performance. Firstly, the performance can be determined for an ideal solar sail to characterise the magnitude of the solar sail loading. It is found that, for a reflectivity of unity, a total solar sail loading of $29.6 \, \mathrm{g \, m^{-2}}$ is required. This is a rather modest performance in comparison to the demands of some interplanetary mission concepts. It is also found that a sail pitch angle of $-0.82°$ is required to provide the necessary side force to displace the equilibrium location away from the Sun–Earth line. As the sail reflectivity is decreased the required solar sail loading increases, as does the required pitch angle, listed in Table 6.1. This increase in pitch angle is needed to compensate for the centre-line effect since the solar radiation pressure force is no longer acting normal to the sail surface. Alternatively, if the solar sail loading is fixed at $29.6 \, \mathrm{g \, m^{-2}}$, the equilibrium point must be displaced closer to the Earth, thus slightly degrading the warning time enhancement, as shown in Fig. 6.2. Although these artificial equilibrium points are relatively easy to achieve, they are in general unstable. However, it can be shown that although unstable, the instability is strictly controllable using feedback to the sail orientation.

Table 6.1. Solar sail loading and pitch angle for
the Geostorm mission for a nominal mission orbit
at $x = 0.98$ au, $y = -0.002$ au.

Reflectivity	σ (g m^{-2})	α (degrees)
1	29.64	−0.82
0.9	28.16	−0.86
0.8	26.68	−0.92

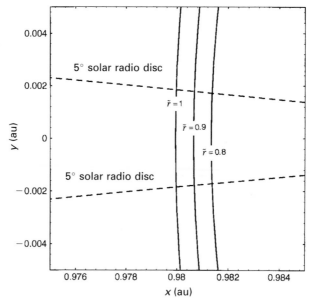

Fig. 6.2. Artificial equilibrium solutions in the x–y plane for a solar sail loading of 29.6 g m^{-2}.

6.2.4 Solar sail design

As noted in section 6.2.2, the Geostorm mission has been designed to capitalise on
the successful demonstration of inflatable structures performed during the IAE
mission on STS-77 in May 1996. The baseline solar sail design therefore centres on
inflatable technology for both deployment and rigidisation of the sail film. Owing to
the modest performance demands on the solar sail, the sail film is manufactured
from commercially available 7.6 μm Kapton, aluminised on one side. The use of off-
the-shelf materials and prior inflatable technology leads to a small, low-cost and,
owing to the back-up mission at L_1, relatively low-risk mission for the first
operational use of a solar sail.

   In order to achieve a total solar sail loading of order 29.6 g m^{-2}, a 67×67 m
square solar sail is required for a total spacecraft mass of order 130 kg, including
approximately 70 kg for the sail film and inflatable structure. The sail structure

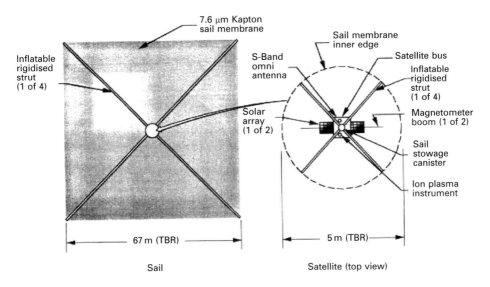

Fig. 6.3. Geostorm solar sail configuration (NASA/JPL).

consists of a set of 47 m long, 0.1 m diameter tubular diagonal spars which inflate for deployment, drawing out the packaged sail film. The spars, along with the sail film, are compactly folded and stored separately in a canister on top of the spacecraft bus for launch and transfer to L_1. Following deployment the spars rigidise in a similar manner to the IAE structure, as described in section 3.4. The bus and payload are located at the centre of the solar sail in a circular cut-out and are attached to the sail spar structure, as shown in Fig. 6.3. Attitude control is provided by conventional hydrazine thrusters mounted on the spacecraft bus. Since solar sail propulsion is not required for either escape from GTO or transfer to L_1, only slow turning rates are required for station-keeping at the artificial Lagrange point.

In addition to returning operational storm-warning data, the Geostorm mission is also conceived as a demonstration of solar sail technology, so that imaging of the sail deployment sequence is required. Images will be obtained using two micro-cameras, one on each face of the sail which are to be deployed at the ends of tension wound wires which unwind automatically on release. The images from the cameras will be stored and slowly returned by embedding compressed image data in the stream of operational storm-warning data. Storm warning is provided by two magnetometers and an ion plasma instrument to characterise the solar wind plasma. The magnetometers provide magnetic field strength and direction every few seconds while the ion plasma instrument provides measurements every few minutes. Knowledge of the magnetic field orientation is important to determine the polarity of the CME, as discussed in section 6.2.1. The entire spacecraft has a design life of three years with a five-year goal, providing storm-warning data until at least 2006. If successful, the mission will form the first in a series of missions providing enhanced solar storm-warning data using solar sail technology.

6.2.5 Other concepts

A similar mission concept (VIGIWIND) has been proposed by the French space agency CNES (Centre National d'Etudes Spatiales) for enhanced warning of geomagnetic storms. In this concept two adjacent slots on an Ariane V auxiliary payload ASAP ring are used to allow up to 200 kg to be delivered to GTO. The CNES solar sail concept has a launch mass of order 160 kg including a 65 kg bus and payload. The sail is manufactured from aluminised 1 μm Mylar with a rigidised inflatable torus structure which is connected to the bus and payload through four inflatable radial booms. A chemical kick-stage is used, firstly, to transfer the stowed solar sail from GTO to an Earth escape trajectory. The sail is then deployed and used to spiral to the operating orbit some 3 million km sunward of the Earth. The mission operating point is again at twice the distance of the L_1 point from the Earth, potentially doubling the storm-warning time available.

While the Geostorm and VIGIWIND mission concepts both operate sunward of the L_1 point, solar sails may also be used to artificially displace the L_2 point closer to the Earth, as demonstrated in section 5.4. A solar sail at such an equilibrium location may be used to monitor the geomagnetic tail *in situ*. Normally investigations of the geomagnetic tail require the use of long elliptical orbits extending along the tail itself. While such orbits may be used to investigate the structure of the geomagnetic tail, it can be difficult to de-convolve spatial and temporal variations in observations. For a solar sail located at an artificial equilibrium point, however, only temporal variations are observed. Such temporal variations can provide an indication of solar-induced disturbances of the geomagnetic field closer to the Earth.

6.3 SOLAR POLAR SAIL MISSION

6.3.1 Background

Images of the Sun obtained from the Earth, or from spacecraft close to the ecliptic plane, only allow good views of the equatorial regions of the solar disc. Images of high-latitude regions towards the solar poles are inevitably foreshortened, hindering the resolution of interesting structures. In addition, little can be inferred about the solar magnetic field and particle environment over the solar poles without direct *in-situ* observations. A solar polar trajectory therefore allows the out-of-plane solar polar magnetic field and solar wind to be fully investigated. When combined with in-plane data obtained from spacecraft orbiting in the ecliptic, the full three-dimensional structure of the solar magnetic field and wind can be inferred. Such unique out-of-plane imaging and solar wind investigations are the goals of the Solar Polar Sail mission. The Solar Polar Sail mission is the result of a response to the February 1996 call for proposals by NASA for new space physics missions. The baseline mission, which has recently been defined by the Jet Propulsion Laboratory Advanced Projects Design Team, is summarised in the next section.

The solar polar mission concept is of course not new. The joint ESA/NASA Ulysses mission used a long elliptical solar polar trajectory, obtained using a gravity assist at Jupiter, to characterise the solar wind and solar magnetic field as a function of solar latitude. The main periods interest for the Ulysses mission occurred when the spacecraft was above 70° latitude at both the south and north solar poles. On its first solar pass Ulysses reached 70° south on 26 June 1994 and spent four months gathering data in this unexplored, high-latitude region. Unexpected findings from the mission included new insights into the global structure of the solar magnetic field and the solar wind. Prior to the Ulysses solar polar mission mission many models of the solar magnetic field assumed that the field was dipolar. However, the Ulysses solar polar mission has demonstrated that the radial component of the solar magnetic field does not increase with latitude, indicating the importance of pressure forces near the Sun for evenly distributing magnetic flux. Similarly, it was found that there are two distinct solar wind states comprising a low-speed equatorial wind and a high-speed polar wind, which only occasionally extends to low latitudes.

Aside from investigating global aspects of the solar magnetic field and solar wind, the Solar Polar Sail mission will also be directed towards imaging coronal mass ejections. As discussed in section 6.2.1, CMEs are primarily responsible for geomagnetic storms which have an adverse effect on terrestrial power distribution networks and satellite communication systems. Imaging the evolution and propagation of CMEs from the Earth is problematic. Firstly, images of the solar disc may reveal indications of CME formation, such as coronal holes. However, propagating CMEs can only be properly imaged for events moving away from the Sun–Earth line. These events are then viewed against dark space, unlike CMEs moving towards the Earth where the bright solar disc prevents effective imaging. Therefore, CMEs which may reach Earth and hence generate geomagnetic storms cannot be adequately imaged. However, using a properly phased solar polar orbit, images may be obtained from a vantage point normal to the Sun–Earth line, allowing CMEs to be tracked from the Sun to the Earth against dark sky. Combing these images with in-plane images of the solar disc will allow both monitoring of changes in the coronal structure and the propagation of CMEs. Such imaging will thus provide a better understanding of the initiation, release and propagation of CME events and provide a useful space weather function.

6.3.2 Mission concept

The Solar Polar Sail spacecraft is launched directly into an Earth escape trajectory to avoid a lengthy spiral orbit escape phase. The mission plan then requires that the solar sail is transferred from escape to a polar orbit 90° to the solar equator, corresponding to an inclination of 83° to the ecliptic plane. In addition, to keep the spacecraft almost normal to the Sun–Earth line the orbit period should be chosen to be resonant with that of the Earth. Therefore, the polar orbit period should be $1/N$ years, for some integer N. This resonance condition will also avoid configurations

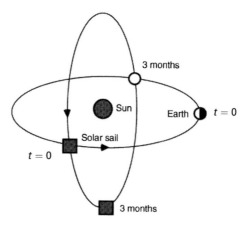

Fig. 6.4. 1 : 1 resonance solar polar orbit.

where the solar sail is occulted by the Sun. The 1 : 1 resonance obtained when $N = 1$ corresponds to a polar orbit at 1 au. This orbit ensures that the solar sail remains almost normal to the Sun–Earth line, providing ideal viewing geometry for CMEs propagating towards Earth, as shown in Fig. 6.4. Although the viewing geometry is good for CMEs, the large orbit radius requires large aperture instruments for effective imaging of the solar disc and solar corona. However, a shorter 3 : 1 resonance obtained with an orbit radius of 0.48 au provides good solar imaging while keeping the spacecraft close to 90° from the Sun–Earth line for much of the mission. This is the orbit selected for the baseline mission concept.

The primary instrument for the Solar Polar Sail mission is a coronograph to image the solar corona. The coronograph provides wide field images of the Sun at a single wavelength with an occulter placed in the instrument field of view to remove the bright solar disc from the image. Views of the solar corona obtained from high above the ecliptic can be combined with in-plane images obtained from Earth to allow the three-dimensional structure of coronal features to be constructed. The second main instrument is an all-sky imager, used primarily for imaging CME propagation. The imager comprises a fish-eye lens with a CCD detector and a system of optical baffles to reduce interference from stray light. The imager can then track the propagation of CMEs against the dark sky background. Again, it is only from a solar polar orbit, normal to the Sun–Earth line that CMEs propagating towards Earth can be adequately viewed against a dark sky.

Aside from the fundamental solar physics to be investigated during the mission, the ability to view CMEs propagating along the Sun–Earth line allows useful early warning of geomagnetic activity. While normal science data can be telemetered in regular sessions, warning data is required in near real-time. A solution to this problem may be provided using beacon-mode technology, currently under development for other deep space missions. During beacon mode only a simple two-level tone signal is telemetered. A change from the quiet tone to an active tone indicates that a CME has been detected. A high-gain antenna from the deep space network (DSN) can then be used to down-link the CME images and supporting data. Since a low bit rate signal is normally telemetered during beacon mode, only small ground stations are required for real-time monitoring. A network of at least three such ground stations is required for full 24-hour coverage.

6.3.3   Mission orbit

The baseline mission requires a cruise duration of order 4.5 years to reach the phased solar polar orbit. The solar sail is firstly injected into an Earth escape trajectory by a chemical kick-stage and uses the cranking orbit scheme described in section 4.3.4.4 to increase orbit inclination, while simultaneously reducing orbit radius. This requires that the solar radiation pressure sail force vector is alternately directed above and below the instantaneous orbit plane, while also maintaining a component of the force vector opposite to the solar sail velocity vector. Recalling from section 4.2.4 the rate of change of orbit inclination with true anomaly for a circular orbit of radius r yields

$$\frac{di}{df} = \frac{r^2}{\mu} W \cos f \qquad (6.4a)$$

$$W = \beta \frac{\mu}{r^2} \cos^2 \alpha \sin \alpha \cos \delta \qquad (6.4b)$$

where W is the out-of-plane component of the solar radiation pressure force experienced by the solar sail. However, ignoring the radial component of solar radiation pressure force, the rate of change of true anomaly for a circular orbit is just $\sqrt{\mu/r^3}$ so that the time rate of change of inclination scales as

$$\frac{di}{dt} \sim r^{-3/2} \qquad (6.5)$$

It can be seen then that cranking orbit manoeuvres become significantly more efficient closer to the Sun since the out-of-plane component of the solar radiation pressure force increases as the inverse square of heliocentric distance. It is this increase in efficiency which gives solar sailing a key advantage and makes it the most attractive propulsion option for the solar polar mission. While solar-electric propulsion systems can also operate more efficiently closer to Sun owing to the increased solar flux for power generation, solar cell efficiency is degraded by elevated operating temperatures. The cranking rate scaling law for such propulsion systems is therefore not nearly as efficient as that for solar sails.

6.3.4   Solar sail design

The baseline concept for the mission is a 150×150 m square solar sail with coilable tubular booms based on the DLR concept detailed in section 3.6.6. In order to minimise the sail assembly loading, a thin $2\,\mu m$ Kapton sail is required with a chromium thermal control coating on the rear surface of the sail film. The bus and

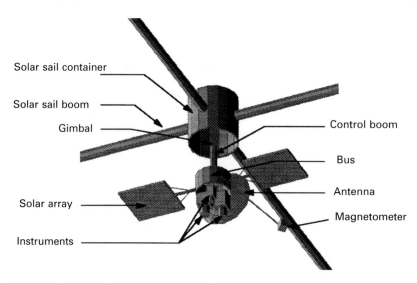

Fig. 6.5. Solar Polar solar sail configuration (NASA/JPL).

payload are mounted at the end of a 10 m deployable boom which is articulated to provide attitude control. Since an Earth escape spiral is not required, attitude control demands are modest, although periodic reversals of the solar sail orientation are required during the cranking orbit phase. The bus and payload are located at the centre of the sail in a circular cut-out which prevents additional thermal input due to solar radiation reflected from the sail film. The overall spacecraft configuration is shown in Fig. 6.5.

   On reaching the mission orbit the solar sail is jettisoned so that the bus and payload become an independent free-flying spacecraft. The bus must therefore fully support the payload, providing attitude control functions for the stringent pointing requirements demanded by payload instruments such as the coronograph. The bus must almost provide propulsion for a clean separation from the solar sail. Based on a suite of field and particle instruments in addition to the coronograph and all-sky imager, the total payload mass is of order 30 kg with a 135 kg bus. Adding the solar sail mass of 150 kg and appropriate contingencies, the total launch mass is some 380 kg, which is within the capacity of the Taurus launch vehicle with a suitable kick-stage. It is quite remarkable that such a high-energy mission, with an effective Δv of order $50 \, \mathrm{km \, s^{-1}}$, can be delivered using a relatively small, low-cost launch vehicle. The mission is truly enabled by solar sailing and makes optimum use of the technology by tackling a high-energy mission which is essentially impossible for conventional propulsion. If the jettisoned solar sail is considered to be the propulsion mass, the effective specific impulse delivered by the solar sail is over 10,500 s.

6.4   MERCURY ORBITER

6.4.1   Background

Although Mercury is a close neighbour in the inner solar system it is a largely unexplored body. Observations from Earth are difficult due to its proximity to the Sun in the sky and *in-situ* observations by spacecraft are difficult due to its location deep within the Sun's gravity well. During its 88-day orbit period Mercury is never more than 28° from the Sun, only allowing observations prior to sunrise and following sunset. In addition, the Δv required for a direct ballistic transfer to a low orbit about Mercury from an Earth-parking orbit is of order $13\,km\,s^{-1}$. Gravity assists can reduce the energy requirements at the cost of an extended mission duration. Owing to these large energy requirements the only Mercury mission to-date has been a series of fly-pasts by the Mariner 10 spacecraft, launched in November 1973. Following a gravity assist from Venus, Mariner 10 passed within 430 miles of Mercury in March 1974 imaging one hemisphere of the planet. The Mercury fly-past resulted in a new orbit for Mariner 10 with an orbit period twice that of Mercury with subsequent fly-pasts in September 1974 and March 1975.

Mercury is somewhat unusual in that there is a $3:2$ resonance between the planetary spin period and orbit period. The spin period of 58.6 days and orbit period of 88 days results in a solar day of 175.9 days' duration. Indeed, due to this $3:2$ resonance, Mariner 10 imaged the same planetary hemisphere on each of its fly-pasts. Therefore, not all of the surface has been imaged, leaving coverage of the planet incomplete. A surprise finding of the Mariner 10 mission was the existence of a weak magnetic field, usually associated with rapid planetary rotation. Although Mercury rotates slowly it does, however, have an iron core, as indicated by its relatively high density. While weak, the magnetic field is strong enough to deflect the solar wind in the vicinity of the planet. Another potential surprise may be the existence of frozen volatiles deep in shaded craters at the poles of Mercury, remnants of cometary impacts early in the history of the solar system. Owing to the lack of basic knowledge of Mercury and such possibilities as ice at Mercury's poles, a Mercury orbiter is a key requirement for future planetary exploration.

6.4.2   Mission concept

Mercury is an ideal candidate for a future solar sail mission since it is extremely difficult to reach using chemical propulsion, even using multiple gravity assists to reduce launch and capture energy requirements. In addition, its close proximity to the Sun results in the efficient use of solar sail propulsion both for orbit transfer, capture and for orbit manoeuvring at Mercury. With a perihelion of only 0.31 au, the solar sail acceleration at Mercury can be over ten times its characteristic acceleration at 1 au. This increased acceleration represents a significant manoeuvring capability once captured in orbit at Mercury. For example, the orbit plane of a polar mapping orbit will remain inertially fixed since Mercury has essentially no polar flattening. This is unlike Earth polar orbits where the harmonics of the geopotential can be used

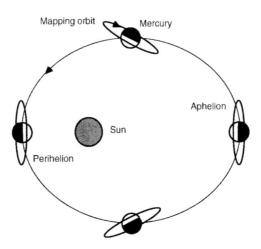

Fig. 6.6. Sun-synchronous orbit conditions at Mercury.

to naturally precess the orbit plane generating Sun-synchronous conditions. A Sun-synchronous condition is obtained when the orbit plane precesses at the same rate as the Sun-line rotates due to the motion of the planet about the Sun. Owing to the high radiation pressure available at Mercury, however, a solar sail may be used to precess the mapping orbit plane, thus generating artificial Sun-synchronous conditions.

Again due to the increased solar flux in the inner solar system, spacecraft in orbit about Mercury face a large thermal input form the Sun and from Mercury through direct reflection and re-radiation of solar energy. For an inertially fixed orbit plane, the orbit will eventually bring the spacecraft over the subsolar point on Mercury, resulting in a large thermal input on both the front and rear surfaces of the spacecraft. For low mapping orbits it can be shown that conventional thermal control methods are not sufficient for such extreme conditions. In addition, an inertially fixed orbit will in general result in regular thermal shocks as the spacecraft transits the terminator through the day–night cycle. In order to avoid these difficulties a Sun-synchronous polar mapping orbit close to the terminator is required, as shown in Fig. 6.6. Such an orbit provides global mapping of the surface of Mercury under constant lighting conditions and limits heating due to reflection and re-radiation from Mercury. As noted earlier, such an orbit can be generated using a solar sail to artificially precess the orbit plane as Mercury orbits the Sun. The DLR mission concept described below has considered such orbits for efficient mapping of Mercury and for maintaining a relatively benign thermal environment.

6.4.3   Mission orbit

Owing to the eccentricity of Mercury's orbit ($e = 0.206$), the rotation rate of the Sun-line observed from Mercury will vary during the 88-day year. It is found that the rotation rate of the Sun-line varies as the inverse square of Mercury's orbit radius so that the Sun-line will appear to rotate quicker at perihelion than at aphelion.

Fortunately, solar radiation pressure also has an inverse square variation so that the precession rate of the polar orbit plane will be automatically adjusted and Sun-synchronous conditions always maintained. This is a unique property of solar sail propulsion which adds to the benefits of using solar sailing for a Mercury orbiter mission.

Although the precession rate is automatically adjusted, a switching strategy is required for circular orbits resulting in a solar sail rotation twice per orbit to direct the solar radiation pressure force normal to the orbit plane. In addition, for the baseline mission characteristic acceleration of $0.25\,\mathrm{mm\,s^{-2}}$, an orbit radius of order 51,000 km is required to achieve the necessary Sun-synchronous condition. Such an orbit is too high for mapping purposes and suffers significant inclination perturbations due to solar radiation pressure and solar gravity. As an alternative, an elliptical polar orbit allows high-resolution mapping at pericentre while avoiding sail rotations to maintain a Sun-synchronous condition. The asymmetry in orbit precession rate at the apocentre and pericentre of the mapping orbit allows the solar sail to be fixed in a Sun-facing attitude while still precessing the orbit plane quickly enough to generate Sun-synchronous conditions. The small regression of the orbit plane at pericentre is more than compensated by the large precession at apocentre.

From section 4.2.4, the precession rate for the elliptical polar mapping orbit is given by the rate of change of the ascending node angle Ω as

$$\frac{d\Omega}{df} = \frac{r^3}{\mu p}\sin(\omega + f)W \tag{6.6}$$

where W is the out-of-plane component of the solar radiation pressure force experienced by the solar sail at the mapping orbit. It can be seen that the precession rate is a strong function of orbit radius so that the greatest precession rate will occur near apocentre while a small regression will occur at pericentre. Substituting for the orbit radius r of the elliptical mapping orbit then yields

$$\frac{d\Omega}{df} = \frac{p^2}{\mu}\frac{\sin(\omega + f)}{(1 + e\cos f)^3}W \tag{6.7}$$

The total change in ascending node angle is now obtained by integrating Eq. (6.7) around one complete elliptical orbit. Integrating from the ascending node, the change in ascending node angle is obtained from

$$\Delta\Omega = \frac{p^2}{\mu}\int_{\omega+f=0}^{2\pi} \frac{\sin(\omega + f)}{(1 + e\cos f)^3}W\,df \tag{6.8}$$

The mean orbit plane precession rate is then obtained from Eq. (6.8) by dividing by the solar sail orbit period T. This operation averages the variation in precession rate during a single orbit due to the asymmetry between the precession experienced on

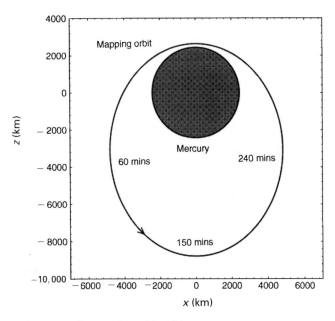

Fig. 6.7. Sun-synchronous polar mapping orbit at Mercury.

each side of the planetary equator. The mean precession rate is then defined as

$$\left\langle \frac{d\Omega}{dt} \right\rangle = \frac{\Delta\Omega}{T} \qquad (6.9)$$

A polar mapping orbit is now selected so that its pericentre is located above the north pole of Mercury with $\omega = \pi/2$. Then, it can be seen from Eq. (6.8) that the precession rate is maximised at apocentre, where the solar sail spends most of its time. Owing to this asymmetry in precession rate between apocentre and pericentre, the solar sail can be fixed in a Sun-facing attitude so that W is constant in both magnitude and sign. As noted earlier, a small retrograde precession is incurred when the solar sail is north of the equator, but this is compensated by the large prograde precession incurred at apocentre. Since the solar sail is always in a Sun-facing attitude, mission operations are greatly simplified.

In order to obtain an elliptical Sun-synchronous mapping orbit a pericentre radius is firstly selected, based on resolution requirements for surface imaging. Then, the required apocentre radius, and so eccentricity, is obtained by integration of Eq. (6.8) so that a Sun-synchronous condition is obtained. For example, a pericentre of 200 km altitude above the north pole of Mercury requires an apocentre of 6350 km for a solar sail with a characteristic acceleration of 0.25 mm s^{-2}, as shown in Fig. 6.7. Such an orbit has a period of 5.08 hours and, owing to the 3 : 2 resonance between the planetary spin period and orbit period, allows full coverage of Mercury in one planetary year. Again, the ability to obtain a Sun-synchronous orbit automatically is a unique property of solar sail propulsion. While a Sun-synchronous orbit may be forced using chemical or solar-electric propulsion, the energy requirements are large. For one planetary year of operation a Δv budget of order $12.5\,\mathrm{km\,s^{-1}}$ is required for the 5.08-hour mapping orbit. This is in addition to the Δv for transfer and propulsive capture at Mercury.

6.4.4   Solar sail design

The DLR Mercury orbiter study identified a baseline mission suitable for delivery using a low-cost launch vehicle, such as the Taurus family with a small kick-stage. The solar sail design utilises the low-mass CFRP coilable boom technology discussed in section 3.6.6 and can deliver a small science payload of 20 kg, excluding the spacecraft bus. The baseline mission requires an $86 \times 86\,\mathrm{m}$ solar sail with a launch mass of 240 kg generating a characteristic acceleration of $0.25\,\mathrm{mm\,s^{-2}}$. The solar sail is injected directly into an Earth escape trajectory using the Taurus launch vehicle and uses solar sail propulsion for orbit transfer, capture and orbit manoeuvres at Mercury. A cruise duration of order 3.5 years is required to reach Mercury with a further 4 weeks for a spiral orbit capture. The capture phase is relatively short owing to the large solar radiation pressure acceleration experience by the solar sail at Mercury. A more advanced option requires a $150 \times 150\,\mathrm{m}$ solar sail with a characteristic acceleration of $0.55\,\mathrm{mm\,s^{-2}}$, resulting in a transfer duration of only 1.8 years. The larger launch mass of order 350 kg would required a more capable member of the Taurus launch vehicle family.

Candidate payload instruments for the mission include a visual band CCD camera for surface mapping, a gamma-ray spectrometer for determining surface composition, a LIDAR instrument for surface elevation mapping derived from laser-ranged altimetry and a magnetometer package for investigating Mercury's magnetic field. Although the solar sail can be used to great advantage for orbit manoeuvres about Mercury, it poses difficulties for instrument design due to the possibility of occultations by the sail. Such viewing difficulties can be alleviated by mounting a selection of instruments on the sail booms, although this poses additional problems for packing and fine pointing control.

6.5   SAMPLE RETURN MISSIONS

6.5.1   Background

Planetary and small body sample return missions represent one of the greatest future challenges for spacecraft engineering. Since the required Δv can be almost double that of a one-way mission, the payload mass fraction of the spacecraft must inevitably be small. Therefore, to place reasonable bounds on the initial spacecraft launch mass, the mass of the returned sample must also be small. The difficulty of a sample return mission scales exponentially, rather than linearly, with the total mission Δv due to the rocket equation. For some initial mass m_0, the final mass

m_f delivered using a propulsion system with specific impulse I_{sp} is

$$m_f = m_0 \exp\left(-\frac{\Delta v}{g I_{sp}} \right) \qquad (6.10)$$

Therefore, defining the ratio of delivered mass to initial launch mass as R, the mass fractions for a one-way mission and return mission scale as

$$R_1 \sim \exp\left(-\frac{\Delta v}{g I_{sp}} \right) \qquad (6.11a)$$

$$R_2 \sim \exp\left(-\frac{2\Delta v}{g I_{sp}} \right) \qquad (6.11b)$$

ignoring staging and the use of propulsion systems with different specific impulses during various mission phases. In general then, for the same initial launch mass the final mass ratio for a one-way and return mission scales as

$$\frac{R_2}{R_1} \sim \exp\left(-\frac{\Delta v}{g I_{sp}} \right) \qquad (6.12)$$

so that sample return becomes increasingly difficult from bodies on high-energy orbits. It is clear then that high specific impulse propulsion can have a significant impact on launch mass, even if used for only part of the mission.

Apart from the Apollo Moon landings, the only sample return missions to date have been the Russian Luna 16, 20 and 24 missions, each of which returned approximately 0.1 kg of lunar soil. The last successful mission, Luna 24, was launched in August 1976. While lunar sample return poses a challenge, a Mars sample return mission is considerably more demanding due to the scaling law described earlier. In addition to the larger Δv required for transfer to and from Mars, Martian surface gravity is 2.3 times that of the Moon, thus leading to increased propellant mass fractions for both descent and ascent. Although Mars is a more massive body, its thin atmosphere allows the possibility of aerocapture to offset the requirements for capture and descent propellant mass. In contrast, small body sample return from asteroids and comets is, in principle, somewhat easier owing to their low surface gravity, although many of these bodies have eccentric, high-energy orbits.

Solar sails can be used to reduce the initial launch mass of a sample return mission and to reduce the total mission duration. Since solar sails do not require propellant, and so are not constrained by the rocket equation, the difficulties associated with conventional chemical propulsion can be overcome. A Mars sample return mission using solar sailing is still challenging and would require a large solar sail to deliver a lander and ascent vehicle to Mars. However, solar sailing is ideally suited to small body sample return missions where the sample can be acquired without a large lander. The escape speed from such small bodies is typically less than a few metres per second, so that the lander does not require a large propellant mass fraction. In addition, solar sailing can enable extremely high energy missions to perform sample returns from comets on highly elliptical orbits which would otherwise be impossible to reach using chemical propulsion. Indeed, the energy required just to reach a comet on a highly elliptical, retrograde orbit was the primary motivation behind the comet Halley rendezvous mission studies more than twenty years ago. Finally, it should be noted that a typical sample return mission requires the sample to be returned to Earth using direct atmospheric entry from an interplanetary trajectory. However, solar sailing offers the possibility of capture and a slow spiral to low altitudes where the sample can be retrieved, and perhaps analysed in an orbital facility. This may be particularly important to reduce terrestrial contamination of samples suspected of containing evidence of past or present biological activity.

6.5.2   Mission concepts

A mission study at the Jet Propulsion Laboratory in 1995 considered the launch of two identical solar sails on a Delta II launch vehicle for a sample return from the two main belt asteroids Vesta and Metis. The use of identical solar sails leads to cost reductions in both development and manufacture while the two target asteroids provide comparative information with which to interpret both samples. The solar sails are launched directly into an Earth escape trajectory and require approximately 3.3 years to rendezvous with their respective targets. A stay time of several months is then used to characterise the asteroids, with their mass distribution determined from the evolution of the solar sail orbits. The mission requires multi-spectral instruments to map the asteroid at several wavelengths to determine surface composition, with very high resolution images used to document the selected landing site. Single 0.5 kg samples are obtained with a small lander using chemical propulsion. The lander then ascends and docks with the solar sail, transferring the sample for return to Earth. During the return phase the solar sail characteristic acceleration is increased from 0.75 to 0.8 mm s^{-2} due to the absence of the lander mass. The samples are retrieved by direct atmospheric entry in a return capsule approximately 7 years after launch.

To achieve two sample returns from a single launch vehicle requires that the lander and sample return capsule mass are minimised. Indeed, there are several missions under development which can provide these technologies, along with other low-mass, miniaturised components. The lander is therefore only expected to mass approximately 50 kg. In addition to a low-mass payload, a thin sail film is required to meet the launch mass and transfer duration requirements for the mission. It is assumed that 1.5 μm Kapton film is available resulting in a 265 × 265 m square solar sail with a launch mass of 668 kg, including a 30% contingency for the sail assembly mass. It can be seen, then, that even projecting the use of low-mass subsystems still results in a rather large solar sail in comparison to other recent mission concepts. Again, this is an indication of the difficulties posed by sample return missions to retrieve even a small body sample in a reasonable mission duration.

Another demanding mission is a sample return from comet Encke, recently studied by DLR. This is a challenging mission since comet Encke has an eccentricity of 0.85, although it is on a prograde orbit, unlike comet Halley which is retrograde. In order to rendezvous with the comet the solar sail uses a solar pass inside the orbit of the Earth in order to increase the orbit eccentricity and so raise the aphelion of the

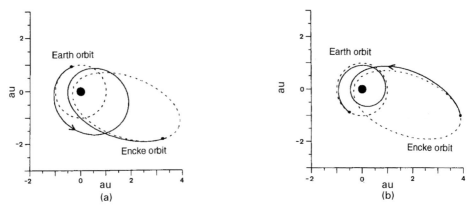

Fig. 6.8. Comet Encke sample return trajectory (DLR): (a) Encke rendezvous; (b) Earth return.

orbit to that of the target body. Similarly, during the return phase another solar pass is required to circularise the orbit for delivery of the sample to Earth, as shown in Fig. 6.8. Using a 150×150 m square solar sail generating a characteristic acceleration of 0.85 mm s^{-2}, a 100-kg bus and payload can be delivered to comet Encke in approximately 3 years. The characteristic acceleration for the return phase increases to 1.0 mm s^{-2} after the lander is discarded, allowing a total mission duration of 5.5 years. Assuming a sail assembly loading of 5 g m^{-2}, the total launch mass is 210 kg so that the mission can be achieved using a launch vehicle from the Taurus family.

6.6   POLAR OBSERVER

6.6.1   Background

Geostationary orbit provides a convenient location for communication satellites, providing a fixed line-of-sight from the satellite to ground terminals. Being located high above a fixed point on the equator, geostationary orbit also provides an ideal vantage point for weather satellites, providing coverage of large geographical regions. While the advantages of geostationary orbit for communications and Earth observation are undisputed, there are, however, some limitations. Owing to their location over the equator, geostationary satellites do not have a good vantage point from which to view high-latitude regions. Imaging of high-latitude regions is degraded by foreshortening effects while the poles are entirely excluded from view. Likewise, communication satellites are extremely difficult to view for users at high latitudes owing to their close proximity to the horizon, and indeed are below the horizon for latitudes above $\pm 81°$.

  High-latitude regions are of importance for a number of military, commercial and environmental interests. Firstly, the high Arctic was a strategically important region during the cold war. While times have changed, there is still a need to provide high-latitude military communications. In addition, the growing interest in the Arctic and Antarctic regions for mineral and oil extraction may lead to a growing demand for communication services. The Arctic and Antarctic are also of great environmental importance and there is a requirement for relaying data from remote weather stations or other automated monitoring platforms. Additional environmental requirements for polar services include continuous imaging of polar weather systems, real-time imaging of aurora – which can cause radar clutter and indicate energy deposition in the magnetosphere – and monitoring of polar ice coverage for climate studies.

The limitations of geostationary orbit for such applications may be overcome to some extent by using satellites in polar orbit. Imaging of high-latitude regions can be accomplished with good viewing geometry and high resolution, although complete mapping relies on the assembly of a mosaic of individual instruments swaths obtained during high-latitude passes. The field of view of the satellite can be broadened by raising the orbit altitude. However, this increases the satellite orbit period and so leads to a longer wait between imaging. Similarly, communication services can be provided using high-inclination Molniya orbits or through constellations of satellites in low Earth orbit. Molniya orbits are inclined, highly elliptical orbits which are oriented so that the orbit apogee is fixed high above the Arctic or Antarctic. When in view, the satellite therefore appears to move slowly across the sky well above the local horizon. At least three satellites are required to provide continuous coverage, with the communication link switching between satellites as each satellite rises and sets. Similarly, low Earth orbit satellite constellations can provide communication links between high-latitude regions and any other point on the surface of the Earth by passing digital data packets between members of the constellation. While such systems are well suited for infrequent mobile communications, they are expensive to use if a continuous, real-time link is required for relaying data to and from remote platforms or other users.

6.6.2   Mission concept

It has been shown in section 5.4 that solar sails may be used to generate artificial Lagrange equilibrium solutions in Sun–planet three-body systems. While in-plane equilibria have applications for missions such as Geostorm, detailed in section 6.2, out-of-plane equilibria may be utilised for continual, low-resolution imaging of the high-latitude regions of the Earth. In fact, if the artificial Lagrange point is located high enough above the ecliptic plane, the solar sail may be stationed almost directly over the north pole, or indeed the south pole, during the appropriate time of year. Such orbits have recently been investigated by the University of Glasgow for NOAA to define mission applications for a solar sail displaced high above the L_1 point. The solar sail can be stationed directly over the north pole at the summer solstice, as shown in Fig. 6.9, but will not remain over the pole during the entire year due to the tilt of the Earth's spin axis. From this unique vantage point a constant daylight view of the north pole is available during summer, however six months later at the winter

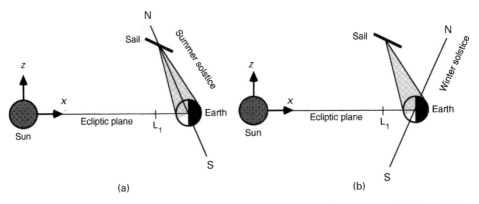

Fig. 6.9. Polar Observer mission orbit illustrating the field of view at the summer and winter solstice.

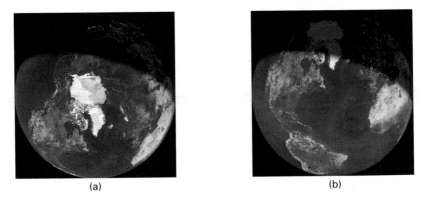

Fig. 6.10. Polar Obserview view at: (a) summer solstice; (b) winter solstice (©*The Living Earth/ Earth Viewer*).

solstice the polar regions are in permanent darkness. An illustration of the view obtained at both the summer and winter solstice is shown in Fig. 6.10.

It will be shown in section 6.6.3 that the required solar sail performance can be minimised by an appropriate selection of polar altitude. It will be found that an equilibrium location some 3.9 million km above the north pole will minimise the demands on the solar sail performance. Closer equilibrium locations are possible using larger, or higher performance solar sails, or indeed selecting a less demanding viewing geometry. At this location 3.9 million km above the Earth, the solar sail is stationed directly over the north pole during the summer solstice. During the winter solstice the solar sail still appears $43°$ above the horizon at the north pole and can be viewed from latitudes down to $49.5°$ at this time.

Although the distance of the solar sail from the Earth is large for imaging purposes, there are potential applications of real-time, low-resolution images for continuous views of large-scale polar weather systems along with Arctic ice and cloud coverage for weather prediction and global climate studies. The minimum

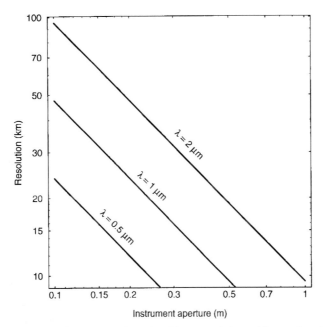

Fig. 6.11. Ground resolution obtained at the Polar Observer mission orbit as a function of instrument wavelength λ.

ground resolution obtained from such imaging is shown in Fig. 6.11 for a range of instrument apertures. For example, a 30 cm aperture instrument stationed at 3.9 million km from the Earth and operating at optical wavelengths provides a minimum ground resolution of order 9 km, assuming near diffraction limited optics. In practice, however, the actual resolution obtained will be degraded due to factors such as the pointing stability of the camera. Higher resolution is possible if an equilibrium location closer to the pole is selected, at the expense of increased demands on the solar sail performance. Space weather applications of these continuous polar orbits are also of interest for real-time imaging of the polar magnetosphere from out-of-the-ecliptic plane and for providing a viewing offset to image Earth-bound coronal mass ejections away from the solar disc.

In addition to imaging missions, direct communication links with high-latitude ground stations are enabled where geostationary satellites appear close to, or indeed below, the local horizon. Such missions have been proposed by physicist Robert Forward as part of the related 'Statite' concept. In this concept equilibrium locations about the L_2 point are envisaged although, as will be seen, these locations do not appear attractive for a realistic solar sail with non-perfect reflectivity. Since the path length to a solar sail high above the L_1 point is much greater than that to geostationary orbit, only moderate bandwidth links appear feasible for a small solar sail and ground segment. Such limited data links do, however, have a number of useful applications. One such application was investigated in some detail by the Canadian Dynacon Corporation for military and civil communication links to northern Canada. A ground station at mid-latitudes would be used to up-link data to a transponder located high above the Arctic using a solar sail. The data would then be down-linked to a ground station in the Arctic for further dissemination to local users. Using a small solar sail delivered by the use of a Pegasus launch vehicle, data rates of up to $1 \, \mathrm{Mbit \, s^{-1}}$ could be sustained. The cost of the continuous link was estimated at less than half of the equivalent cost of procuring time on a commercial satellite constellation, or launching a dedicated series of conventional polar orbiters.

A similar communication scheme could be used to obtain continuous, real-time data from automated polar weather or science stations. Indeed, such a scheme may have applications for data returns from landers or surface rovers at the polar regions of Mars. While a polar orbiter relay satellite can also perform the same function, the orbiter is only visible at most once per orbit. A solar sail could be used to deliver the landers or rovers to Mars and then act as a transponder to enable continuous data returns.

Lastly, from a vantage point high above the Arctic, a solar sail could also be utilised as a real-time solar physics platform. The solar sail would have continuous visibility of both the Sun and a single high-latitude ground station for continuous data returns. This is in contrast to conventional Lagrange point solar physics missions which require a complex and costly network of at least three low-latitude ground stations to ensure continual links to the satellite. The solar sail could be used as an instrument platform, or merely as a transponder for a number of conventional Lagrange point satellites.

6.6.3   Mission orbit

The solar sail loading required to achieve the Polar Observer mission orbit can be determined from the analysis used to evaluate the Geostorm mission artificial Lagrange point, detailed in section 6.2.3. Both the Geostorm and Polar Observer orbits belong to the same connected family of equilibrium solutions in the vicinity of the L_1 Lagrange point. It will be assumed that the solar sail is non-ideal and has a realistic reflectivity of 0.85. Then, using Eq. (6.3b) it is found that for a given solar sail lightness number there is a large surface attached to L_1 on which the solar sail will remain in equilibrium, with the gravitational, light pressure and centripetal forces all in balance. This surface extends sunward, providing the equilibrium location for Geostorm, and above the ecliptic plane, providing the equilibrium location for the Polar Observer mission. A section of the resulting surfaces of solar sail lightness number is shown in Fig. 6.12 (see also Table 6.2). The Polar Observer solutions high above L_1 can be seen along with a second family of solutions near L_2. However, these solutions are not attractive for polar applications due to their restricted viewing geometry at the summer solstice. The surfaces S_1 and S_2 define the volume of space within which equilibrium is possible.

It can be shown that for a given orbit geometry there is an optimum distance at which to station the solar sail from the Earth in order to minimise the required solar sail performance. For the orbit geometry described earlier, with the solar sail stationed directly over the north pole during the summer solstice, this optimum

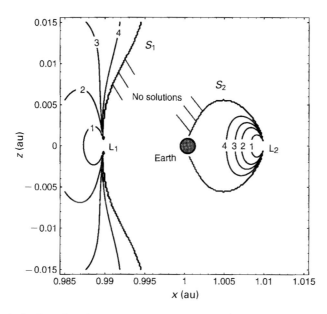

Fig. 6.12. Required solar sail performance contours for out-of-plane equilibria (see Table 6.2 for values).

Table 6.2. Solar sail lightness number, characteristic acceleration and loading for Lagrange point equilibrium solutions in the vicinity of the Earth for a solar sail with reflectivity 0.85.

	Contour			
	1	2	3	4
β	0.02	0.04	0.06	0.1
a_0 (mm s^{-2})	0.12	0.24	0.36	0.59
σ (g m^{-2})	76.50	38.25	25.50	15.30

distance is found to be 0.0261 au (3.9 million km) with a required solar sail lightness number of 0.099 (15.4 g m^{-2}). The variation of the required solar sail performance with polar altitude is shown in Fig. 6.13, with the optimum equilibrium location clearly seen. While a higher performance solar sail will provide an equilibrium point stationed closer to the Earth, it should be noted that for this particular orbit geometry and sail reflectivity the closest possible polar distance is of order 2.5 million km. Orbits below this distance are interior to the surface S_1 indicated on Fig. 6.12 and so are forbidden. However, for a less-demanding orbit geometry this minimum distance can be reduced at the expense of poorer polar viewing.

If a single instrument is to be delivered to the mission orbit, for example a CCD optical imager, the bus and payload may be limited to a mass of order 150 kg. Then,

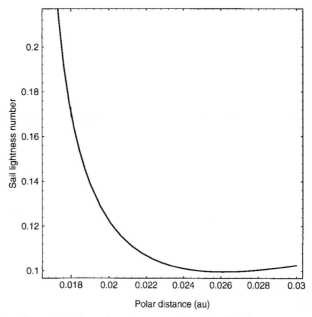

Fig. 6.13. Variation of required solar sail performance with polar altitude.

for a moderate sail assembly loading of $6.0\,\mathrm{g\,m}^{-2}$, a $126 \times 126\,\mathrm{m}$ square solar sail is required. A sail assembly loading of $6.0\,\mathrm{g\,m}^{-2}$ requires a relatively thin sail film and a low-mass boom structure, representing an improvement in technology from the Geostorm in-plane Lagrange point mission. The total spacecraft launch mass of 245 kg is, however, well within the Earth escape capacity of the Taurus family of launch vehicles. A larger payload with an array of imaging devices would of course require a larger solar sail and launch vehicle. Similarly, a larger solar sail would allow an equilibrium location closer to the lower limit of 2.5 million km, providing a significant improvement in ground resolution.

It is appropriate to note here the recent interest in whole Earth imaging from the natural L_1 point, some 1.5 million km from the Earth. Studies have investigated a small L_1 spacecraft with a total mass of order 220 kg carrying a three-colour CCD camera. A 25 cm telescope will provide a ground resolution of order 14 km, providing an Earth image every three minutes and requiring a down-link data rate of $50\,\mathrm{kbit\,s}^{-1}$. Such missions can exploit some of the advantages of the Polar Observer mission for continuous imaging of large geographical regions, for example weather systems and cloud cover, but from an in-plane location the poles are of course not in continual view.

6.7 MICRO-SOLAR SAIL CONSTELLATIONS

6.7.1 Background

The possibility of scaling down conventional solar sail concepts to accommodate extremely low-mass payloads has been investigated in section 3.5.7. It was demonstrated that if developments in microelectromechanical systems (MEMS) technology continue to a level where self-contained payloads of order 0.2 kg are feasible, square solar sails only a few metres in diameter appear possible. With a launch mass of order 0.25 kg, many such micro-solar sails could be deployed from a dispenser on a single launch to an Earth escape trajectory. Owing to the small area of these solar sails, packing and deployment are likely to be less complex tasks than for larger sails. In addition, while planetary or small body rendezvous missions require complex and precise attitude manoeuvring, some mission applications may only require a fixed, or slowly varying sail attitude, which can be easily controlled by passive means for a simple micro-solar sail. Some of these speculative possibilities for future missions will now be explored.

6.7.2 Mission concepts

A micro-solar sail constellation could be formed by the deployment of tens or even hundreds of small, autonomous solar sails to perform continuous *in-situ* measurements of, for example, the solar wind. Such distributed observations would allow the extended structure and time evolution of the solar wind to be inferred. These microspacecraft can then be thought of as 'pixels' of a synthetic image generated by convolving sampled data from each micro-spacecraft. If the individual spacecraft are deployed from a single dispenser, the solar sails may be used to disperse the orbits of the members of the constellation. For example, if the shape of each sail is designed to trim to some different fixed pitch attitude, then individual members of the constellation will either spiral inwards towards the Sun or outwards away from the initial Earth escape trajectory. In doing so, the resulting change in semi-major axis will induce an azimuthal drift in the relative positions of the individual micro-solar sails. Over a period of time the constellation will then have an almost uniform azimuthal distribution, while slowly diffusing inwards and outwards in the radial direction. An example of this process is shown in Fig. 6.14 using only a four-member constellation for clarity. If the micro-solar sails are agile enough to generate an out-of-plane component of solar radiation pressure force, then cranking manoeuvres may also be used to induce a uniform distribution in inclination. Such a constellation would then allow a continuous three-dimensional map of the solar wind and solar magnetic to be generated.

Perhaps the greatest difficulty associated with such missions is returning data from micro-spacecraft which are distant from the Earth. However, rather than returning data directly, the constellation members may pass packets of data between themselves. When a packet of data reaches a constellation member close to the Earth, the data packet may then be directly retrieved. In this way the communica-

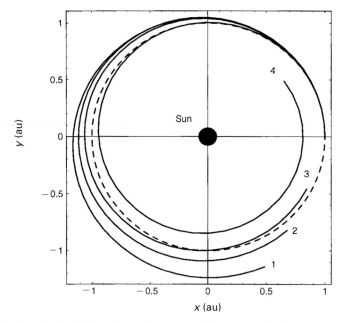

Fig. 6.14. Solar sail constellation dispersal with a characteristic acceleration of 0.25 mm s^{-2}.

tions architecture can be designed so that the link budget of each spacecraft is sufficient to pass data across the expected mean spacing of constellation members. Appropriate time tagging and identification of the data packets may then allow a complete map to be reconstructed from discrete packets of data. Although such mission concepts are at present speculative, the possibilities offered by micro-solar sails in combination with MEMS technology may enable totally new forms of exciting space science missions in the future.

6.8   NON-KEPLERIAN ORBITS

6.8.1   Sun-centred non-Keplerian orbits

Large families of Sun-centred non-Keplerian orbits were presented in section 5.2, where it was demonstrated that an advanced solar sail could be displaced high above the ecliptic plane in a unique, displaced orbit. The orbit radius, out-of-plane displacement distance and orbit period were found to be functions of the solar sail pitch angle and lightness number. It was also demonstrated that individual orbits could be patched together, resulting in additional families of new orbits. These families of displaced non-Keplerian orbits have a diverse range of applications for solar physics and space weather missions. For example, the Ulysses mission demonstrated the value of field and particle observations from out of the ecliptic plane, but provided relatively short passes of the solar poles. Displaced solar sail

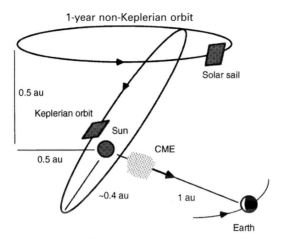

Fig. 6.15. Sun-centred non-Keplerian orbit.

orbits could in principle provide a continuous out-of-plane view. Other advantages of out-of-plane observations of the Sun have been discussed for the Solar Polar Sail Mission in section 6.3.1.

As an example mission, a 1-year Earth-synchronous non-Keplerian orbit will be considered with an orbit radius of 0.5 au and out-of-plane displacement distance of 0.5 au. Such an orbit requires a solar sail with a lightness number of 0.88. From this vantage point high above the ecliptic plane, coronal mass ejections moving towards Earth can be clearly imaged against a dark sky, as shown in Fig. 6.15. Such imaging is of importance for space weather studies where CMEs propagating directly towards Earth cannot be adequately imaged from the Earth against the bright solar disc, as discussed in section 6.3.1. The design of a high-performance solar sail for such a mission has been discussed in section 3.5.6. A solar sail with a bus and payload mass of order 35 kg requires a 100 m radius rotating disc solar sail with a thin metallic film and sublimating substrate. The mass of the deployed sail film is only 16 kg, although an additional 90 kg is required for a 2 μm sublimating substrate. The total launch mass is then approximately 140 kg, although additional mass would be required for deployment and the initial spin-up of the solar sail. Once on-station at the non-Keplerian orbit the solar sail would have a unique vantage point from which to perform continuous observations. In addition, if the solar sail spin axis is precessed such that the sail is rotated edgewise to the Sun, a patch to an elliptical Keplerian orbit is obtained with a perihelion of 0.387 au and an inclination of 45°, allowing further direct investigation of the Sun, as shown in Fig. 6.15. On return to the aphelion of the ellipse at 0.5 au the solar sail can be patched back to a non-Keplerian orbit. Alternatively, the solar sail can be patched directly to another non-Keplerian orbit normal to the first, allowing observations from above and below the ecliptic plane. Although a bus and payload mass of order 35 kg presents a challenge for a high-performance interplanetary spacecraft, advances in micro-spacecraft technologies can enable such capabilities.

For a sail film with a specialised thermal coating, operation at solar distances inside 0.15 au may also be possible. Of particular interest are orbits with a period of approximately 25 days which are synchronous with the solar rotation. A Keplerian orbit with a 25-day period requires a semi-major axis of 0.167 au. However, a solar sail can have a synchronous 25-day orbit period, but at a closer solar distance, either in-plane or displaced out of the ecliptic plane. Such an orbit would allow the formation, evolution and dissipation of solar features to be continuously imaged during a complete solar rotation. Also of interest are static equilibrium locations directly over the solar poles using a solar sail of unit lightness number. Since both solar gravity and solar radiation pressure have an inverse square variation (other than close to the Sun, see section 2.5) an equilibrium condition can be established at any distance over the solar poles.

6.8.2   Planet-centred non-Keplerian orbits

A similar family of non-Keplerian orbits was discussed in section 5.3 for planet-centred orbits where it was demonstrated that near polar orbits could be displaced along the Sun–planet line. For this family of orbits the required solar sail perform-ance is dependent on the planetary body about which the orbit is established. For example, it was shown that the requirements for displaced orbits about Mercury are significantly less demanding than for those about the Earth, since the ratio of the solar radiation pressure force to the local planetary gravitational force experienced by the solar sail is much larger for Mercury. For Earth-centred orbits mission applications include investigation of the full three-dimensional structure of the geomagnetic tail. For conventional missions, long elliptical orbits are required to gather data along the tail during each complete orbit. However, it then becomes difficult to deconvolve spatial and temporal variations in observations owing to the orbital motion of the spacecraft. Using a displaced orbit, instruments are maintained at a fixed distance along the geomagnetic tail. For example, a 17.5-day orbit with a radius of 20 Earth radii and a displacement distance of 40 Earth radii requires a solar sail lightness number of 0.74. By orienting the solar sail edgewise to the Sun a 12-day elliptical Keplerian orbit with a perigee of 25 Earth radii is achieved, allowing motion along the geomagnetic tail for further investigations, as shown in Fig. 6.16. In addition, patching the initial displaced orbit to other displaced orbits can generate interesting new families of orbits, as detailed in section 5.3.5.

6.9   OUTER SOLAR SYSTEM MISSIONS

6.9.1   Background

Delivering spacecraft to the outer solar system using chemical propulsion inevitably requires a long mission duration and a high launch energy. While gravity assists can be used to reduce the mission duration and launch energy, mission opportunities become somewhat limited. For example, the Voyager 1 and 2 spacecraft launched in

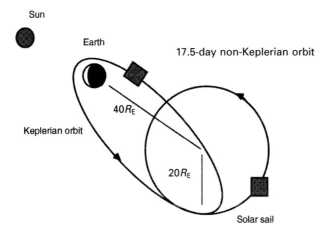

Fig. 6.16. Earth-centred non-Keplerian orbit.

September and August 1977 respectively provided a unique tour of the outer solar system. Since the final encounter of Neptune by Voyager 2 in 1989, the spacecraft have also been providing valuable insights into the structure of the outer heliosphere. The heliosphere is the bubble of plasma generated by the solar wind as it expands from the Sun into interstellar space. Eventually the supersonic solar wind merges with interstellar plasma thus defining the heliopause, the beginning of true interstellar space. Both spacecraft are moving in the general direction of the solar apex, the direction of motion of the Sun with respect to the interstellar plasma, at speeds of order 3 au per year above and below the ecliptic plane. Although the location of the heliopause is uncertain, it is expected that it will be detected directly at a distance of order 100 au. Already the spacecraft have detected low-frequency radio emissions from this region. Assuming the ageing spacecraft survive beyond the heliopause, the interstellar medium will be sampled directly for the first time, perhaps thirty years after launch.

Missions beyond the heliopause are also of interest for a number of novel applications. Firstly, the focus of the gravitational lens of the Sun extends outwards from approximately 550 au. The lensing effect predicted by general relativity could in principle provide spectacular observations at, for example, interstellar hydrogen radio frequencies or optical wavelengths. Although 550 au represents the minimum distance required to obtain interesting observations, travel beyond this point out to 1000 au will still provide additional information. Candidate directions for such a mission include the galactic anti-centre to view the galactic core. In addition to optical and radio frequency observations, a 1000 au class mission would also provide an extremely long baseline for accurate measurements of stellar parallax and experiments in gravitational wave detection.

In order to reach such comparatively vast distances in a reasonable duration, techniques other than gravity assists must be considered. As noted above, the two Voyager spacecraft are in cruise at speeds of order 3 au per year. Clearly, this is inadequate for any future missions to 100 au or beyond. Therefore, to obtain a cruise speed of order 50 au per year or greater will require the use of a high specific impulse propulsion system. One candidate system is nuclear electric propulsion using radioisotope power sources. Such systems can deliver a large payload mass while the radioisotope power unit can also provide electrical power for the payload once the required heliocentric distance has been reached. Solar sails can also generate a high cruise speed by using a close solar approach. However, in order to obtain a high solar sail lightness number a large payload mass fraction is not desirable. In addition, the solar sail will also require a separate radioisotope power source for use by the payload once on station.

6.9.2   Mission orbit

The type of trajectory used to reach the outer solar system is dependent on the performance of the solar sail. While a moderate performance solar sail, with a characteristic acceleration of order $1 \, \mathrm{mm \, s^{-2}}$, can use a close solar approach to reach solar system escape speeds, fast missions beyond 100 au will require high performance solar sails. The design of high-performance solar sails has been discussed in section 3.5.6 and solar system escape trajectories briefly identified in section 4.3.2. In particular, it was shown that a solar sail with a lightness number greater than equal to 0.5 could escape directly from the solar system merely by selecting a Sun-facing attitude. If the solar sail lightness number is greater than unity a nett repulsive force is exerted by the Sun, leading to a hyperbolic orbit with the Sun at the empty focus of the hyperbola. For a high-performance solar sail with a lightness number greater than unity, such a Sun-facing attitude may in fact be desirable to avoid rotating such a large, gossamer structure.

The cruise speed attainable using a hyperbolic escape trajectory will now be investigated. It will be assumed that a stowed solar sail is delivered onto an elliptical transfer orbit using a high-energy upper-stage or a series of gravity assists, as shown in Fig. 6.17. Then, the solar sail will be deployed at the perihelion of the transfer orbit in a Sun-facing attitude. If the sail shape is slightly conical, with the apex of the cone directed towards the Sun, passive Sun pointing can then be achieved. Any rotations away from the Sun-line will generate a restoring torque, thus maintaining the Sun-facing attitude. The elliptical transfer orbit will have an aphelion radius of R_{E}, taken to be 1 au, and perihelion radius R_{P} so that the solar sail speed prior to deployment at perihelion is given by

$$v_{\mathrm{P}}^2 = \frac{2\mu R_{\mathrm{E}}}{(R_{\mathrm{E}} + R_{\mathrm{P}})R_{\mathrm{P}}} \tag{6.13}$$

Following deployment of the sail, the solar sail is transferred to a hyperbolic orbit since the effective solar gravity now changes sign. Since the sail is oriented in a Sun-facing attitude, energy conservation requires that

$$\tfrac{1}{2}v^2 - \frac{\mu(1-\beta)}{r} = \tfrac{1}{2}v_{\mathrm{P}}^2 - \frac{\mu(1-\beta)}{R_{\mathrm{P}}} \tag{6.14}$$

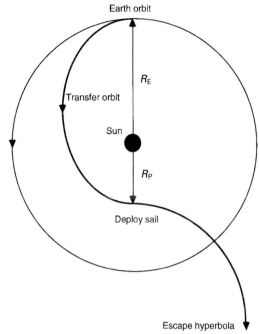

Fig. 6.17. Hyperbolic escape trajectory.

where the solar sail has speed v at orbit radius r. The asymptotic speed of the solar sail v_∞ is now obtained when $r \to \infty$. Since the asymptotic speed is quickly approached, this is effectively the solar sail cruise speed. In addition, owing to the inverse square variation of solar radiation pressure with heliocentric distance, the sail becomes ineffective after several astronomical units and so can be jettisoned. From Eqs (6.13) and (6.14) the cruise speed is now obtained as

$$v_\infty = \left[\frac{2\mu R_E}{(R_E + R_P)R_P} - \frac{2\mu(1 - \beta)}{R_P} \right]^{1/2} \qquad (6.15)$$

Therefore, since R_E is fixed the cruise speed can be determined as a function of the perihelion radius and the solar sail lightness number, as shown in Fig. 6.18. It can be seen that a high cruise speed can be obtained for a high-performance solar sail if a close solar approach is possible. Indeed, there is a significant advantage to be obtained from decreasing the sail deployment distance, rather than increasing the solar sail lightness number, even if this requires an increase in sail mass through the addition of thermal coatings.

As an alternative to using a hyperbolic escape orbit, another class of escape trajectory can be utilised which does not require a ballistic transfer for the close solar approach. In this scheme the sail attitude is fixed relative to the Sun-line such that the solar sail initially loses orbital angular momentum. Then, as the solar sail falls sunward, it rapidly accelerates as gravitational potential energy is converted to

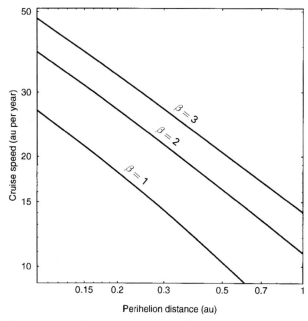

Fig. 6.18. Solar sail cruise speed obtained using a close solar pass.

kinetic energy. After the solar pass this energy would normally be converted back to potential energy and lost. However, using a solar sail of unit lightness number solar gravity can be effectively switched off, thus allowing the spacecraft to escape without losing the energy gained during the infall to the Sun. The trajectory therefore efficiently extracts energy from the Sun's gravity well. The trajectory shown in Fig. 6.19 is obtained for a solar sail of unit lightness number and a fixed pitch angle of $-40°$ relative to the Sun-line. The asymptotic cruise speed attained is approximately 18 au per year, although this can be greatly increased by allowing a closer solar pass. It can be seen that the magnitude of the solar sail angular momentum diminishes and eventually reverses as the solar sail falls sunward. This angular momentum reversal (H-reversal) trajectory was first proposed by Giovanni Vulpetti for a 550 au mission to the solar gravitational lens. Since the sail attitude is fixed relative to the Sun-line, attitude control is simplified since such an orientation can be obtained passively through an appropriate choice of solar sail geometry.

6.9.3   Outer planet missions

Owing to the inverse square variation of solar radiation pressure with heliocentric distance, a solar sail cannot be used as the sole means of propulsion for rendezvous missions to the outer planets. The solar radiation pressure in the outer solar system is too weak to allow either capture or a planet-centred spiral to a low planetary orbit. For such rendezvous missions the solar sail can only be used to accelerate the

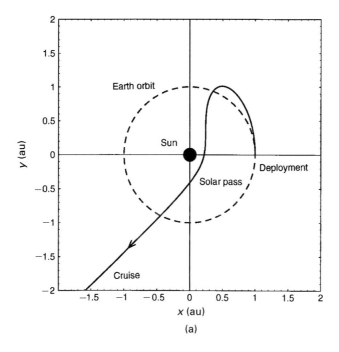

(a)

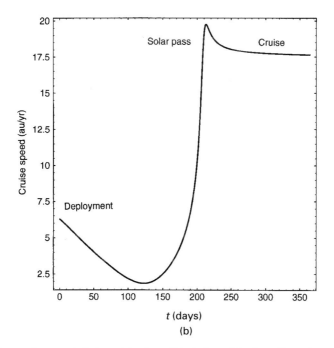

(b)

Fig. 6.19. *H*-reversal escape trajectory: (a) solar sail trajectory; (b) solar sail speed.

payload while in the inner solar system to obtain a hyperbolic-like transfer trajectory. While this limits the utility of solar sails for outer solar system missions, the use of a high-energy launcher can be avoided, thus significantly reducing mission costs. In addition, intermediate gravity assists are not required, thus increasing flexibility in the mission planning process. For orbit insertion the payload must carry its own retro-propulsion system with space storable propellants, or use aerobraking if appropriate.

If the transfer duration to the target body is decreased by using a close solar pass, the arrival speed at the target increases, requiring a larger braking manoeuvre. For example, a solar sail with a characteristic acceleration of $1 \, \mathrm{mm \, s^{-2}}$ can reach Pluto in approximately 13 years using a single loop with a solar pass at 0.49 au. The arrival speed of $10 \, \mathrm{km \, s^{-1}}$ relative to Pluto, however, necessitates a fly-past mission. In fact the same mission can be performed using a double loop with two close solar approaches. A solar sail with a characteristic acceleration of only $0.7 \, \mathrm{mm \, s^{-2}}$ requires approximately 10 years to reach Pluto with solar passes of 0.6 au and 0.32 au. The use of a sail film with specialised thermal coatings may allow solar passes closer than 0.2 au and so lead to a significant reduction in transfer duration. If the payload mass for outer solar system missions can be kept small, solar sails offer the possibility of reaching the outer planets, starting from an Earth escape trajectory using only a small, low-cost launch vehicle.

6.9.4 550 au and beyond

The delivery of payloads to the focus of the solar gravitational lens at 550 au has been the long-standing aim of an international study led by Claudio Maccone and Giovanni Vulpetti. The Aurora mission is designed to deliver a radio telescope to the focus of the solar gravitational lens for a range of astrophysics applications. In particular, the galactic anti-centre is seen as a valuable target direction in order to view the galactic core in great detail. In addition, the recent Solar Gravitational Telescope study conducted by NASA/JPL has considered the delivery of an infra-red telescope to the focus of the solar gravitational lens. In order to limit the mission duration to 10 years, the H-reversal trajectory described in section 6.9.2 is used with a high-performance solar sail. If the resulting solar pass is limited to 40 solar radii (0.186 au) – determined as the minimum heliocentric distance for a thermally limited ultra-thin sail film – a solar sail with a lightness number greater than 5 is required. This corresponds to a total solar sail loading of $0.31 \, \mathrm{g \, m^{-2}}$ and presents a significant challenge, even if a perforated sail film is used. For a large 1 km diameter solar sail the total launch mass would then be approximately 250 kg, including the payload mass. However, it has been determined that for the solar gravitational telescope mission the payload mass alone may be of order 2500 kg.

It is clear then that in order for solar sailing to be a competitive propulsion choice for this mission a solar sail with a lower lightness number will be required, allowing a larger payload mass fraction to be delivered. In order to meet the 10-year flight time constraint a closer solar approach is then required, necessitating the development of advanced thermal coatings. A secondary issue to the rapid delivery of the payload is the achievable navigation accuracy at 550 au. In order to image a desired target object, the telescope must be positioned to an accuracy of order 100 m owing to the large effective gain provided by the Sun. This is clearly an extremely challenging requirement. Unfortunately the solar sail is unable to provide useful trajectory control beyond several astronomical units so that an auxiliary propulsion system would be required to meet the stringent navigation requirements at 550 au.

In addition to the primary mission goals, a number of additional science goals appear feasible. Indeed, these goals may be enough to justify a dedicated outer solar system mission using solar sailing without the demands of observations at 550 au. Firstly, the cruise phase from the inner solar system presents a valuable opportunity to investigate the large-scale structure of the heliosphere out to and beyond the heliopause. Once outside the heliopause the solar sail will reach true interstellar space, allowing direct investigation of the interstellar space environment. Also, the extending baseline between the Earth and the solar sail may be used to obtain accurate stellar distances using parallax measurements. There may be similar opportunities for investigations of gravitational waves. A less certain scientific goal is the observation of comets in the Kuiper belt, extending to several hundred astronomical units. Since the density of comets in this region is uncertain, random encounter observation rates may be low.

It is also of interest to note that 550 au corresponds to a distance of approximately 3 light days, well short of the 4.3 light years to the Alpha Centauri star system. In order to reach across true interstellar distances in a reasonable duration, mildly relativistic speeds of order 10% of the speed of light will be required. It seems likely that laser-driven sails will be required for such missions. Although laser-driven light sails present formidable engineering challenges, advances in micro-spacecraft technologies mean the challenges are perhaps less daunting than at first appear. Such advanced light sail systems will be discussed in Chapter 7.

6.10 SUMMARY

It has been shown that solar sail propulsion opens up a diverse range of new and exciting mission opportunities. The most attractive are those high-energy missions which are truly enabled by solar sailing. While some are clearly high-energy missions, such as the Solar Polar Sail mission and Sun-synchronous Mercury orbiter, others do not so obviously belong to this class. For example, the Geostorm mission requires only a moderate performance solar sail. However, the time-integrated acceleration required to maintain an artificial Lagrange point orbit over an extended mission duration also places this mission within the high-energy class. It should therefore be remembered that the use of a low or moderate performance solar sail does not constitute a low-energy mission if the solar sail is used for an extended duration.

A further consideration is the time-line of development of solar sail technology for future mission applications. Historically, solar sail development has been viewed as a series of evolutionary steps, beginning with a small demonstration mission in Earth orbit and culminating in high-energy interplanetary missions. However, recent mission concepts allow the development of useful mission applications in step with technology flight testing and validation. Again, the Geostorm mission requires only a relatively small, moderate performance solar sail. In the past such a solar sail would have been viewed as an engineering demonstration for more ambitious missions. However, the Geostorm mission concept allows the utilisation of such a solar sail for an extremely exciting and valuable mission application. Similarly, the Polar Observer mission requires a solar sail with approximately double the performance of the Geostorm design, allowing evolutionary development and again a novel mission application. Both missions are in fact utilising the same class of artificial Lagrange point orbits and are high-profile demonstrations of the novel orbits and mission applications enabled by solar sailing.

6.11   FURTHER READING

Solar storm missions

West, J.L., 'NOAA/DOD/NASA Geostorm Warning Mission', *JPL D-13986*, Jet Propulsion Laboratory, Pasadena, October 1996.

Prado, J.Y., Perret, A., Pignolet, G. & Dandouras, I., 'Using a Solar Sail for a Plasma Storm Early Warning System', IAA-96-IAA.3.3.06, *47th International Astronautical Congress*, October 1996.

McInnes, C.R., 'Solar Sail Performance Requirements for the Geostorm Warning Mission', *DR-9623*, Dept of Aerospace Engineering, University of Glasgow, Glasgow, October 1996.

McInnes, C.R., 'Solar Sail Force Model and Up-dated Performance Requirements for the Geostorm Warning Mission', DR-9805, Dept of Aerospace Engineering, University of Glasgow, Glasgow, May 1998.

Solar polar missions

Neugebauer, M. *et al.*, 'A Solar Polar Sail Mission', Internal Report, Space Physics Research Element (3239), Jet Propulsion Laboratory, Pasadena, October 1996 [available from http://spacephysics.jpl.nasa.gov/spacephysics/SolarPolarSail/].

Leipold, M., 'To the Sun and Pluto with Solar Sails and Micro-Sciencecraft', IAA-L98-1004, *3rd IAA International Conference on Low-Cost Planetary Missions*, Pasadena, April 1998.

Mercury orbiter missions

French, J.R. & Wright, J., 'Solar Sail Missions to Mercury', *Journal of the British Interplanetary Society*, **40**, 543–550, 1987.

Leipold, M., Borg, E., Lingner, S., Pabsch, A., Sachs, R. & Seboldt, W., 'Mercury Orbiter with a Solar Sail Spacecraft', *Acta Astronautica*, **35**, Suppl., 635–644, 1995.

Leipold, M., Lingner, S., Borg, E. & Brueckner, J., 'Mercury Imaging from a Sun-Synchronous Solar Sail Orbiter', *Journal of the British Interplanetary Society*, **49**, 105–112, 1996.

Leipold, M. & Wagner, O., 'Mercury Sun-Synchronous Polar Orbits using Solar Sail Propulsion', *Journal of Guidance, Control and Dynamics*, **19**, 6, 1337–1341, 1996.

Non-Keplerian orbits

McInnes, C.R., 'Solar Sail Halo Trajectories: Dynamics and Applications', IAF-91-334, *42nd International Astronautical Congress*, Montreal, October 1991.

Forward, R.L., 'Statite: A Spacecraft That Does Not Orbit', *Journal of Spacecraft and Rockets*, **28**, 5, 606–611, 1991.

Carroll, K.A., 'POLARES Solar Sail Feasibility Study', *TM 39-306*, Dynacon Enterprises Ltd, Ontario, October 1993.

McInnes, C.R., 'Advanced Trajectories for Solar Sail Spacecraft', *International Conference Space Missions and Astrodynamics III*, Turin, June 1994.

McInnes, C.R., 'An Examination of the Constant Polar Orbit: Discussion Document', DR-9809, Dept of Aerospace Engineering, University of Glasgow, Glasgow, April 1998.

McInnes, C.R., 'Mission Applications for High Performance Solar Sails', IAA-L.98-1006, *3rd IAA International Conference on Low-Cost Planetary Missions*, Pasadena, April 1998.

Mulligan, P., 'Solar Sails for the Operational Space Community', *Space Views*, August 1998 [available from http://www.spaceviews.com/1998/08/article3a.html].

Outer solar system missions

Eshleman, V., 'Gravitational Lens of the Sun: Its Potential for Observations and Communications over Interstellar Distances', *Science*, **205**, 1133–1135, September 1979.

Maccone, C., 'Space Missions Outside the Solar System to Exploit the Gravitational Lens of the Sun', *Journal of the British Interplanetary Society*, **47**, 45–52, 1994.

Uphoff, C., 'Very Fast Solar Sails', *International Conference Space Missions and Astrodynamics III*, Turin, June 1994.

Vulpetti, G., 'The AURORA Project: Flight Design of a Technology Demonstration Mission', *1st IAA Symposium on Realistic Near-Term Advanced Scientific Space Missions*, Turin, June 1996.

Vulpetti, G., '3D High-Speed Escape Heliocentric Trajectories by All-Metallic-Sail Low-Mass Sailcraft', *Acta Astronautica*, **39**, 1–4, 161-170, 1997.

Leipold, M., 'To the Sun and Pluto with Solar Sails and Micro-Sciencecraft', IAA-L98-1004, *3rd IAA International Conference on Low-Cost Planetary Missions*, Pasadena, April 1998.

West, J.L., 'Design Issues for a Mission to Exploit the Gravitational Lensing Effect at 550 AU', *2nd IAA Symposium on Realistic Near-Term Advanced Scientific Space Missions*, Aosta, Italy, June 1998.

Miscellaneous

Kiefer, J.W., 'Feasibility Considerations for a Sail-Powered Multi-Mission Solar Probe', *Proceedings of the 15th International Astronautical Federation Congress*, Warsaw, Vol. 1, Gauthier-Villars, Paris, 1964.

Wright, J.L. & Warmke, J.M., 'Solar Sail Mission Applications' AIAA-76-808, AIAA/AAS Astrodynamics Conference, San Diego, August 1976.

McInnes, C.R., 'Solar Sailing: A New Tool for Solar System Research', *Vistas in Astronomy*, **34**, 4, 369–408, 1991.

McInnes, C.R.: 'The Historical Development and Mission Applications of Solar Sail Spacecraft', *Interdisciplinary Science Reviews*, **20**, 4, 289–301, 1995.

Frisbee, R. H. & Brophy, J. R., 'Inflatable Solar Sails for Low-Cost Robotic Mars Missions', AIAA-97-2762, *33rd AIAA/ASME/SAE/ASEE Joint Propulsion Conference*, Seattle, July 1997.

Gershman, R. & Seybold, C., 'Propulsion Trades for Space Science Missions', IAA-L.98-1001, *3rd IAA International Conference on Low-Cost Planetary Missions*, Pasadena, April 1998.

Prado, J.Y., 'Low Cost Mission Opportunities Using a Solar Sail in Addition to Ariane V', COSPAR 98, Nagoya, July 1998.

7

Laser-driven light sails

7.1 INTRODUCTION

In order to overcome the inverse square variation of solar radiation pressure with heliocentric distance, a long-standing concept for outer solar system and interstellar missions centres on transporting momentum to a sail using a collimated beam of light from a space-based laser. This method was first proposed by physicist Robert Forward in 1962 and was later re-invented by a number of authors. Unfortunately the physics of the problem shows that only 6.7 newtons of force can be generated for each Gigawatt of laser power used. Therefore, vast quantities of power are required to accelerate a light sail with a mass of even a few hundred kilograms. In addition, to accelerate the light sail across interstellar distances requires a large aperture collimator to focus a diffraction-limited spot onto the light sail. While the engineering challenges posed by laser-driven light sails are great, analysis shows that in comparison to other schemes they do in fact provide an attractive means of propulsion for interstellar and interstellar precursor missions.

Firstly, the fundamental physics of laser-driven light sails will be investigated and their efficiency compared with conventional reaction propulsion. It will be shown that light sails only become efficient for extremely high-energy missions, such as rapid outer solar system and interstellar missions. The main benefit of laser-driven light sails for interstellar missions is that the source of motive force is fixed, while only the light sail and payload are accelerated. Therefore, the mass of the interstellar vehicle is relatively small in comparison to other schemes which require reaction mass to be transported. A further benefit is that the laser, the main component of the propulsion system, remains in the solar system where it can be maintained. These benefits are offset, however, by the need to direct the laser energy in a collimated beam across interstellar distances.

Following the discussion of light sail efficiency, the dynamics of laser-driven light sails will be considered with both a classical and relativistic analysis. The classical analysis is adequate for the more modest speeds required for outer solar system or interstellar precursor missions. However, for true interstellar missions a relativistic analysis is required. This analysis accounts for the fundamental effects of relativistic mechanics and effects such as the Doppler shift of laser photons, as viewed by the light sail. Consideration of the Doppler shift of the laser photons leads to interesting conclusions regarding the overall efficiency of light sail systems.

In addition to light sail mechanics, a discussion of the engineering requirements for light sail system design is presented. Although aluminium is generally considered a suitable material for the light sail film, other high-temperature materials offer potential advantages. For example, although niobium is denser than aluminium, its high melting point allows a significantly higher thermally limited acceleration. Following the discussion of light sail design, candidate laser devices are also considered. High peak power is attainable at present using single-shot chemical lasers. However, for sustained operation closed loop solar-pumped lasers are required. The energy from the laser system must then be directed, requiring a collimator such as a Fresnel zone lens. This lens system uses annular rings of thin film spaced by vacuum to provide a low-mass lens tuned to the laser wavelength. Although the lens is simple in design, it is physically large in scale and requires accurate control of the lens plane to project a diffraction-limited spot onto the light sail over interstellar distances.

Lastly, a number of potential mission applications for laser-driven light sails are investigated. These applications are classified as interstellar precursor missions, which range up to 1000 au, and true interstellar probes traversing several light years or more. Although true interstellar missions require vast quantities of energy and large-scale space engineering, precursor missions are somewhat less daunting, particularly if the light sail mass is minimised by assuming a miniaturised payload. The benefits of miniaturisation are also discussed for interstellar missions where the advantages of multiple, low-mass light sails are discussed. Rather than accelerating a single large light sail, a series of small lights are sails accelerated in turn to eventually form a chain from the target star back to the solar system.

7.2   LIGHT SAIL PHYSICS

The light sail propulsion system requires a laser and optical collimating system, as shown in Fig. 7.1. Since the flux of energy directed onto the light sail from the laser is in the form of electromagnetic radiation (or photons in the particulate view of light), momentum will be transported from the laser to the sail. For a laser of power output P, the rate of momentum transport by the directed energy flux is just P/c, where c is the speed of light. Therefore, for a sail with reflectivity $\tilde{r}$ and mass m the acceleration experienced by the light sail is given by

$$a = \frac{(1+\tilde{r})P}{mc} \tag{7.1}$$

For a real light sail, the sail reflectivity will be less than unity so that the sail will absorb a small faction of the incident laser energy. However, if a high-power laser is used, this absorbed energy may be considerable. For a light sail of reflectivity $\tilde{r}$, a

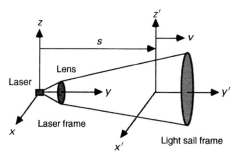

Fig. 7.1. Laser-driven light sail system configuration.

fraction $1 - \tilde{r}$ of the incident laser energy will be absorbed. This absorbed energy will then be dissipated by re-radiation on both sides of the light sail. If the light sail is a thin metallic film with area A and uniform emissivity ε on both sides, the sail equilibrium temperature may be obtained from the thermal balance equation, viz.

$$(1 - \tilde{r})P = 2A\varepsilon\tilde{\sigma}T^4 \tag{7.2}$$

where $\tilde{\sigma}$ is the Stefan–Boltzmann constant and T is the sail equilibrium temperature. Then, using Eq. (7.1) the thermally limited acceleration of the light sail is obtained as

$$\tilde{a} = 2\varepsilon\frac{1 + \tilde{r}}{1 - \tilde{r}}\frac{\tilde{\sigma}\tilde{T}^4}{\sigma c} \tag{7.3}$$

where $\tilde{T}$ is the maximum allowable sail film temperature and σ is the total mass per unit area of the light sail. It can be seen that the light sail acceleration is extremely sensitive to the maximum film temperature. A high-temperature film will therefore sustain a high acceleration, but at the expense of greater laser power. In addition, for a power-limited system Eq. (7.1) gives the maximum acceleration of the light sail as

$$\tilde{a} = \frac{4}{\pi d^2}\frac{1 + \tilde{r}}{\sigma c}\tilde{P} \tag{7.4}$$

where $\tilde{P}$ is the maximum power output of the laser and d is the diameter of the light sail. In this case it is assumed that the sail film is operating below its maximum temperature and that the technology constraint is the available laser power.

In order to direct a collimated beam of laser energy across interstellar distances, a large aperture lens is required. For a lens of diameter D and a laser operating at wavelength λ, the diffraction-limited divergence angle of the laser beam θ is defined as

$$\theta \approx \frac{2.44\lambda}{D} \tag{7.5}$$

This angle arises from the first null point of the Bessel function for a circular aperture and defines a cone which contains 84% of the laser energy. The remaining laser energy is dispersed outside this cone within the diffraction pattern formed by the aperture. Therefore, using Eq. (7.5) the laser beam spot diameter d_S at distance s

from the lens is given by

$$d_S = \frac{2.44\lambda s}{D}$$
(7.6)

If the laser spot diameter is constrained to the diameter of the light sail d, then the maximum distance $\tilde{s}$ at which the light sail will experience full acceleration is defined as

$$\tilde{s} = \frac{dD}{2.44\lambda}$$
(7.7)

Beyond this distance the light sail acceleration decreases as the sail fails to capture all of the available laser energy. Consideration of the geometry of the system shows that the light sail acceleration diminishes as the inverse square of the distance of the light sail from the lens beyond the diffraction limit $\tilde{s}$.

7.3 LIGHT SAIL MECHANICS

7.3.1 Light sail efficiency

In order to assess the potential benefits of laser-driven light sails, the relative energy efficiency of conventional reaction propulsion and light sail propulsion will be compared. Firstly, a conventional rocket will be considered with exhaust speed u and propellant mass flow rate $\dot{m}$. Then, assuming that energy is produced purely in the form of kinetic energy in the exhaust gas, the energy E_R expended by the rocket after some duration t is given by

$$E_R = \int_0^t \tfrac{1}{2}\dot{m}u^2 \, dt$$
(7.8)

For a total change in speed Δv, the rocket equation may be used to relate the final mass m_f of the rocket to the initial mass m_0 as

$$m_f = m_0 \exp\left(-\frac{\Delta v}{u}\right)$$
(7.9)

Then, integrating Eq. (7.8) and substituting for m_0 from Eq. (7.9) it is found that

$$E_R = \tfrac{1}{2}m_f u^2 \left[\exp\left(\frac{\Delta v}{u}\right) - 1\right]$$
(7.10)

For a light sail, the total energy expended is obtained purely from the laser. Since no reaction mass is expelled, the mass of the light sail is constant. Then, for a laser of power output P, the force exerted on an ideal light sail is given by

$$m_f \frac{dv}{dt} = \frac{2P}{c}$$
(7.11)

where the Doppler shift of the laser photons with light sail speed v is neglected. Integrating Eq. (7.11) it can be seen that

$$m_f \Delta v = \frac{2E_L}{c}$$
(7.12)

where E_L is the time-integrated energy output from the laser during the acceleration of the light sail through a change in speed Δv. Therefore, the energy expended by the laser is given by

$$E_L = \tfrac{1}{2} m_f c \, \Delta v \tag{7.13}$$

The ratio of rocket energy to light sail energy R can now be obtained from Eqs (7.10) and (7.13) as

$$R = \frac{u^2}{c \, \Delta v} \left[\exp\left(\frac{\Delta v}{u} \right) - 1 \right] \tag{7.14}$$

It can be seen that this ratio is only a function of the rocket exhaust speed u and the required total change in speed Δv. In the lower limit where $\Delta v / u \ll 1$ the ratio may be approximated to

$$R \approx \frac{u}{c} \tag{7.15a}$$

so that the rocket requires much less energy to accelerate through some change in speed Δv than the light sail since $u/c \ll 1$. However, in the upper limit where $\Delta v / u \gg 1$ the ratio can also be approximated as

$$R \approx \frac{u^2}{c \, \Delta v} \exp\left(\frac{\Delta v}{u} \right) \tag{7.15b}$$

which grows exponentially with the required Δv. Therefore, for high-energy missions the light sail is significantly more energy efficient than a conventional rocket. The regime where the transition from rocket to light sail efficiency occurs can be determined by setting $R = 1$. For example, a rocket engine using cryogenic hydrogen and oxygen has a typical exhaust speed of $4.5 \, \text{km s}^{-1}$. Then, it is found that $R > 1$ when the required Δv is greater than $62 \, \text{km s}^{-1}$. This corresponds to a Δv equivalent to five times the solar system escape speed from 1 au. The variation of the energy ratio is shown in Fig. 7.2 where the exponential increase above $R = 1$ can be seen. It is therefore concluded that laser-driven light sails are only suitable for extremely high-energy missions, such as interstellar and interstellar precursor missions.

7.3.2 Classical light sail mechanics

7.3.2.1 Terminal speed

As discussed in section 7.2, a light sail operating beyond the diffraction limit of the lens system will experience a decrease in acceleration as the sail fails to capture all of the available laser energy. Therefore, for a given laser, lens and light sail system the light sail will achieve some terminal speed in the limit $s \to \infty$ as the laser-induced acceleration diminishes. With this in mind the equation of motion for a light sail which experiences a thermally limited acceleration is defined in two domains by

$$\frac{d^2 s}{dt^2} = \tilde{a}, \quad s < \tilde{s} \tag{7.16a}$$

$$\frac{d^2 s}{dt^2} = \tilde{a} \left(\frac{\tilde{s}}{s} \right)^2, \quad s > \tilde{s} \tag{7.16b}$$

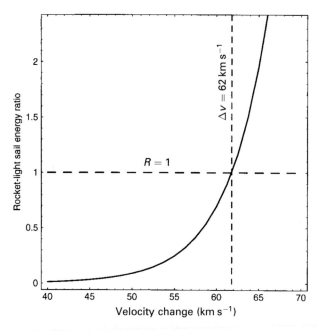

Fig. 7.2. Light sail energy efficiency compared to reaction propulsion.

where $\tilde{s}$ is again the distance at which the laser spot diameter is equal to the light sail diameter. Therefore, within the diffraction limit of the lens the light sail experiences constant acceleration, while beyond the diffraction limit the acceleration decreases as an inverse square law. Integrating Eq. (7.16a) from some initial speed v_0 at distance s_0 yields the light sail speed as

$$v^2 = v_0^2 + 2\tilde{a}(s - s_0), \qquad s < \tilde{s} \tag{7.17}$$

Then, Eq. (7.16b) may also be integrated to obtain the light sail speed beyond the diffraction limit as

$$v^2 = \tilde{v}^2 + 2\tilde{a}\tilde{s}^2 \left(\frac{1}{\tilde{s}} - \frac{1}{s} \right), \qquad s > \tilde{s} \tag{7.18}$$

where $\tilde{v}$ is the speed of the light sail at the diffraction limit of the lens system, obtained from Eq. (7.17). It will now be assumed that the light sail begins from rest so that $v_0 = 0$ at $s_0 = 0$. Then, the terminal speed of the light sail v_∞ may be obtained directly from Eq. (7.18) in the limit $s \to \infty$ as

$$v_\infty = 2\sqrt{\tilde{a}\tilde{s}} \tag{7.19}$$

It is interesting to note that $v_\infty = \sqrt{2}\,\tilde{v}$ so that an increment in speed can be obtained by accelerating the light sail beyond the diffraction limit of the lens system. However, since laser energy is continually being lost, this does not represent an efficient use of the light sail system. The phases of acceleration up to and beyond the diffraction

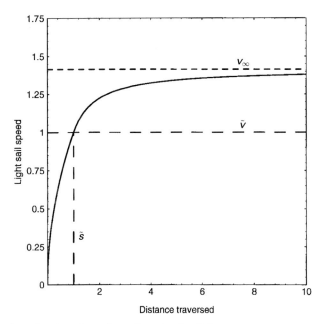

Fig. 7.3. Light sail acceleration phases (non-dimensional units).

limit are shown in Fig. 7.3 using non-dimensional units, where $\tilde{v}$ and $\tilde{s}$ are both taken as unity. From Eqs (7.19) and (7.3) the terminal speed of a thermally limited sail is found to be

$$ v_\infty = \sqrt{\frac{3.28}{\lambda} Dd\left(\frac{1+\tilde{r}}{1-\tilde{r}}\right)\frac{\tilde{\sigma\varepsilon}}{\sigma c}\tilde{T}^4} \tag{7.20a}$$

so that the terminal speed is limited by both the maximum sail temperature and the product Dd. However, for a power-limited system Eq. (7.4) shows that

$$ v_\infty = \sqrt{\frac{2.09}{\lambda}\frac{D}{d}\frac{1+\tilde{r}}{\sigma c}\tilde{P}} \tag{7.20b}$$

so that the terminal speed is limited by D/d.

For the thermally limited case the light sail acceleration is independent of the sail area and the laser power is unconstrained. Therefore, the terminal speed scales as the product of the lens diameter and the sail diameter to maximise the diffraction limited distance over which the light sail is accelerated. However, for the power-limited case a large collimator and small light sail are required to minimise the sail mass and so maximise the light sail acceleration. In both cases it can be seen that the light sail terminal speed scales inversely as the wavelength of the laser. A shorter wavelength laser will lead to a narrower beam so that the light sail can be accelerated over a longer distance. Similarly, the terminal speed scales inversely as the light sail mass per unit area, as expected. A thin, low-density metallic film is therefore required for a high-performance light sail. However, owing to the sensitivity of the thermally limited acceleration to sail temperature, the metallic film mass properties may be traded-off against the film thermal properties. These considerations will be discussed in section 7.4.1.

7.3.2.2   Boost–coast

For outer solar system or interstellar mission applications, a short acceleration phase is required followed by a relatively long cost phase at high speed. The short boost phase will minimise the required diffraction-limited $\tilde{s}$ and so will minimise the scale of the engineering required for the lens system. Owing to the loss of energy beyond the diffraction limit of the lens, it will be assumed that the boost phase is terminated at the diffraction limit. Then, for such a two phase scheme the speed of the light sail is given by

$$v = \tilde{a}t, \quad t < \tilde{t} \tag{7.21a}$$

$$v = \tilde{v}, \quad t > \tilde{t} \tag{7.21b}$$

where $\tilde{t}$ is the time to reach the diffraction limit $\tilde{s}$. Therefore, the distance traversed by the light sail at time t is given by

$$s = \tilde{s} + \tilde{v}(t - \tilde{t}), \quad t > \tilde{t} \tag{7.22}$$

where the time to reach the diffraction limit of the lens system is obtained from

$$\tilde{t} = \sqrt{\frac{2\tilde{s}}{\tilde{a}}} \tag{7.23}$$

Then, noting that $\tilde{v} = \tilde{a}\tilde{t}$, Eq. (7.22) yields the total time required for the light sail to traverse distance s as

$$t = \sqrt{\frac{2\tilde{s}}{\tilde{a}}} + \frac{s - \tilde{s}}{\sqrt{2\tilde{a}\tilde{s}}}, \quad t > \tilde{t} \tag{7.24}$$

It can be seen from Eq. (7.24) that the coast time scales inversely as the product of the light sail acceleration $\tilde{a}$ and the diffraction limit $\tilde{s}$. Therefore, for a fixed coast time the performance of the light sail can be traded directly with the performance of the lens system to minimise the engineering requirements for the entire light sail system.

7.3.3   Relativistic light sail mechanics

The earlier classical analysis of light sail mechanics is adequate for speeds which are non-relativistic. However, since one of the main applications of laser-driven light sails is for interstellar missions, where relativistic speeds are necessary, a fully relativistic analysis is appropriate. The relativistic analysis firstly transforms the frame of reference to that of the light sail, as shown in Fig. 7.1. The apparent laser power is then obtained in this frame of reference to form an equation of motion. In order to obtain the true trajectory of the light sail, the frame of reference must then be transformed back to that of the laser. Apart from the usual mechanical relativistic effects, the main phenomena of interest is the Doppler shift of laser photons viewed from the light sail frame of reference. As the light sail accelerates, the laser light appears increasingly red shifted, thus reducing the momentum imparted to the light sail.

For a light sail of rest mass m_0, the one-dimensional equation of motion in the rest frame of the light sail is given by

$$m_0 \frac{dv^\mu}{d\tau} = \begin{bmatrix} 0 \\ f \end{bmatrix} \tag{7.25}$$

where $v^\mu = \gamma(c, v)$ is the velocity 4-vector of the light sail with γ the Lorentz factor, to be defined later. The proper time τ is defined as the time measured in a frame of reference co-moving with the light sail. The light pressure force f exerted on the light sail by the laser must now be obtained in the rest frame of the sail to account for the observed red shift of laser photons. For a laser of power P, the laser power P' viewed from the light sail frame of reference may be written as

$$P' = \frac{dE'}{dt'} = P \left[\frac{dE'}{dE} \right] \left[\frac{dt}{dt'} \right] \tag{7.26}$$

Then, using special relativity it can be shown that the laser energy dE and time element dt transform from the laser frame to the light sail frame as

$$\frac{dE'}{dE} = \gamma(1 - \beta) \tag{7.27a}$$

$$\frac{dt'}{dt} = \gamma(1 + \beta) \tag{7.27b}$$

where $\beta = v/c$ is the normalised speed of the light sail. For a laser of power P, the force exerted on a perfectly reflecting light sail in the rest frame of the sail is therefore given by

$$f = \frac{2P}{c} \left(\frac{1 - \beta}{1 + \beta} \right) \tag{7.28}$$

It can be seen from Eq. (7.28) that as $\beta \to 1$ the light pressure force vanishes. This is due to the momentum transferred to the light sail diminishing with increasing photon red shift. It should be noted that when viewed from the light sail frame of reference, photons reflected from the sail will be observed to have the same wavelength as the incident photons. However, viewed from the laser frame of reference, the reflected photons will appear red shifted due to the light sail motion relative to the laser. This difference in energy between the incident and reflected photons lies in the kinetic energy added to the light sail. It can therefore be concluded that the efficiency of momentum transfer is greatest when the light sail is at rest, but that the efficiency of energy transfer is greatest as $\beta \to 1$.

The equation of motion in the rest frame of the light sail must now be transformed to the frame of reference of the laser to calculate the light sail trajectory. In order to simplify the analysis, it will be assumed that the light sail is always within the diffraction limit of the lens system. Then, using a standard Lorentz transformation it is found that in the laser frame of reference

$$m_0 \frac{dv^\mu}{d\tau} = \gamma \begin{bmatrix} 1 & \beta \\ \beta & 1 \end{bmatrix} \begin{bmatrix} 0 \\ f \end{bmatrix} \tag{7.29a}$$

$$\gamma = \frac{1}{\sqrt{1 - \beta^2}} \tag{7.29b}$$

where γ is again the Lorentz factor. This transformed equation of motion can now be written in scalar form with two components

$$m_0 c \frac{d\gamma}{d\tau} = \frac{2\gamma\beta P}{c} \left(\frac{1-\beta}{1+\beta} \right) \tag{7.30a}$$

$$m_0 c \frac{d(\gamma\beta)}{d\tau} = \frac{2\gamma P}{c} \left(\frac{1-\beta}{1+\beta} \right) \tag{7.30b}$$

For ease of illustration Eq. (7.30a) will be solved by firstly noting that by the definition of the Lorentz factor

$$\frac{d\gamma}{d\tau} = \frac{\gamma\beta}{1 - \beta^2} \frac{d\beta}{d\tau} \tag{7.31}$$

Therefore, Eq. (7.30a) may be written as

$$\frac{d\beta}{d\tau} = \frac{(1-\beta)^2}{\tau_0} \tag{7.32}$$

where $\tau_0 = m_0 c^2 / 2P$ is a time constant. Using the mass–energy equivalence $E = mc^2$ it can be seen that this time constant represents the time required for the laser to deliver an energy output equivalent to half of the light sail rest mass. Assuming the light sail begins at rest, Eq. (7.32) can be integrated to yield the light sail speed as

$$\beta(\tau) = \frac{\tau/\tau_0}{1 + \tau/\tau_0} \tag{7.33}$$

It can be seen that the constant τ_0 forms a natural timescale for the dynamics of the system. In particular, when an energy output equivalent to the light sail rest mass has reached the light sail, its speed is $\frac{2}{3}c$. Having obtained the light sail speed, the distance s traversed may be obtained by noting that

$$\frac{ds}{d\tau} = \frac{dt}{d\tau} \frac{ds}{dt} \tag{7.34}$$

Then, by further noting that $dt/d\tau = \gamma$ and that $ds/dt = \beta c$, an additional equation of motion can be formulated as

$$\frac{1}{c} \frac{ds}{d\tau} = \frac{\tau/\tau_0}{\sqrt{2\tau/\tau_0 + 1}} \tag{7.35}$$

Again assuming that the light sail begins at rest, this equation may be integrated to yield the distance traversed in proper time τ as

$$\frac{1}{c}s(\tau) = \frac{\tau_0}{6}\left(\frac{2\tau}{\tau_0}+1\right)^{3/2} - \frac{\tau_0}{2}\left(\frac{2\tau}{\tau_0}+1\right)^{1/2} + \frac{\tau_0}{3} \qquad (7.36)$$

Now Eqs (7.33) and (7.36) both define the motion of the light sail as it accelerates, driven by laser pressure. However, both of these equations are referenced to the proper time τ measured in the light sail frame of reference. In order to obtain the trajectory of the light sail as viewed from the laser frame of reference, the relationship between proper time and laser frame time must be obtained. Again noting that $dt/d\tau = \gamma$ and writing the Lorentz factor in terms of τ from Eq. (7.33), it is found that

$$\frac{dt}{d\tau} = \frac{\tau/\tau_0 + 1}{\sqrt{2\tau/\tau_0 + 1}} \qquad (7.37)$$

This equation may then be integrated to yield

$$t(\tau) = \frac{\tau_0}{6}\left(\frac{2\tau}{\tau_0}+1\right)^{3/2} + \frac{\tau_0}{2}\left(\frac{2\tau}{\tau_0}+1\right)^{1/2} - \frac{2\tau_0}{3} \qquad (7.38)$$

Therefore, Eqs (7.33), (7.36) and (7.38) can now be used to determine the light sail trajectory viewed from the laser frame of reference.

In order to quantify the effect of relativistic dynamics, a 1000 kg light sail will be accelerated for three years with a laser power of 65 GW. This case is typical of a large interstellar fly-past probe and will be discussed in detail in section 7.5.2. The light sail trajectory is shown in Fig 7.4, where it can be seen that the light sail acceleration diminishes with increasing speed. This is principally due to the red shift of laser photons viewed from the light sail frame of reference, leading to less efficient momentum transport. However, viewed from the laser frame of reference, reflected photons are increasingly red shifted so that the efficiency of energy extraction from the laser photons in fact increases, as discussed earlier. It can be seen that the correction due to relativistic mechanics is not too great since the maximum light sail speed is of order 10% of the speed of light. However, for higher speeds the relativistic light sail will lag significantly behind the predictions from classical mechanics.

7.4 LIGHT SAIL DESIGN

7.4.1 Light sail films

The design of a laser-driven light sail is similar in many respects to the design of a high-performance solar sail in that the light sail must have a low mass per unit area. More importantly, however, the sail film must have a high reflectivity and high operating temperature to allow the use of high laser power levels. If the laser power is a free parameter, then the light sail reflectivity, emissivity and operating temperature limit the maximum achievable acceleration, as discussed in section 7.2.

For metallic films the light sail acceleration will clearly increase as the film

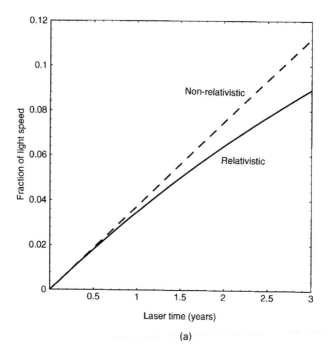

(a)

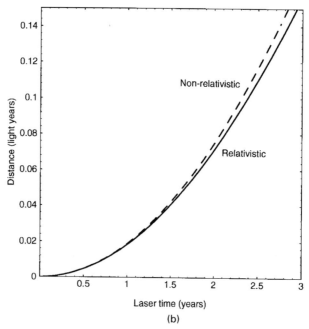

(b)

Fig. 7.4. 1000 kg light sail accelerated by a 65 GW laser: (a) light sail speed (relativistic/non-relativistic); (b) light sail distance traversed (relativistic/non-relativistic).

becomes thinner. However, for ultra-thin light sails the transmittance of the metallic film also begins to increase. Therefore, there is a trade-off between reducing the film mass and maintaining its opacity. For example, the reflectivity of aluminium falls to 0.5 for a thickness of order 0.01 μm as it becomes semi-transparent. In addition, although a bulk metal may have a relatively high melting point, in thin film form the light sail may fail due to agglomeration. This effect occurs since a thin film has a large surface to volume ratio. The surface energy of the film can fall to a lower energy state by forming into small droplets. In addition to aluminium, there are a number of other candidate light sail materials. A particularly attractive candidate material is niobium, a highly reflective metal commonly used in various industrial processes. Although the bulk density of niobium is $8.85\,\mathrm{g\,cm}^{-3}$ compared to $2.7\,\mathrm{g\,cm}^{-3}$ for aluminium, it has a significantly higher melting point at 2741 K allowing a higher thermally limited acceleration.

More exotic possibilities for light sails include multi-layer thin films, optically tuned to the wavelength of the laser. Such multi-layer films can, in principle, have extremely high reflectivities approaching unity at a single wavelength. The films are manufactured from quarter wavelength thickness layers of dielectrics with alternating films of high and low refractive index film. Alternatively, a single dielectric medium can be separated at quarter wavelength intervals in vacuum using spacer posts. Using such optically tuned thin films, fabricated using diamond-like materials manufactured artificially from carbon and nitrogen, extremely high reflectivities appear possible. In principle the light sail absorbance may be as low as 10^{-6}, greatly increasing the thermally limited acceleration of the light sail. However, since a high reflectivity is only obtained at a single wavelength these concepts may not be suitable for relativistic light sails with a fixed laser wavelength. As the light sail accelerates, the benefit of a tuned thin film will be lost owing to the Doppler shift of laser photons. The wavelength of some laser devices, such as the free-electron laser, can however be altered to compensate for the red-shifted laser photons viewed from the light sail frame of reference.

7.4.2 Laser systems

Although laser technology has advanced rapidly in recent years, the power levels required for interstellar laser-driven light sails are not yet in sight. While there have been significant developments in high-power lasers for military and laser fusion applications, the technology has centred on high-power, short-duration pulses rather than quasi-continuous output.

In general, laser systems can be either open-loop or closed-loop, indicating whether the lasing medium is recycled in the laser device. For long-duration acceleration, open-loop chemical lasers are clearly unsuitable. Although relatively high power levels are possible owing to the efficiency of these devices, the chemical reaction is irreversible, requiring the laser to be refuelled before another pulse is delivered. Closed-loop systems are possible using electrical discharge lasers where carbon dioxide or carbon monoxide atoms are pumped to energetic states using electrical energy. However, the most promising lasers for space use are directly solar-pumped lasers and free-electron lasers. Solar-pumped lasers use solar photons to directly excite the lasing medium. A large collector is required to capture solar photons with the lasing medium pumped through a cavity at the focal point of the collector. The cavity has a window to allow solar photons to enter and a window to allow laser energy to escape. A lasing medium must be selected which has a wide absorption band for solar photons and a high conversion efficiency. This minimises the waste heat to be rejected by the device and optimises use of the available spectrum of solar energy.

While lasers which use a chemical medium operate at a single wavelength, free-electron lasers can be tuned. These devices use relativistic electrons to transfer energy to a light wave in a similar manner to a travelling wave amplifier. The electron beam is constrained by an alternating magnetic field which can tune the electrons to resonate with the light wave. Since no fluids or gases are required, high operating powers and conversion efficiencies of order 50% are possible. Unlike directly solar-pumped lasers, however, solar energy must firstly be converted to electrical energy to drive the laser. It can be seen then that a closed-loop solar-pumped laser will be required with either a direct solar-pumped system, or an indirectly pumped free-electron laser. Owing to the vast power requirements for laser-driven light sails, the lasers and their solar collectors would be in Earth orbit for precursor missions, or for advanced missions, in orbit around Mercury to utilise the enhanced solar flux close to the Sun. In addition, many individual lasers are likely to be required with the laser outputs phase locked with each other for coherence.

7.4.3   Optical collimating systems

In order to accelerate the light sail over large distances the laser energy must be focused by a suitable optical collimating system. As noted in section 7.2, any optical collimator is diffraction limited. Therefore, either a short-wavelength laser must be used or extremely large aperture optics must be contemplated. While x-ray lasers could in principle be beneficial in reducing the aperture of the collimating system, the sail mass must then be increased in order to stop the x-ray photons and extract momentum from the beam. Unfortunately, a more conventional laser operating closer to optical wavelengths at 1 μm requires a large aperture collimator.

A candidate collimating system is a large Fresnel zone lens with a diameter of up to 1000 km for large interstellar probes. These lenses are formed from annular rings of thin plastic film spaced on radial spokes, as shown in Fig. 7.5. The thickness of the plastic film is selected so that the excess path length through the film is one half of the laser wavelength. The lens geometry is designed such that the path length from any annular zone of the lens to the focal point is an integer number of half wavelengths. Then, the wave fronts from the even zones of the lens will arrive in phase, while those from the odd zones also arrive in phase due the excess path length through the plastic film. Although the lens must be extremely large it has a relatively low mass due to the use of thin plastic film which is spaced by vacuum. However, in order to project a focused spot of energy onto the light sail the lens geometry must be carefully controlled in the radial plane. In addition to control of the lens plane, the relative

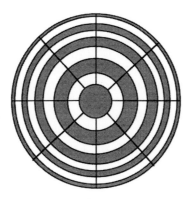

Fig. 7.5. Schematic illustration of a Fresnel zone lens.

geometry of the lens and laser system must be carefully considered. For advanced systems with lasers in orbit close to the Sun, the lens may be levitated by laser pressure in the outer solar system to provide a fixed line-of-sight to the light sail. Whatever the configuration selected, the sustained operation of the laser and lens system poses exacting engineering challenges for the future.

Lastly, it is of interest to note that it is not feasible to use an optical collimator to directly focus solar energy onto a light sail. In order to generate a power density at the sail greater than the solar intensity at the lens would required a lens aperture greater than the diameter of the Sun. It is also of some interest to note that a large Fresnel zone lens used for collimating laser energy would make a quite remarkable telescope, albeit operating at a single wavelength.

7.4.4 Impact damage and interstellar drag

Although interstellar space is essentially a hard vacuum, there is a low-density population of hydrogen atoms. Therefore, due to the large cross-section and high speed of the light sail there is the possibility of significant back-pressure, limiting the light sail to some terminal speed where the light pressure and gas back-pressure balance. In addition, there is also a possibility of accumulated impact damage producing defects in the light sail film which would ultimately degrade its structural integrity. However, it will be shown that for thin light sail films, the film is essentially transparent to particles with interstellar hydrogen passing through the light sail virtually unimpeded.

Firstly, the energy and momentum flux incident on the light sail will be obtained before the transport efficiency to the sail film is considered. Owing to relativistic effects, the number density of interstellar hydrogen atoms n measured from the light sail frame of reference is observed to be

$$n = \gamma n_0 \tag{7.39}$$

where n_0 is their proper number density. Therefore, for a light sail of cross-section A

moving with speed $c\beta$, the rate at which hydrogen atoms impinge on the light sail is given by

$$\dot{n} = c\beta\gamma n_0 A \tag{7.40}$$

Again using relativistic mechanics, the kinetic energy E of a hydrogen atom of rest mass m_H viewed from the light sail frame of reference is defined by

$$E = m_H c^2(\gamma - 1) \tag{7.41}$$

Therefore, using the impact rate of hydrogen atoms from Eq. (7.40) the energy flux $\dot{Q}$ (energy per unit time per unit area) at the light sail is found to be

$$\dot{Q} = \xi_E n_0 m_H c^3 \beta\gamma(\gamma - 1) \tag{7.42}$$

where only a fraction ξ_E of the incident energy is transported to the light sail film. Similarly, the momentum p of a hydrogen atom viewed from the light sail frame of reference is defined by

$$p = m_H c\sqrt{\gamma^2 - 1} \tag{7.43}$$

Again, using the impact rate obtained from Eq. (7.40), it is found that the back-pressure P exerted on the light sail is given by

$$P = \xi_P n_0 m_H c^2 \beta\gamma\sqrt{\gamma^2 - 1} \tag{7.44}$$

where ξ_P is the efficiency of momentum transfer. Therefore, Eqs (7.42) and (7.44) define the energy and momentum transport to the light sail by impinging interstellar hydrogen atoms.

Now that the heating and back-pressure of the light sail film have been obtained, the efficiency of energy transport from the impinging particles to the sail film will be assessed. Since the light sail will be extremely thin, it will be shown that particles will in fact pass through the film largely unimpeded. If the light sail film has stopping power dE/dx, then an incident particle of energy E has final kinetic energy $\tilde{E}$ on exit from a light sail, given by

$$\tilde{E} \approx E - \left(\frac{dE}{dx}\right)l \tag{7.45}$$

where l is the light sail film thickness. From impact calculations it is found that for non-relativistic particles the stopping power is inversely proportional to the incident particle energy, viz.

$$-\left(\frac{dE}{dx}\right) \propto \frac{1}{E} \tag{7.46}$$

The efficiency of energy transport from the impinging particles to the light sail ξ_E can now be defined as $(E - \tilde{E})/E$ so that

$$\xi_E \propto \frac{1}{E^2} \tag{7.47}$$

Therefore, as the light sail accelerates the sail film becomes increasingly transparent to interstellar hydrogen with little energy being deposited. At relativistic speeds the stopping power is modelled by the Bethe–Bloch equation which shows that particle energy loss reaches a minimum at a Lorentz factor of $\gamma \approx 2\text{–}3$ before undergoing a modest rise to $\gamma \approx 10$. For a typical transport efficiency of order 10^{-5} at non-relativistic speeds, both the energy and momentum transfer to the thin light sail film will be small. It is also of interest to note that damage to denser structural elements is likely to be rapidly annealed since any vaporised material will quickly cool before all but a small fraction of it escapes.

Lastly, in the extreme relativistic regime it is found that the light sail does in fact have a terminal speed imposed not by interstellar gas, but by the 3 K cosmic microwave background radiation. As the light sail accelerates, the background photons incident on the front surface of the sail become blue shifted while the laser photons incident on the rear surface become red shifted. In addition, relativistic aberration leads to the background photons arriving almost opposite to the direction of motion of the light sail, thus increasing the back-pressure further. The terminal speed of the light sail is attained when the driving pressure of the laser is balanced by the back-pressure from the blue-shifted background photons. However, it is found that such effects are limited to speeds extremely close to the speed of light with a Lorentz factor $\gamma \approx 10^2$.

7.5   MISSION APPLICATIONS

7.5.1   Interstellar precursor

In this section a brief analysis of laser-driven light sail missions, ranging from 100 to 1000 au, will be presented. As discussed in section 6.9.1, the lower bound of 100 au is of interest since this is estimated to be the location of the heliopause, the boundary at which the solar wind merges with interstellar plasma, so defining the beginning of true interstellar space. The upper bound of 1000 au is also of interest for a number of applications. Firstly, the focus of the gravitational lens of the Sun extends outwards from approximately 550 au, allowing spectacular astronomical observations. While such a mission presents great challenges, 550 au corresponds to a distance of only 3 light days, well short of the 4.3 light years to Alpha Centauri, the closest star in the solar neighbourhood. In addition to gravitational lens observations, a 1000 au class mission would provide a vast base-line for measurements of stellar parallax and gravitational waves.

Using the boost–coast analysis from section 7.3.2.2 the requirements for the interstellar precursor missions described above will be investigated. In keeping with current advances in micro-spacecraft technologies only low-mass light sails will be considered. The light sail mass is assumed to be divided equally between the sail film, supporting structure and payload. The light sail design would be similar to a high-performance solar sail with a total light sail loading of $1\,\mathrm{g\,m^{-2}}$ and a reflectivity of 0.85. In order to limit the scale of the engineering required, the laser power will be set at 10 GW for a wavelength of 1 μm and the Fresnel zone lens diameter will be set at 1 km. For a 30 kg light sail, the sail has a diameter of 195 m and is accelerated at approximately $2\,\mathrm{m\,s^{-2}}$ for 3.3 days, reaching the diffraction limit of the lens at a

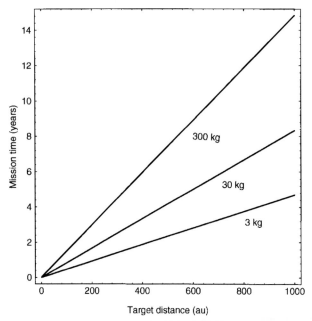

Fig. 7.6. Interstellar precursor mission duration as a function of light sail mass.

distance of 0.54 au. The cruise speed of the light sail is then 570 km s^{-1}, which corresponds to only 0.2% of the speed of light. At this cruise speed the light sail reaches the heliopause at 100 au in 0.8 years, the solar gravitational lens at 550 au in 4.6 years and 1000 au in 8.3 years, as shown in Fig. 7.6. It can be seen that trip times can be significantly reduced by lowering the total light sail mass through payload miniaturisation. Lastly, although a 10 GW continuous output laser is a major challenge, the requirements for a solar-pumped laser of this magnitude are perhaps more modest than expected. Assuming a conversion efficiency of 50%, only a 4.3 km diameter solar collector is required in Earth orbit to drive the laser.

7.5.2 Interstellar probe

The engineering challenges for a true interstellar mission are formidable. Even for a relatively low-mass light sail, enormous quantities of energy are required in comparison with contemporary spacecraft propulsion. Following the analysis of Robert Forward, a 1000 kg light sail will be considered for a fly-past mission to Alpha Centauri, 4.3 light years distant from the Sun. Again, the light sail mass is assumed to be divided equally between the sail film, supporting structure and payload. It will also be assumed that the sail is manufactured from ultra-thin 0.016 μm aluminium film with a total light sail loading of 0.1 g m^{-2}, resulting in a sail diameter of 3.6 km. At this thickness the sail film becomes semi-transparent, has a reflectivity of 0.82 and a maximum operating temperature of 600 K. The thermally

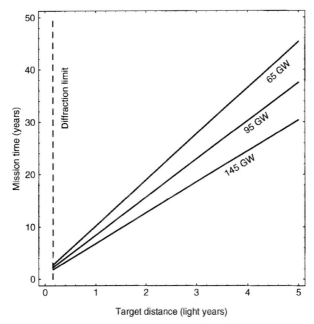

Fig. 7.7. Interstellar boost–coast mission duration as a function of laser power.

limited acceleration of the light sail is then $0.36\,\mathrm{m\,s}^{-2}$, requiring a laser power of 65 GW with an assumed operating wavelength of 1 μm. For a 1000 km diameter Fresnel zone lens, the boost–coast analysis in section 7.3.2.2 shows that the light sail reaches the diffraction limit of the lens after 2.7 years at a distance of 0.16 light years. The light sail then coasts at 10% of the speed of light, arriving at Alpha Centauri some 40 years later. The total trip time can be reduced with a higher cruise speed, as shown in Fig. 7.7, at the expense of a higher temperature sail film and increased laser power. Similarly, while the engineering challenges for the light sail system can be alleviated by reducing the light sail mass, the need for a communication system with a link budget capable of traversing interstellar distances limits the minimum mass of the probe.

Although 65 GW of laser power is large, the requirements for a solar-pumped laser are again more modest than expected. Assuming a conversion efficiency of 50%, an 11 km diameter solar collector is required in Earth orbit to capture enough energy to drive the laser. The collector area will clearly decrease if the laser is stationed closer to the Sun. For example, a laser system in orbit about Mercury would only required a collector diameter of 4.6 km. The greatest challenge for this mission is, however, the mass of the Fresnel zone lens. Even for a thin film lens manufactured from 1 μm Kapton, the total lens mass is some 560,000 metric tons, requiring almost 111,500 rings. The lens mass can be reduced by increasing the light sail acceleration so that it can attain its cruise speed over a shorter distance with a smaller lens. Current developments in micro-spacecraft technologies indicate that,

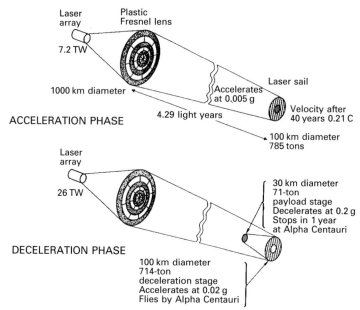

Fig. 7.8. Interstellar rendezvous mission (Robert Forward).

rather than increasing the laser power, the light sail acceleration may be increased more readily by reducing the mass of the light sail and its payload.

Rather than accelerate a single light sail to cruise speeds, a more efficient scheme may be to use a large number of much smaller light sails. The light sails would be rapidly accelerated in turn, resulting in a chain stretching from the solar system to the target star. As noted above, the lower limit to the mass of a single light sail is bounded by the need for a communication system capable of returning data across interstellar distances. For the multiple light sail concept, data would be returned from the target by passing data back along the chain, then each individual light sail would only be required to transfer data to its nearest neighbour. In this way a continuous stream of data would be returned as each light sail passed through the target star system. This is in stark contrast to a fly-past with a large, single light sail where data is gathered at the target star on a timescale of hours. Data returned from the first light sails passing through the target star system would also allow re-targeting of subsequent light sails in order to view planets or other targets of interest.

The laser-driven light sail scheme has been extended to include the possibility of rendezvous, or even return missions. This extended scheme hinges on a segmented light sail which is used in stages to accelerate the payload to the target star, decelerate the payload and finally return to Earth. The challenges for this class of mission are even more impressive than for a fly-past mission. The scenario for a rendezvous mission is shown in Fig. 7.8. Initially both sail stages are accelerated to the diffraction limit of the lens and then coast to the target star. On approach to the target the outer segment of the light sail detaches and is used as a reflector to direct laser energy onto the inner segment. With active control of the outer segment geometry, the inner segment is decelerated and brought to rest as it passes through the target star system. During this phase of the mission some $26\,\mathrm{TW}$ (26×10^{12} watts) of laser power are required. After deceleration the inner segment may be used as a conventional solar sail to explore the target star system. The staged light sail concept can be extended further by assuming a three-stage sail. In this scenario the outer segment decelerates the inner two segments which are used to explore the target star system, as described above. For the return trip the laser is activated at the appropriate time and the light sail positioned to intercept the long pulse of laser energy leaving the solar system. The remaining outer segment of the light sail then detaches and directs the laser energy onto the central return stage, which is finally decelerated directly by the laser on return to the solar system.

Lastly, it is again noted that laser-driven light sails are inefficient in terms of energy utilisation at low speeds. Reflected photons will have only a modest Doppler shift indicating the small fraction of energy transferred to the light sail. After the laser photons are reflected from the light sail their remaining energy is lost, resulting in very poor energy utilisation. In order to increase the efficiency of the light sail at low speeds a secondary reflector placed behind the laser may, in principle, be beneficial. Using a secondary reflector laser, photons would be recycled many times, thus increasing the effective laser intensity incident on the light sail. Analysis shows that such a large optical cavity greatly enhances the initial performance of the light sail system, although the benefits are quickly reduced as the light sail accelerates.

7.6 SUMMARY

The fundamental physics of laser-driven light sails have been presented and their efficiency compared to conventional reaction propulsion. It has been demonstrated that light sails offer an attractive means of propulsion for very high energy missions such as interstellar and interstellar precursor missions. For low-energy missions, and hence low speeds, momentum transfer from the laser photons to the light sail is efficient while energy efficiency is poor. However, at relativistic speeds the laser photons reflected from the light sail are strongly red shifted, indicating efficient energy transfer.

An analysis of both classical and relativistic light sail mechanics has also been presented. The analysis shows that if the laser power is not constrained, the cruise speed of the light sail is a strong function of the operating temperature of the sail film. Since the coast phase of a light sail mission is long relative to the boost phase, the cruise speed of the light sail is clearly a key mission design parameter. A relativistic analysis shows that for mildly relativistic speeds of order 10% of the speed of light, the predictions of classical mechanics still give reasonable agreement.

While metallic films are candidates for light sail manufacture, optically tuned multi-layer thin films offer many advantages. In particular, such films can have an extremely low absorbance at a single wavelength allowing high laser powers to be used. However, optically tuned films require tuneable lasers to compensate for the red shift of laser photons as viewed from the light sail frame of reference. Tuning of the laser is possible using free-electron lasers which can be indirectly pumped by solar photons using an intermediate electrical generating stage.

Along with the challenges of engineering high-power space-based lasers, a large optical collimating system is required to project a diffraction limited spot of laser energy onto the light sail. While Fresnel zone lenses appear attractive, vast structures are required for large interstellar light sails. The entire scale of the light sail system can be greatly reduced by assuming a small, low-mass light sail with a miniaturised payload. For interstellar probes, the minimum mass of the light sail is dictated by the need to return data across interstellar distances. A possible solution to this limitation has been discussed by using multiple low-mass light sails accelerated in turn to form a chain from the target star back to the solar system.

7.7 FURTHER READING

Forward, R.L., 'Pluto –The Gateway to the Stars', *Missiles and Rockets*, **10**, 26–28, April 1962.

Marx, G., 'Interstellar Vehicle Propelled by Terrestrial Laser Beam', *Nature*, **211**, 22–23, July 1966.

Redding, J.L., 'Interstellar Vehicle Propelled by Terrestrial Laser Beam', *Nature*, **213**, 588–589, February 1967.

Moeckel, W.E., 'Propulsion by Impinging Laser Beams', *Journal of Spacecraft and Rockets*, **9**, 942–944, December 1972.

Rather, J.D.S., Zeiders, G.W. & Vogelsang, K.R., 'Laser Driven Light Sails – An Examination of the Possibilities for Interstellar Probes and Other Missions', W.J. Schafer and Associates Inc., *NASA CR 157362*, December 1976.

Forward, R.L., 'Roundtrip Interstellar Travel using Laser-Pushed Light Sails', *Journal of Spacecraft and Rockets*, **21**, 187–195, March 1984.

Meyer, T.R., McKay, C.P., McKenna, P.M. & Pryor, W.R., 'Rapid Delivery of Small Payloads to Mars', The Case for Mars II, *American Astronautical Society Science and Technology Series*, **62**, 419–431, 1985.

Forward, R.L., 'Laser Weapons Target Practice with Gee-Whiz Targets', *Proceedings of SDIO/DARPA Workshop on Laser Propulsion*, Lawrence Livermore National Laboratory, July 1986.

McInnes, C.R. & Brown, J.C., 'Terminal Velocity of a Laser-Driven Light Sail', *Journal of Spacecraft and Rockets*, **27**, 48–52, January 1990.

Simmons, J.F.L. & McInnes, C.R., 'Was Marx Right? or How Efficient are Laser Driven Interstellar Spacecraft?', *American Journal of Physics*, **61**, 205–207, March 1993.

Maccone, C., 'Space Missions Outside the Solar System to Exploit the Gravitational Lens of the Sun', *Journal of the British Interplanetary Society*, **47**, 45–52, February 1994.

Landis, G.A., 'Small Laser-Propelled Interstellar Probe', Paper 95-IAA.4.1.102, *46th International Astronautical Federation Congress*, Oslo, October 1995.

Index